AF581868

Rainfall Thresholds and Other Approaches for Landslide Prediction and Early Warning

Rainfall Thresholds and Other Approaches for Landslide Prediction and Early Warning

Editors

Samuele Segoni
Stefano Luigi Gariano
Ascanio Rosi

MDPI • Basel • Beijing • Wuhan • Barcelona • Belgrade • Manchester • Tokyo • Cluj • Tianjin

Editors
Samuele Segoni
University of Firenze
Italy

Stefano Luigi Gariano
Italian National Research Council (CNR-IRPI)
Italy

Ascanio Rosi
University of Firenze
Italy

Editorial Office
MDPI
St. Alban-Anlage 66
4052 Basel, Switzerland

This is a reprint of articles from the Special Issue published online in the open access journal *Water* (ISSN 2073-4441) (available at: https://www.mdpi.com/journal/water/special_issues/Rainfall_Thresholds).

For citation purposes, cite each article independently as indicated on the article page online and as indicated below:

LastName, A.A.; LastName, B.B.; LastName, C.C. Article Title. *Journal Name* **Year**, *Volume Number*, Page Range.

ISBN 978-3-0365-0930-3 (Hbk)
ISBN 978-3-0365-0931-0 (PDF)

Cover image courtesy of Stefano Luigi Gariano.

Contents

About the Editors

Samuele Segoni graduated in Geological Sciences at the University of Firenze, Italy, and in 2008, he obtained a PhD in Earth Sciences, also at the University of Firenze. Since 2005, he has been carrying out research and teaching activities at the Department of Earth Sciences of the University of Firenze, where he is currently enrolled as Assistant Professor. Since 2019, he has been Chair Associate at the UNESCO Chair on the Prevention and Sustainable Management of Geo-hydrological Hazards. His main research topics are: the prediction and mapping of landslide hazards; application of physically based models for the triggering of shallow landslides at regional scale; landslide susceptibility assessment, statistical rainfall thresholds for landslide triggering, regional scale landslide early warning systems; civil protection, land planning, landslide risk assessment. Within these topics, he authored more than 50 peer-reviewed articles in international journals, and he is the member of Editorial Boards of various international journals.

Stefano Luigi Gariano, Ph.D., is a research scientist at the Research Institute for Geo-Hydrological Protection of the Italian National Research Council (CNR IRPI) in Perugia, Italy. He graduated in Environmental Engineering from the University of Calabria, Italy, and achieved a Ph.D. in Geology with a thesis on the impact of climate change on landslides from the University of Perugia, Italy. His main research topics include: (i) analysis of rainfall time series and of series of landslide occurrences; (ii) objective and reproducible reconstruction of rainfall conditions responsible for landslide occurrence; (iii) definition and validation of empirical rainfall thresholds for the possible initiation of landslides; (iv) evaluation of the impact of climate and environmental changes on landslides; (v) national and regional early warning systems for rainfall-induced landslides, in different climatic and physiographic environments. He is involved in several projects, mostly focused on the operational prediction of rainfall-induced landslides. He has authored more than 100 publications, including more than 40 peer-reviewed articles in international journals. He is organizing and convening scientific sessions on landslide early warning systems and on climate change effects on landslides at various international conferences. He is a founding member of "LandAware", the international network on Landslide Early Warning Systems and he is an Alternative Board Member for CNR IRPI at the International Consortium on Landslides.

Ascanio Rosi is currently working as a fixed-term research scientist at the Department of Earth Sciences of the University of Florence, where he obtained his PhD in Earth Sciences in 2013, with a thesis on the use of satellite radar data for subsidence and landslide mapping. From 2019, he has been a Chair Associate at the UNESCO Chair on Prevention and Sustainable Management of Geo-hydrological Hazards. His main research topics are the use of new technologies to create landslide databases; the identification of rainfall condition associated with the triggering of landslides; the use of artificial intelligence techniques to identify the geological and meteorological condition associated with the initiation of landslides; the development of landslide early warning system based on empirical and statistical analyses and landslide risk assessment. From 2012, he authored and co-authored over 20 articles in international journals; he also served as a reviewer and an editor for several international journals.

 water

Editorial

Preface to the Special Issue "Rainfall Thresholds and Other Approaches for Landslide Prediction and Early Warning"

Samuele Segoni [1,*], Stefano Luigi Gariano [2] and Ascanio Rosi [1]

[1] Department of Earth Sciences, University of Firenze, Via La Pira 4, 50121 Florence, Italy; ascanio.rosi@unifi.it
[2] CNR-IRPI—Research Institute for Geo-Hydrological Protection of the Italian National Research Council, 06127 Perugia, Italy; stefano.luigi.gariano@irpi.cnr.it
* Correspondence: samuele.segoni@unifi.it

Citation: Segoni, S.; Gariano, S.L.; Rosi, A. Preface to the Special Issue "Rainfall Thresholds and Other Approaches for Landslide Prediction and Early Warning". *Water* **2021**, *13*, 323. https://doi.org/10.3390/w13030323

Academic Editor: Y. Jun Xu
Received: 22 January 2021
Accepted: 26 January 2021
Published: 28 January 2021

Landslides are frequent and widespread destructive processes causing casualties and damage worldwide [1,2]. The majority of the landslides are triggered by intense and/or prolonged rainfall [3]. Therefore, the prediction of the occurrence of rainfall-induced landslides is an important scientific and social issue. To mitigate the risk posed by rainfall-induced landslides, landslide early warning systems (LEWS) can be built and applied at different scales as effective non-structural mitigation measures [4]. Usually, the core of a LEWS is constituted of a mathematical model that predicts landslide occurrence in the monitored areas [5–7]. In the last decades, rainfall thresholds have become a widespread and well-established technique for the prediction of rainfall induced landslides, and for the setting up of prototype or operational LEWS at regional scale [8–11]. A rainfall threshold expresses, with a mathematical law, the rainfall condition that, when reached or exceeded, is likely to trigger one or more landslides in a given area. Rainfall thresholds can be defined with relatively few parameters and are very straightforward to operate, because their application within LEWS is usually based only on the comparison of monitored and/or forecasted rainfall with the identified critical conditions. Because of these advantages, the technique of rainfall thresholds has received growing attention from the early 1980s of the last century to present. To date, rainfall thresholds have become the most widespread method to develop (operational or prototypal) regional scale warning systems irrespective of physical settings, landslide characteristics, and technological level of the countries financing research programs and applications [10,11].

Despite that, the technique is still affected by some limitations, making the topic a prolific research field for the landslide community. Among the most cogent research trends: the evaluation and reduction of possible sources of uncertainties [12,13]; the reduction of the false alarm rate committed by the models [14]; the strife for improving quantity and quality of input data [15]; the definition of standardized and objective methods of analysis [16,17]; the comparison between different possible rainfall parameters to identify the optimal ones for each case of study [18]; the attempts to enhance the performances of the thresholds by the joint use of instrumental monitoring [19]; the combination of rainfall thresholds into more complex forecasting systems combining different techniques, among which landslide susceptibility zonation [20,21] and antecedent soil conditions analyses [22]; the tests with hydrological parameters instead of the classical rainfall parameters [23,24]; the experiments on the exportability of consolidated models to completely different test sites [25].

In this wide panorama of open research questions, the present special issue can contribute to the advancement of the state of the art, as some of the aforementioned criticalities are tackled in the papers collected. Indeed, this special issue collects contributions about recent research advances or well-documented applications of rainfall thresholds, as well as other innovative methods for landslide prediction and early warning. All contributions are focused on the development of LEWS or are preparatory studies on forecasting models with the perspective of future operational implementations.

Moreover, besides scientific advances, the development of the recent literature highlights the interest, by an international audience, of new case studies, new approaches, new objectives (reliable results before establishing an operational LEWS). In this regard, the special issue collects case studies from three continents and a wide range of countries: Bhutan, China, India, Italy, Slovenia, Taiwan, and a site across Democratic Republic of Congo, Uganda, Rwanda, and Burundi. This allows accounting for very different climatic and geological settings, two relevant factors in the definition of critical rainfall conditions for landslide initiation. Moreover, the papers account for scales of application ranging from the local scale to the national scale. An interesting advance, useful especially in data-scarce regions, is represented by the use of satellite-based rainfall estimates and freely available global landslide catalogues in the calculation of the thresholds. Interestingly, contributions focused on different approaches useful in landslide analyses (e.g., numerical modeling, susceptibility and hazard analysis) are also proposed in this special issue to cover a broad spectrum of studies.

To better address the readers towards the content of the special issue, a short summary of each published paper is provided hereafter.

– In the paper by Yang and co-authors [26], the authors presented the Runout modeling of the Yining landslides (China), made using DAN-W software. Triggering factors of the landslide have been identified in a combination of snow melt and geological setting of the slopes. The numerical model was calibrated using field survey and laboratory tests results and allowed the researchers to estimate the velocity of the landslide, which reached a maximum of 20.5 m/s and to estimate the duration of the paroxistic event in 22 s. The outcomes of this paper showed the importance of slope monitoring, since landslide triggering can be a quick event, leaving no time for countermeasure operations once the landslide started its mobilisation.
– Dikshit and co-authors [27] investigated the rainfall conditions that can lead to landslide triggering in the Chukha Dzongkhag area (Buthan) and defined a rainfall threshold based on E-D (cumulative rainfall-duration) relationship. They also discovered that 10 days and 30 days antecedent rainfall play an important role in the occurrence of landslides in the investigated area.
– Abraham and co-authors [28] try to define empirical rainfall thresholds for the Idukki area in India, to set the first step to establish a landslide early warning system. Two types of thresholds have been defined: (i) classical I-D (mean rainfall intensity-duration) thresholds, (ii) threshold based on short (1 day) and long duration rainfall (from 3 to 40 days). One of the main outcomes of the paper is the clear importance of antecedent rainfall (30 and 40 days before failure) in the triggering of landslides for the investigated area.
– Using satellite-based rainfall estimates from TMPA 3B42 Real-Time v.7 and information on 184 dated landslides in the period 2001–2019, Monsieurs and co-authors [29] applied the modified antecedent rainfall–susceptibility threshold approach (previously proposed by the same authors [30]) to calculate and validate regional rainfall thresholds in a data-scarce region: the western branch of the East African Rift. The method was here tested and improved by means of newly available regional-scale susceptibility data: a regional model and a continental model. The main methodological novelty is the stratified selection of data linked to the lowest landslide-triggering antecedent rainfall values. A statistical analysis on the effect of outliers in small datasets on the estimation of parameter uncertainties with bootstrapping statistical technique is a valid methodological corollary to this work.
– The contraposition between empirical and physically-based thresholds includes different methods (the first ones are defined using past rainfall and landslide data, the latter integrate stability analyses and hydrogeological modeling) and applications (the first ones are mostly applied at a regional scale, while the second ones are mainly used at a local scale). Bordoni and co-authors [31] present a comparison between thresholds defined with the two methods using landslide and rainfall data collected in the period 2000–2018 in the Oltrepò Pavese, in Northern Italy. They used the CTRL-T tool [17] to define the empirical thresholds and the TRIGRS model [32] to calculate the physically-based thresholds. After validating both

thresholds against an independent dataset, the authors observed that the physically-based thresholds discriminate better than empirical thresholds the landslide triggering and non-triggering rainfall events. This is due mostly to the fact that the adopted physically-based model considers the antecedent soil hydrological conditions, which are known to have a primary role in slope instability.

– Lin and co-authors [33] presented the definition of SWI-D (soil water index-duration) thresholds to define the condition of landslide triggering in Taiwan. In this paper, besides the classical rainfall thresholds, the authors proposed an approach based on the definition of soil water content, calculated by the use of a 3-layers tank model, where each tank represents a soil layer, from ground surface to the bedrock. Results of the work highlighted that the water content of the deeper layer is more relevant in the triggering of large landslides and therefore that their initiation is more related to long rainfall events rather than shorter ones.

– This study proposed by Dikshit and co-authors [34] presents a landslide hazard assessment in a 180 km long road corridor in Bhutan, combining (i) rainfall thresholds based on daily rainfall amount and 30-days antecedent rainfall; (ii) temporal probability analysis of landslide triggering using a Poisson probability model; (iii) landslide susceptibility map developed with the AHP (Analytical Hierarchy Process) method. The study gains relevant knowledge for the strategic infrastructure analyzed, and poses the basis for further developments of the research towards an operational landslide warning system in the area.

– He and co-authors [35] defined four groups of national rainfall thresholds for landslide occurrence in China based on 771 landslide events occurred in the period 1998–2017. In particular, they used the satellite precipitation product produced by the NOAA's (National Oceanic and Atmospheric Administration) Climate Prediction Center Morphing technique (CMORPH) and calculated both rainfall event–duration (E–D) and normalized (by mean annual precipitation) (EMAP–D) rainfall thresholds. Moreover, they defined thresholds for rainy season and non-rainy season, and thresholds for short (<48 h) and long (≥48 h) durations. The main findings retrieved from the results are that: (i) the slope of the thresholds for long durations is larger than that for short durations, and (ii) the thresholds in the non-rainy season are generally lower than those in the rainy season.

– The study proposed by Abraham and co-authors [36] faces the operational difficulties encountered when trying to establish a regional scale I-D threshold in an area monitored by a sparse rain gauge network at daily temporal resolution. The paper investigates the sensitivity of the results to different model configurations adopted in selection of the rain gauges, in defining the rainfall intensity and in dividing the area into smaller sub-zones. After a comparative validation, the authors conclude that in their case of study, selecting the rain gauge based on maximum average intensity performs better than choosing the nearest rain gauge.

– Abraham and co-authors [37] applied in a sub-Himalayan test site in India a state-of-the-art rainfall threshold model called SIGMA [38,39], which is based on statistical anomalies observed in varying time-windows of antecedent rainfall to account for both shallow and deep-seated landslides. The application is interesting because SIGMA was purposely developed for an Italian test site affected by both kinds of landslides and was conceived to be operated using rainfall measurements at daily temporal resolution: this is the first reported attempt to apply it in other geographical climatic settings. Results are encouraging since a quantitative and comparative validation shows that the effectiveness of the model is higher than other approaches based on I-D and E-D thresholds.

– Given that a recent validation of the prototype landslide early warning system in Slovenia highlighted the need to define new reliable rainfall thresholds, Jordanova and co-authors [40] addressed this task taking advantage of a consolidated tool [17] that allows the automated calculation and validation of empirical, frequentist thresholds at different non-exceedance probabilities. Other than new national thresholds (compared with other regional, national, and global thresholds), the authors determined additional thresholds for two different environmental classifications: the first based on three classes of mean annual rainfall and the second based on four lithological units. Through these additional analyses, two findings are observed: (i) the area with the highest mean annual rainfall has the highest thresholds, which indicates the landscape adaptation to higher average rainfall; (ii) the areas characterized by rocks prone to weathering have the lowest thresholds signal that the lithology influences landslide occurrence conditions.

The contributions collected in the special issue "Rainfall Thresholds and Other Approaches for Landslide Prediction and Early Warning" provide interesting understanding and new perspectives on the very wide topic of rainfall thresholds for landslide prediction. The different aspects covered in this special issue demonstrate that the definition, validation, and application of rainfall thresholds are complex tasks which require detailed data and rigorous methods. The research contributions deal with both empirical and physically-based approaches, use different sources for landslide and rainfall data and are implemented in different study areas with diverse temporal scales.

Some important aspects were not covered in this special issue: the topic of landslide initiation is still open for new ideas and innovations. However, we think that this collection of manuscripts could be useful for the community involved in operational prediction of landslides and landslide early warning at all levels [41], from the academic sector to the practitioners and end-users.

Author Contributions: Conceptualization, investigation, writing—original draft preparation, and writing—review and editing, S.S., S.L.G., A.R. All authors have read and agreed to the published version of the manuscript.

Funding: This research received no external funding.

Institutional Review Board Statement: Not applicable.

Informed Consent Statement: Not applicable.

Conflicts of Interest: The authors declare no conflict of interest.

References

1. Pereira, S.; Zêzere, J.L.; Quaresma, I.; Santos, P.P.; Santos, M. Mortality patterns of hydro-geomorphologic disasters. *Risk Anal.* **2015**, *36*, 1188–1210. [CrossRef] [PubMed]
2. Froude, M.J.; Petley, D.N. Global fatal landslide occurrence from 2004 to 2016. *Nat. Hazards Earth Syst. Sci.* **2018**, *18*, 2161–2181. [CrossRef]
3. Sidle, R.C.; Ochiai, H. Landslides: Processes, prediction, and land use. *Water Resour. Monogr.* **2006**. [CrossRef]
4. Segoni, S.; Piciullo, L.; Gariano, S.L. Preface: Landslide early warning systems: Monitoring systems, rainfall thresholds, warning models, performance evaluation and risk perception. *Nat. Hazards Earth Syst. Sci.* **2018**, *18*, 3179–3186. [CrossRef]
5. Calvello, M. Early warning strategies to cope with landslide risk. *Rivista Italiana di Geotecnica* **2017**, *2*, 63–69. [CrossRef]
6. Intrieri, E.; Gigli, G.; Casagli, N.; Nadim, F. Brief communication "Landslide Early Warning System: Toolbox and general concepts". *Nat. Hazards Earth Syst. Sci.* **2013**, *13*, 85–90. [CrossRef]
7. Intrieri, E.; Gigli, G.; Mugnai, F.; Fanti, R.; Casagli, N. Design and implementation of a landslide early warning system. *Eng. Geol.* **2012**, *147–148*, 124–136. [CrossRef]
8. Guzzetti, F.; Peruccacci, S.; Rossi, M.; Stark, C.P. The rainfall intensity-duration control of shallow landslides and debris flows: An update. *Landslides* **2008**, *5*, 3–17. [CrossRef]
9. Segoni, S.; Piciullo, L.; Gariano, S.L. A review of the recent literature on rainfall thresholds for landslide occurrence. *Landslides* **2018**, *15*, 1483–1501. [CrossRef]
10. Piciullo, L.; Calvello, M.; Cepeda, J.M. Territorial early warning systems for rainfall-induced landslides. *Earth-Sci. Rev.* **2018**, *179*, 228–247. [CrossRef]
11. Guzzetti, F.; Gariano, S.L.; Peruccacci, S.; Brunetti, M.T.; Marchesini, I.; Rossi, M.; Melillo, M. Geographical landslide early warning systems. *Earth-Sci. Rev.* **2020**, *200*, 102973. [CrossRef]
12. Gariano, S.L.; Melillo, M.; Peruccacci, S.; Brunetti, M.T. How much does the rainfall temporal resolution affect rainfall thresholds for landslide triggering? *Nat. Hazards* **2020**, *100*, 655–670. [CrossRef]
13. Peres, D.J.; Cancelliere, A.; Greco, R.; Bogaard, T.A. Influence of uncertain identification of triggering rainfall on the assessment of landslide early warning thresholds. *Nat. Hazards Earth Syst. Sci.* **2018**, *18*, 633–646. [CrossRef]
14. Rosi, A.; Segoni, S.; Canavesi, V.; Monni, A.; Gallucci, A.; Casagli, N. Definition of 3D rainfall thresholds to increase operative landslide early warning system performances. *Landslides* **2020**, 1–13. [CrossRef]

15. Battistini, A.; Rosi, A.; Segoni, S.; Lagomarsino, D.; Catani, F.; Casagli, N. Validation of landslide hazard models using a semantic engine on online news. *Appl. Geogr.* **2017**, *82*, 59–65. [CrossRef]
16. Segoni, S.; Rossi, G.; Rosi, A.; Catani, F. Landslides triggered by rainfall: A semi-automated procedure to define consistent intensity–duration thresholds. *Comput. Geosci.* **2014**, *63*, 123–131. [CrossRef]
17. Melillo, M.; Brunetti, M.T.; Perruccacci, S.; Gariano, S.L.; Roccati, A.; Guzzetti, F. A tool for the automatic calculation of rainfall thresholds for landslide occurrence. *Environ. Model. Softw.* **2018**, *105*, 230–243. [CrossRef]
18. Leonarduzzi, E.; Molnar, P. Deriving rainfall thresholds for landsliding at the regional scale: Daily and hourly resolutions, normalisation, and antecedent rainfall. *Nat. Hazards Earth Syst. Sci.* **2020**, *20*, 2905–2919. [CrossRef]
19. Abraham, M.T.; Satyam, N.; Bulzinetti, M.A.; Pradhan, B.; Pham, B.T.; Segoni, S. Using Field-Based Monitoring to Enhance the Performance of Rainfall Thresholds for Landslide Warning. *Water* **2020**, *12*, 3453. [CrossRef]
20. Segoni, S.; Tofani, V.; Rosi, A.; Catani, F.; Casagli, N. Combination of rainfall thresholds and susceptibility maps for dynamic landslide hazard assessment at regional scale. *Front. Earth Sci.* **2018**, *6*, 85. [CrossRef]
21. Palau, R.M.; Hürlimann, M.; Berenguer, M.; Sempere-Torres, D. Influence of the mapping unit for regional landslide early warning systems: Comparison between pixels and polygons in Catalonia (NE Spain). *Landslides* **2020**, *17*, 2067–2083. [CrossRef]
22. Wicki, A.; Lehmann, P.; Hauck, C.; Seneviratne, S.I.; Waldner, P.; Stähli, M. Assessing the potential of soil moisture measurements for regional landslide early warning. *Landslides* **2020**, 1–16. [CrossRef]
23. Bogaard, T.; Greco, R. Invited perspectives: Hydrological perspectives on precipitation intensity–duration thresholds for landslide initiation: Proposing hydro-meteorological thresholds. *Nat. Hazards Earth Syst. Sci.* **2018**, *18*, 31–39. [CrossRef]
24. Marino, P.; Peres, D.J.; Cancelliere, A.; Greco, R.; Bogaard, T.A. Soil moisture information can improve shallow landslide forecasting using the hydrometeorological threshold approach. *Landslides* **2020**. [CrossRef]
25. Rosi, A.; Canavesi, V.; Segoni, S.; Dias Nery, T.; Catani, F.; Casagli, N. Landslides in the mountain region of Rio de Janeiro: A proposal for the semi-automated definition of multiple rainfall thresholds. *Geosciences* **2019**, *9*, 203. [CrossRef]
26. Yang, L.; Wei, Y.; Wang, W.; Zhu, S. Numerical Runout Modeling Analysis of the Loess Landslide at Yining, Xinjiang, China. *Water* **2019**, *11*, 1324. [CrossRef]
27. Dikshit, A.; Sarkar, R.; Pradhan, B.; Acharya, S.; Dorji, K. Estimating Rainfall Thresholds for Landslide Occurrence in the Bhutan Himalayas. *Water* **2019**, *11*, 1616. [CrossRef]
28. Abraham, M.T.; Pothuraju, D.; Satyam, N. Rainfall Thresholds for Prediction of Landslides in Idukki, India: An Empirical Approach. *Water* **2019**, *11*, 2113. [CrossRef]
29. Monsieurs, E.; Dewitte, O.; Depicker, A.; Demoulin, A. Towards a Transferable Antecedent Rainfall—Susceptibility Threshold Approach for Landsliding. *Water* **2019**, *11*, 2202. [CrossRef]
30. Monsieurs, E.; Dewitte, O.; Demoulin, A. A susceptibility-based rainfall threshold approach for landslide occurrence. *Nat. Hazards Earth Syst. Sci.* **2019**, *19*, 775–789. [CrossRef]
31. Bordoni, M.; Corradini, B.; Lucchelli, L.; Valentino, R.; Bittelli, M.; Vivaldi, V.; Meisina, C. Empirical and Physically Based Thresholds for the Occurrence of Shallow Landslides in a Prone Area of Northern Italian Apennines. *Water* **2019**, *11*, 2653. [CrossRef]
32. Baum, R.L.; Savage, W.Z.; Godt, J.W. *TRIGRS—A Fortran Program for Transient Rainfall Infiltration and Grid-Based Regional Slope-Stability Analysis, Version 2.0*; Open-File Report, 1159; US Geological Survey: Reston, VA, USA, 2008.
33. Lin, G.W.; Kuo, H.L.; Chen, C.W.; Wei, L.W.; Zhang, J.M. Using a Tank Model to Determine Hydro-Meteorological Thresholds for Large-Scale Landslides in Taiwan. *Water* **2020**, *12*, 253. [CrossRef]
34. Dikshit, A.; Sarkar, R.; Pradhan, B.; Jena, R.; Drukpa, D.; Alamri, A.M. Temporal Probability Assessment and Its Use in Landslide Susceptibility Mapping for Eastern Bhutan. *Water* **2020**, *12*, 267. [CrossRef]
35. He, S.; Wang, J.; Liu, S. Rainfall Event–Duration Thresholds for Landslide Occurrences in China. *Water* **2020**, *12*, 494. [CrossRef]
36. Abraham, M.T.; Satyam, N.; Rosi, A.; Pradhan, B.; Segoni, S. The Selection of Rain Gauges and Rainfall Parameters in Estimating Intensity-Duration Thresholds for Landslide Occurrence: Case Study from Wayanad (India). *Water* **2020**, *12*, 1000. [CrossRef]
37. Abraham, M.T.; Satyam, N.; Kushal, S.; Rosi, A.; Pradhan, B.; Segoni, S. Rainfall Threshold Estimation and Landslide Forecasting for Kalimpong, India Using SIGMA Model. *Water* **2020**, *12*, 1195. [CrossRef]
38. Martelloni, G.; Segoni, S.; Fanti, R.; Catani, F. Rainfall thresholds for the forecasting of landslide occurrence at regional scale. *Landslides* **2012**, *9*, 485–495. [CrossRef]
39. Segoni, S.; Rosi, A.; Fanti, R.; Gallucci, A.; Monni, A.; Casagli, N. A regional-scale landslide warning system based on 20 years of operational experience. *Water* **2018**, *10*, 1297. [CrossRef]
40. Jordanova, G.; Gariano, S.L.; Melillo, M.; Peruccacci, S.; Brunetti, M.T.; Jemec Auflič, M. Determination of Empirical Rainfall Thresholds for Shallow Landslides in Slovenia Using an Automatic Tool. *Water* **2020**, *12*, 1449. [CrossRef]
41. Calvello, M.; Devoli, G.; Freeborough, K.; Gariano, S.L.; Guzzetti, F.; Kirschbaum, D.; Nakaya, H.; Robbins, J.; Stähli, M. LandAware: A new international network on Landslide Early Warning Systems. *Landslides* **2020**, *17*, 2699–2702. [CrossRef]

Technical Note

Estimating Rainfall Thresholds for Landslide Occurrence in the Bhutan Himalayas

Abhirup Dikshit [1,2], Raju Sarkar [1,3,*], Biswajeet Pradhan [2,4], Saroj Acharya [1] and Kelzang Dorji [1]

1 Center for Disaster Risk Reduction and Community Development Studies, Royal University of Bhutan, Rinchending 21101, Bhutan
2 Centre for Advanced Modelling and Geospatial Information Systems (CAMGIS), University of Technology Sydney, NSW 2007, Australia
3 Department of Civil Engineering, Delhi Technological University, Bawana Road, Delhi 110042, India
4 Department of Energy and Mineral Resources Engineering, Choongmu-gwan, Sejong University, 209, Neungdongro Gwangjin-gu, Seoul 05006, Korea
* Correspondence: rajusarkar.cst@rub.edu.bt

Received: 22 June 2019; Accepted: 31 July 2019; Published: 5 August 2019

Abstract: Consistently over the years, particularly during monsoon seasons, landslides and related geohazards in Bhutan are causing enormous damage to human lives, property, and road networks. The determination of thresholds for rainfall triggered landslides is one of the most effective methods to develop an early warning system. Such thresholds are determined using a variety of rainfall parameters and have been successfully calculated for various regions of the world at different scales. Such thresholds can be used to forecast landslide events which could help in issuing an alert to civic authorities. A comprehensive study on the determination of rainfall thresholds characterizing landslide events for Bhutan is lacking. This paper focuses on defining event rainfall–duration thresholds for Chukha Dzongkhag, situated in south-west Bhutan. The study area is chosen due to the increase in frequency of landslides during monsoon along Phuentsholing-Thimphu highway, which passes through it and this highway is a major trade route of the country with the rest of the world. The present threshold method revolves around the use of a power law equation to determine event rainfall–duration thresholds. The thresholds have been established using available rainfall and landslide data for 2004–2014. The calculated threshold relationship is fitted to the lower boundary of the rainfall conditions leading to landslides and plotted in logarithmic coordinates. The results show that a rainfall event of 24 h with a cumulated rainfall of 53 mm can cause landslides. Later on, the outcome of antecedent rainfall varying from 3–30 days was also analysed to understand its effect on landslide incidences based on cumulative event rainfall. It is also observed that a minimum 10-day antecedent rainfall of 88 mm and a 20-day antecedent rainfall of 142 mm is required for landslide occurrence in the area. The thresholds presented can be improved with the availability of hourly rainfall data and the addition of more landslide data. These can also be used as an early warning system especially along the Phuentsholing–Thimphu Highway to prevent any disruptions of trade.

Keywords: rainfall thresholds; Bhutan; shallow landslides

---

1. Introduction

Rainfall or earthquake triggered landslides are common in some parts of the world, causing loss of human life and property [1]. A global dataset of landslide disasters [2] showed that three-quarters of all landslide events occurred in the Himalayan arc between 2004 and 2016. Bhutan is one of the highly susceptible landslide zones in the Himalayan region [3]. The majority of landslides in Bhutan are initiated due to heavy monsoon precipitation and aggravated due to the increase in human activities. The increasing number of landslide events in Bhutan can be attributed to complex geological conditions, steep slopes, climate change, type of soil, and tectonic activity. Landslides in Bhutan mostly occur during the monsoons during which the torrential rainfall leads to several flash floods and landslide triggering, cutting off many parts of south Bhutan from the rest.

The relationship between the amount of rainfall associated with landslide occurrences is generally studied using either an empirical or physical based approach [4–7]. Physical process models are based on numerical models which study the relationship between rainfall, pore water pressure, soil type, and volumetric water content that can lead to slope instability. Such a study is usually site specific due to variation in soil properties. It is a challenge to extend this approach to large areas, as the extensive data that is required are usually not available, and their use for an early warning system is either experimental or prototype based [8–10]. On the other hand, empirical methods study the landslides that are caused by rainfall events—both massive downpour that triggers instantaneous landslides and the low but continuous antecedent rain that destabilizes the slope and triggers the landslide. Although, there are many factors like rainfall, earthquake, geology, soil type etc. involved for landslide triggering, in the present study, precipitation rates have been considered as this is the primary cause of several changes in soil properties, pressure variations, etc. The minimum quantity of precipitation requisite for landslide occurrences known as thresholds can be determined using empirical models. The limit is defined by lower-bound lines to the precipitation conditions causing landslides and plotted in Cartesian, semi-logarithmic, or logarithmic coordinates [11]. Contingent upon the kind of available rainfall data, empirical thresholds can be summarized as follows: (1) thresholds which combine rainfall data obtained from specific rainfall events [12], (2) thresholds involving antecedent parameters [6], and (3) alternating thresholds, like hydrological thresholds [13]. Therefore, several works can be found depicting rainfall thresholds based on empirical techniques [4,7,14–18]. The present study highlights the importance of antecedent rainfall along with the determination of cumulative event-rainfall–duration thresholds for an operational early warning system. A recent review on rainfall thresholds [19] showed that the thresholds could be used to predict landslide events at various geographical extents and also in a broad spectrum of physical settings [10]. The study also found that, for setting up an early warning system using empirical rainfall thresholds, various factors needs to be taken care of: (i) collection of reliable and large rainfall and landslide datasets, (ii) selecting threshold parameters depending on landslide characteristics and precipitation data, (iii) defining the events and using an objective and standardized methodology, (iv) validation of the thresholds determined. The recent development is on defining objective thresholds using semi-automated algorithms [20–23].

In the context of Bhutan, it is difficult to study the relationship using the physical based approach as the data required is not available. Therefore, this study is lying on an empirically based approach, which defines thresholds using available records of daily rainfall and landslide data in the period 2004–2014. The available scientific literature for landslides in Bhutan is very sparse with no operating landslide early warning system [24,25]. Only one study has been carried out to determine site-specific rainfall thresholds [25], which determined thresholds using an algorithm-based approach, CTRL-T (Calculation of Thresholds for Rainfall-Induced Landslides Tool). However, more extensive work can be found for the Indian Himalayas [11,26–28] and Nepal Himalayas [29]. The study also discusses the significance of antecedent rainfall and the possibility to use the output as a decision tool for landslide determination. The effect of antecedent rainfall on landslides was analysed for various time durations: 3, 7, 10, 20, and 30 days. Further, the analysis has been validated using a statistical indicator, i.e., threat score for 2015 rainfall data. The paper is presented in five sections; the first two sections are listed as 'Study Area' and 'Rainfall and Landslide data'. The third section is listed as 'Methodology' where explanation on threshold calculation and the uncertainty associated with it are described. The fourth heading is marked as 'Results and Discussions' summarizing threshold determination along with understanding the importance of antecedent rainfall. The findings of the present study are summarized in section five 'Conclusions'.

2. Study Area

Bhutan is situated in the eastern section of the Himalaya with an area of 38,394 km^2 (Figure 1). The nation is enclosed with the Tibetan Plateau in the north and Indian states Arunachal Pradesh in the east, Bengal and Assam in the south, and the Darjeeling–Sikkim Himalayas in the west. Bhutan is divided into 20 Dzongkhags and has elevations varying from 150 m to 7570 m [30]. The elevation of the present study area Chukha Dzongkhag ranges from 1000 m to 4200 m. Most of the people in the region depend on agriculture and livestock for their livelihood.

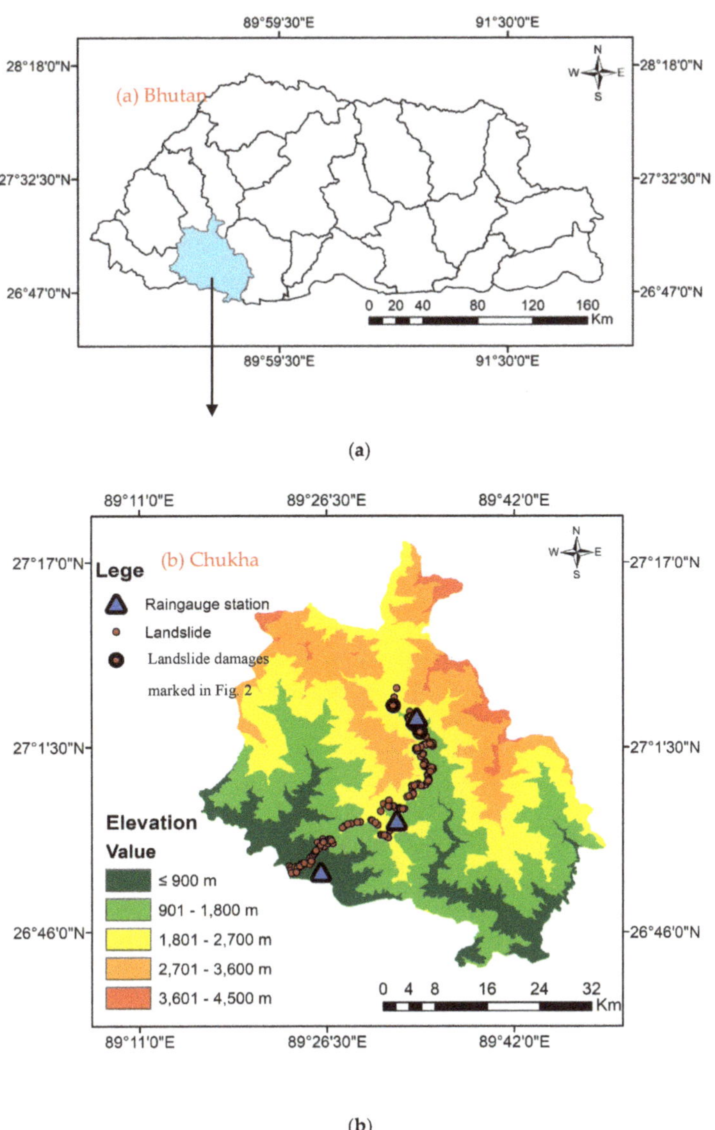

Figure 1. Location of (**a**) Bhutan; and (**b**) elevation map of Chukha Dzongkhag depicting rain gauge and landslide locations.

Geologically, the Chukha region belongs to the Lesser Himalayan formation which includes a wide variety of sedimentary and low-grade metamorphic rocks. The study area is mainly comprised of tectonically active metasedimentary rocks such as phyllite, schist, quartzite, and limestone. The northern part belongs to the Higher Himalayan crystalline rocks comprising mainly of garnetiferous mica-shist, quartzite, and gneiss. The soils in the region are mostly moderate to high weathered and are comprised of weaker fracture phyllites, which make the soil texture very fine, and slopes are very unstable [25]. Based on the practical experiments, the soil can be classified as poorly graded sand (SP) mixed with pebbles. The geology of the region described in [31]. The tectonic setting of the region is very similar to Nepal and Indian Himalayas [25]. The morphology of the unstable slopes leading to failures are complex and managed by several factors including lithology and rock type.

The region receives yearly rainfall of 4000–6000 mm with heavy bursts reaching up to 800 mm/day [18]. More than three-quarters of the annual rainfall occurs during the monsoon period of June to September [25]. Figure 1b depicts the landslide occurrences between 2004–2014 and the rain gauge coordinates used for the analysis. The monsoon rainfall leads to erosion of rocks causing widespread slope instability and mass movements. The increase in anthropogenic activities has escalated deforestation leading to slope instability. The landslide typology in the region can be described as rock fall, rockslide, debris flow, debris slide, and earth slide [25]. The major effects due to this monsoonal rainfall of this region are roadblocks. As mentioned earlier, the Phuentsholing-Thimphu highway (also known as Asian Highway) situated in Chukha Dzongkhag is one of the major road connections of Bhutan, which is key to the trade, passes through these landslides affected areas. The landslide events along the highway pose serious logistic problems as it affects the economy of the region. Figure 2a,b illustrates the damage caused by landslides along the highway.

(a)

(b)

Figure 2. Landslide damages in Chukha Dzongkhag along Phuentsholing-Thimphu Highway after (**a**) 2016 and (**b**) 2017 monsoon.

3. Rainfall and Landslide Data

The daily rainfall data were collected from the three rain gauges from Malbase, Gedu, and Chukha. These three rain gauges are maintained by the National Center for Hydrology and Meteorology, Royal Government of Bhutan. The presence of a properly distributed rain gauge network is essential for determining rainfall thresholds. Figure 3 depicts the cumulative rainfall distribution for the study period (2004–2014). The rainfall data collected for 2004–2014 shows that about 78% of the annual rainfall falls during the monsoon (June–September) with pre-monsoon (March–May) and post-monsoon (October–December) contributing 18% and 4%, respectively.

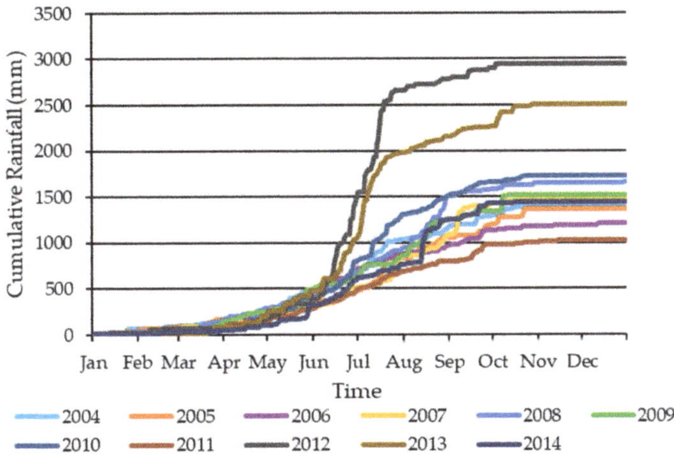

Figure 3. Variation of cumulative daily rainfall with time (2004–2014).

The landslide data were provided by project Dantak of Border Road Organization (BRO), of the Govt. of India. The landslide locations are confined along the Phuentsholing–Thimphu highway, most prominent trade route, which connects the capital city Thimphu with neighbouring countries by road. The total number of landslides during the study period was 248, out of which 63, 81, and 105 landslides occurred in Malbase, Gedu, and Chukha regions respectively. The landslides in the region are mostly shallow with depths ranging from few decimetres to metres [25]. The catalogue of landslides included coordinates and date of failures. Figure 4 depicts the year-wise landslide occurrences.

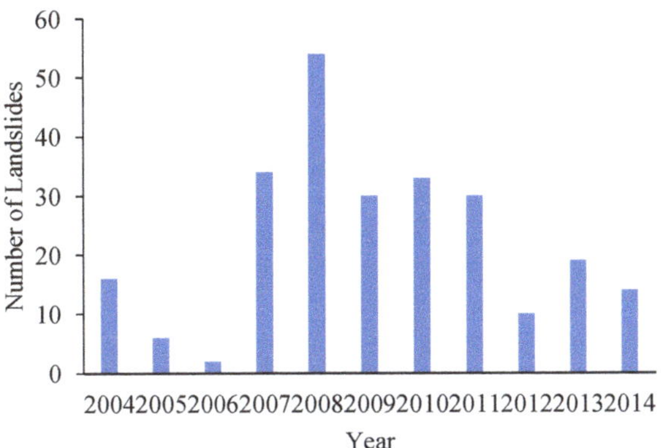

Figure 4. Year-wise landslide occurrences in Chukha Dzongkhag (2004–2014).

4. Method

This section explains the procedure to determine thresholds. The determination of thresholds using empirical methods can be divided into two primary steps. The first step is the collection of rainfall and landslide data, and the next step is to apply empirical models to determine the thresholds. An event rainfall is determined by the number of consecutive days of rainfall before the landslide incident. This helps in calculating the number of days of rainfall before landslide and the total rainfall. Thereafter, the rainfall events leading to landslides are plotted in a log(E) vs log(D) graph and the distribution fitted to the power law equation. The equation of the threshold is $E = (\alpha \pm \Delta\alpha).\ D^{(\gamma \pm \Delta\gamma)}$, where E is cumulated event rainfall (mm), D is duration (h), α is intercept, and γ is the slope of threshold curve [4]. The uncertainties $\Delta\alpha$ and $\Delta\gamma$ are determined using a bootstrap nonparametric statistical technique [32]. This uncertainties measure the variation of the threshold around a central tendency line, which depends on multiple factors, but primarily on the number and the distribution of the empirical data points representing different rainfall conditions that have resulted in landslides [33]. The distribution of rainfall conditions which have resulted to triggering of landslides is fitted the power law equation in a log-log graph. The thresholds were determined using the methodology proposed by [4] and further modified by [32] for various exceedance probabilities ranging from 1% to 50%.

The equation involves the discrete and continuous maximum likelihood function estimation to fit the data in agreement with the equation. There are two assumptions involved with the use of power law equation. (1) With the increase in the cumulated rainfall, there is a nonlinear increase in the probability of landslides. It asserts that the possibility of landslide decreases when threshold reduces and vice versa. (2) With the increase in rainfall duration, the occurrences of slope failure reduce [27].

The study of antecedent rainfall for landslide incidences is important as it may lead to an increase in soil moisture content leading to slope instability. The impact of antecedent rainfall should be a site-specific study and may not always hold good for other regions with similar geological and rainfall conditions [34]. The variation in soil moisture content across an area is difficult to accurately determine as it depends on various factors like the variation in soil type, depth, climatic variation, etc. [7]. Various authors have used different periods to determine the correlation between antecedent rainfall and number of days for landslide triggering. [35–38] examined for 3, 4, 18, and 180 days respectively. [14] used 7, 10, and 15 days, whereas [39] assessed 2, 5, 15, and 25 days based on a trial and error basis. In this study, we considered 3, 7, 10, 15, 20, and 30 days.

5. Results and discussion

In this study, an event rainfall–duration threshold has been determined using available rainfall and landslide data [32,40]. The definition of rainfall and landslide event to determine any kind of threshold is very critical [18]. In the present study, to determine the thresholds, the landslides after the initial failure were not considered, i.e., if, on a particular day, five landslides occur, they will be recognized as one landslide event. This approach is similar to other works [40] and reduces the number of rainfall events with landslides. After that, landslide events which initiated due to very low rainfall values (lower than 25th percentile) were discarded to determine more accurate thresholds [25] as such landslide incidences may not be solely initiated by rainfall. Thus, the number of landslides reduced to 51 landslide events. With the method explained in the previous section, the threshold came out to be $E = (5.68 \pm 1.80).\ D^{(0.70 \pm 0.04)}$. Figure 5a depicts the 51 rainfall conditions, which led to landslides, threshold at 5% ($T_{5,B}$) exceedance probability and the associated uncertainty (Figure 5b,c). The relative uncertainties for the γ parameter are less than the acceptable value of 10% [32]. However, the uncertainty for α is slightly higher at 31.7%, which can be attributed to limited number of empirical data used to determine thresholds [25]. The power-law function and maximum likelihood estimations are used to get the best fitted curve by utilization the present set of data. The lower boundary points are determined using regression analysis and a standard deviation is used to understand the distribution of normally distributed data from the mean value [27]. The graph is drawn on a logarithmic scale. [25] also defined ED thresholds for Chukha region using a semi-automatic empirical approach and defined threshold as E = (7.3 ± 2.0). $D^{(0.71 \pm 0.06)}$ using 43 landslide data points for 2004–2014. The thresholds defined for Kalimpong region, India by [11] depicted ED relationship as E = 3.52. $D^{0.41}$.

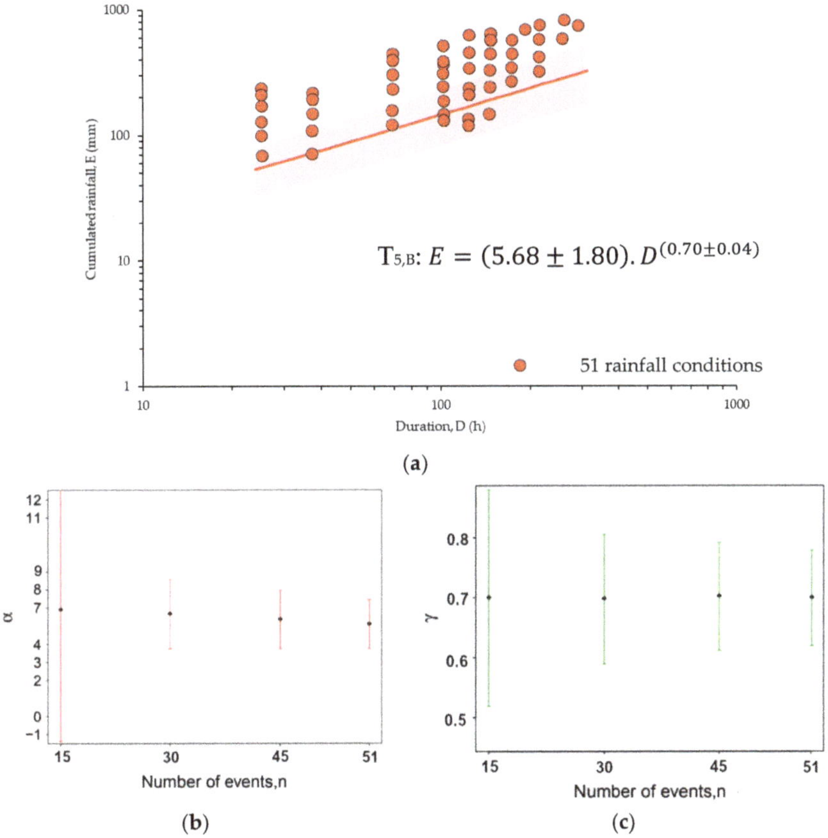

Figure 5. (**a**) Event rainfall–duration (ED) threshold for 51 rainfall conditions in log–log coordinates (**b**) Variation in parameter α, and associated uncertainty Δα, as a function of the number of events. (**c**) Variation in parameter γ, and associated uncertainty Δγ, as a function of the number of events.

To evaluate the importance of antecedent rainfall the rainfall events during the monsoon of year 2012 were analyzed, a total of 50 rainfall events occurred which led to eight landslide events. As there were more than 1000 rainfall events during 2004–2014, the landslide events of 2012 were selected to understand the significance of antecedent rainfall. Figure 6 depicts the daily (R_{day}) and 30-day ($R_{30\text{-}day}$) antecedent precipitation values for 2012. The daily rainfall values were comparatively higher on the day of landslide occurrence. Adopting the calculated threshold for triggering landslide, the number of daily rainfall events exceeding the reference period is high which depicts that antecedent precipitation plays an important role in landslide initiation [41]. The landslide events during this period were characterized by an $R_{30\text{-}day}$ value of 350 mm. Considering the $R_{30\text{-}day}$ value of 350 mm, the number of rainfall events exceeding it is 14. This analysis resulted in fewer data points, which depict the significance of antecedent rainfall, and thus its effect on the distribution of soil moisture during the initiation of triggering rainfall [41].

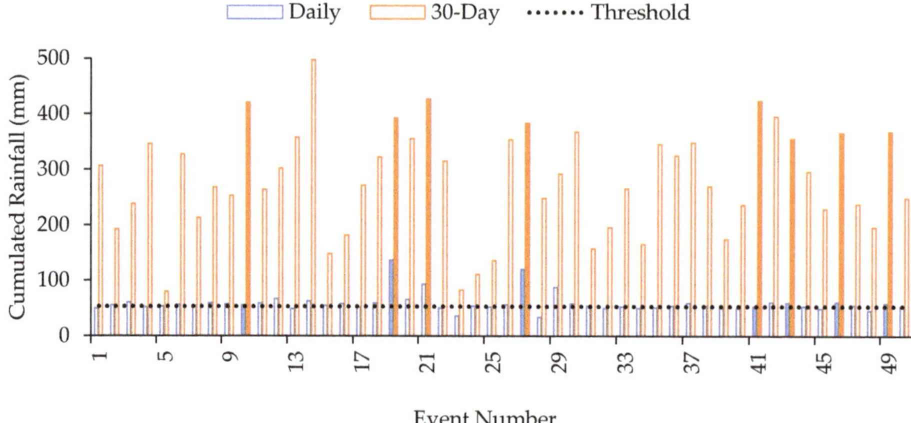

Figure 6. Cumulated rainfall during 2012 monsoon period.

To determine the number of antecedent rainfall days significant for landslide triggering, a trial and error approach considering various days was used [11,27]. The daily rainfall on the day of rainfall for landslide events was plotted against antecedent rainfall for various time durations of 3, 7, 10, 20, and 30 days (Figure 7). The diagonal line divides the graph to differentiate the scattering bias of daily rainfall (ordinate) and antecedent rainfall (abscissa). The diagonal depicts that the daily precipitation data on the day of failure and the antecedent precipitation prior to failure are same [27]. Figure 7 shows graphs of daily rainfall corresponding to antecedent rainfall of various periods. As observed in Figure 7, the majority of the landslide events are biased towards antecedent rainfall as compared to daily rainfall. In the case of 3-day antecedent rainfall, 19.6% of the landslides are biased towards daily rainfall and the remaining 80.4% (41 of the 51 landslides) are biased towards 3-day antecedent rainfall prior to failure. Similarly, for 7-day antecedent conditions, the biasness of daily rainfall towards landslide initiation decreases to 13.7%. In the remaining cases (10-day, 20-day, and 30-day) the effect decreases to 1.9%. Such a comparison between the daily and antecedent rainfall data would give more clarity with the availability of hourly data. The observation of plots for similar bias antecedent conditions shows that a greater number of points in the case of 10 days is concentrated along the abscissa in comparison to other cases where a scattering of points along the abscissa is prevalent [27]. Therefore, it can be concluded that an antecedent rainfall of 88.35 mm for a minimum of 10 days provides the best correlation for triggering of landslides in the region.

The use of thresholds for an operational early warning system can be justified by validating it with an independent dataset which is lacking in various studies conducted, as mentioned in the review article by [19]. The present study validates the determined thresholds using the rainfall data of 2015 by determining the threat score [3,42]. Threat score (TS) is defined as the number of true positive cases (TP) divided by the summation of true positive (TP), false negative (FN), and false positive (FP) cases [43].

$$TS = \frac{TP}{TP + FN + FP} \tag{1}$$

During this period, eight landslide events occurred due to a total of 46 rainfall events. As determined earlier, a $R_{30\text{-day}}$ precipitation value of 350 mm could be used as a pre-filter, and it was found that only 11 of the 46 rainfall events exceeded the value. Thereafter, the biasness of the landslide events with respect to daily and antecedent rainfall was carried out and the results show that only one case was slightly biased towards the daily rainfall whereas the rest were biased towards antecedent rainfall. This analysis shows that a $R_{30\text{-day}}$ value could be used as a pre-filter for determining thresholds and antecedent rainfall plays a significant part for landslide initiation in the region. Finally, the threat score for 2015 landslide event was calculated using the ED threshold value of 53.3 mm. The results determined were TP equal to six, FN and FP were two and three respectively, and so TS was found to be 0.54. The result shows that the rainfall thresholds can be used the first step and eventually the threshold effectiveness will be improved in time when additional data will be collected, as shown in other long-term projects [12]. However, when using as an early warning system the effect of daily as well as antecedent rainfall needs to be considered.

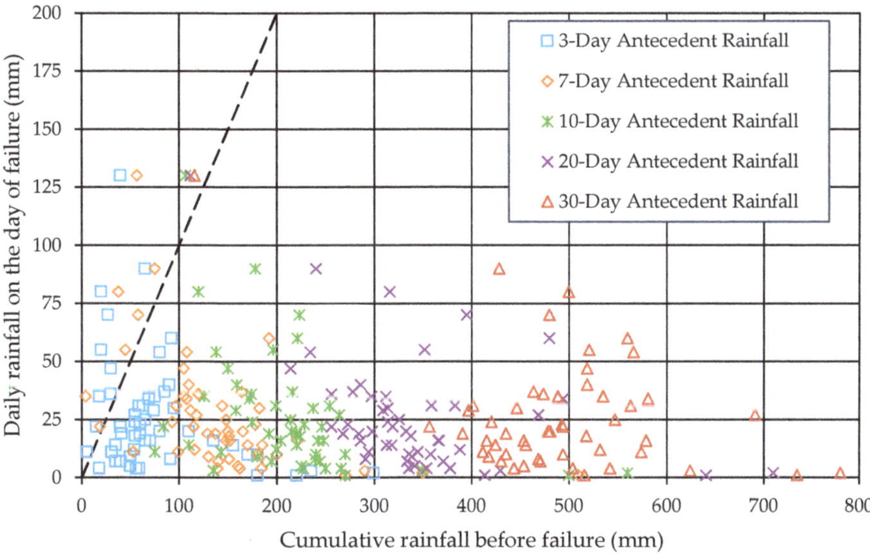

Figure 7. Relation between antecedent rainfall before failure (3, 7, 10, 20, and 30 days) and daily rainfall for landslide occurrences.

6. Conclusions

Rainfall-induced landslides are one of the major destructive natural disasters in Bhutan. However, minimal study has been conducted to develop a landslide early warning system, either regionally or locally. This paper attempts to determine rainfall thresholds in terms of event rainfall–rainfall duration and antecedent rainfall for landslide incidences using daily rainfall data for Chukha region located in the south-western part of Bhutan Himalayas. The analysis was carried out from the available rainfall and landslide occurrences between 2004 to 2014. The majority of the landslide occurrences are along the Phuentsholing–Thimphu highway, which is an important road connection for the country as it connected the capital with India and used for trade purposes. The thresholds determined for the region depicts a minimum event rainfall of 55 mm for short duration of 24 h can trigger landslides. Thereafter, the significance of antecedent rainfall was carried out using the rainfall events of 2012 monsoon. The analysis revealed the importance of antecedent rainfall and therefore it was necessary to determine the antecedent time window necessary for landslide occurrence. The biasness of the rainfall events which resulted in landslides revealed that a 10-day antecedent rainfall would provide the best correlation for landslide occurrences in the study region. Further, it can be stated that $R_{30\text{-day}}$ value of 350 mm could be used as a pre-filter before the use of ED thresholds. The derived rainfall thresholds can be improved with the availability of hourly rainfall data to use for an effective warning system.

Author Contributions: A.D. and R.S. carried out the analysis and wrote the article, B.P. provided technical assistance and contributed in writing the article, S.A. carried out the GIS work and K.D. provided the data for carrying out analysis.

Funding: The research was funded from BRACE project (NERC/GCRF NE/P016219/1) granted to Raju Sarkar.

Acknowledgments: The authors are thankful to National Center for Hydrology and Meteorology, Royal Government of Bhutan for providing rainfall data and the Border Roads Organization (Project DANTAK), Government of India for providing landslide data. Authors are also thankful to staff of College of Science and Technology, Royal University of Bhutan, who helped directly or indirectly while carrying out the present study. The authors acknowledge the two anonymous reviewers for their useful comments and suggestions.

Conflicts of Interest: The authors declare no conflict of interest.

References

1. Petley, D. Global patterns of loss of life from landslides. *Geology* **2010**, *40*, 927–930. [CrossRef]
2. Froude, M.J.; Petley, D.N. Global fatal landslide occurrence from 2004 to 2016. *Nat. Hazards Earth Syst. Sci.* **2018**, *18*, 2161–2181. [CrossRef]
3. Sarkar, R.; Dorji, K. Determination of the Probabilities of Landslide Events—A Case Study of Bhutan. *Hydrology* **2019**, *6*, 52. [CrossRef]
4. Brunetti, M.T.; Peruccacci, S.; Rossi, M.; Luciani, S.; Valigi, D.; Guzzetti, F. Rainfall thresholds for the possible occurrence of landslides in Italy. *Nat. Hazards Earth Syst. Sci.* **2010**, *10*, 447–458. [CrossRef]
5. Baum, R.L.; Savage, W.Z.; Godt, J.W. *TRIGRS—A Fortran Program for Transient Rainfall Infiltration and Grid-Based Regional Slope-Stability Analysis*; Open-File Report; US Geological Survey: Reston, VA, USA, 2002; Volume 02–424, 61p.
6. Baum, R.L.; Savage, W.Z.; Godt, J.W. *TRIGRS—A Fortran Program for Transient Rainfall Infiltration and Grid-Based Regional Slope-Stability Analysis, Version 2.0*; Open-File Report; US Geological Survey: Reston, VA, USA, 2008; Volume 2008–1159, 75p.
7. Guzzetti, F.; Peruccacci, S.; Rossi, M.; Stark, C.P. Rainfall thresholds for the initiation of landslides in central and southern Europe. *Meteorol. Atmos. Phys.* **2007**, *98*, 239–267. [CrossRef]
8. Mercogliano, P.; Segoni, S.; Rossi, G.; Sikorsky, B.; Tofani, V.; Schiano, P.; Catani, F.; Casagli, N. Brief communication "A prototype forecasting chain for rainfall induced shallow landslides". *Nat. Hazards Earth Syst. Sci.* **2013**, *13*, 771–777. [CrossRef]
9. Thiebes, B.; Bell, R.; Glade, T.; Jäger, S.; Mayer, J.; Anderson, M.; Holcombe, L. Integration of a limit-equilibrium model into a landslide early warning system. *Landslides* **2014**, *11*, 859–875. [CrossRef]
10. Huggel, C.; Khabarov, N.; Obersteiner, M.; Ramírez, J.M. Implementation and integrated numerical modeling of a landslide early warning system: A pilot study in Colombia. *Nat. Hazards* **2010**, *52*, 501–518. [CrossRef]
11. Dikshit, A.; Satyam, D.N. Estimation of rainfall thresholds for landslide occurrences in Kalimpong, India. *Innov. Infrastruct. Solut.* **2018**, *3*, 24. [CrossRef]
12. Segoni, S.; Rosi, A.; Fanti, R.; Gallucci, A.; Monni, A.; Casagli, N. A Regional-Scale Landslide Warning System Based on 20 Years of Operational Experience. *Water* **2018**, *10*, 1297. [CrossRef]
13. Reichenbach, P.; Cardinali, M.; De Vita, P.; Guzzetti, F. Regional hydrological thresholds for landslides and floods in the Tiber River Basin (Central Italy). *Environ. Geol.* **1998**, *35*, 146–159. [CrossRef]
14. Aleotti, P. A warning system for rainfall-induced shallow failures. *Eng. Geol.* **2004**, *73*, 247–265. [CrossRef]
15. Rosi, A.; Canavesi, V.; Segoni, S.; Nery, T.D.; Catani, F.; Casagli, N. Landslides in the Mountain Region of Rio de Janeiro: A Proposal for the Semi-Automated Definition of Multiple Rainfall Thresholds. *Geosciences* **2019**, *9*, 203. [CrossRef]
16. Segoni, S.; Rossi, G.; Rosi, A.; Catani, F. Landslides triggered by rainfall: A semiautomated procedure to define consistent intensity-duration thresholds. *Comput. Geosci.* **2014**, *63*, 123–131. [CrossRef]
17. Rosi, A.; Peternel, T.; Jemec-Auflič, M.; Komac, M.; Segoni, S.; Casagli, N. Rainfall thresholds for rainfall-induced landslides in Slovenia. *Landslides* **2016**, *13*, 1571–1577. [CrossRef]
18. Gariano, S.L.; Brunetti, M.; Iovine, G.; Melillo, M.; Peruccacci, S.; Terranova, O.; Vennari, C.; Guzzetti, F. Calibration and validation of rainfall thresholds for shallow landslide forecasting in Sicily, southern Italy. *Geomorphology* **2015**, *228*, 653–665. [CrossRef]
19. Segoni, S.; Piciullo, L.; Gariano, S.L. A review of the recent literature on rainfall thresholds for landslide occurrence. *Landslides* **2018**, *15*, 1483–1501. [CrossRef]

20. Lagomarsino, D.; Segoni, S.; Rosi, A.; Rossi, G.; Battistini, A.; Catani, F.; Casagli, N. Quantitative comparison between two different methodologies to define rainfall thresholds for landslide forecasting. *Nat. Hazards Earth Syst. Sci.* **2015**, *15*, 2413–2423. [CrossRef]
21. Melillo, M.; Brunetti, M.T.; Peruccacci, S.; Gariano, S.L.; Guzzetti, F. An algorithm for the objective reconstruction of rainfall events responsible for landslides. *Landslides* **2015**, *12*, 311–320. [CrossRef]
22. Iadanza, C.; Trigila, A.; Napolitano, F. Identification and characterization of rainfall events responsible for triggering of debris flows and shallow landslides. *J. Hydrol.* **2016**, *541*, 230–245. [CrossRef]
23. Staley, D.M.; Kean, J.W.; Cannon, S.H.; Schmidt, K.M.; Laber, J.L. Objective definition of rainfall intensity–duration thresholds for the initiation of post-fire debris flows in southern California. *Landslides* **2013**, *10*, 547–562. [CrossRef]
24. Piciullo, L.; Calvello, M.; Cepeda, J.M. Territorial early warning systems for rainfall-induced landslides. *Earth Sci. Rev.* **2018**, *179*, 228–247. [CrossRef]
25. Gariano, S.L.; Sarkar, R.; Dikshit, A.; Dorji, K.; Brunetti, M.T.; Peruccacci, S.; Melillo, M. Automatic calculation of rainfall thresholds for landslide occurrence in Chukha Dzongkhag, Bhutan. *Bull. Eng. Geol. Environ.* **2018**. [CrossRef]
26. Sengupta, A.; Gupta, S.; Anbarasu, K. Rainfall thresholds for the initiation of landslide at Lanta Khola in North Sikkim, India. *Nat. Hazards* **2010**, *52*, 31–42. [CrossRef]
27. Kanungo, D.P.; Sharma, S. Rainfall thresholds for prediction of shallow landslides around Chamoli-Joshimath region, Garhwal Himalayas, India. *Landslides* **2014**, *11*, 629–638. [CrossRef]
28. Dikshit, A.; Sarkar, R.; Satyam, N. Probabilistic approach toward Darjeeling Himalayas landslides—A case study. *Cogent Eng.* **2018**, *5*, 1537539. [CrossRef]
29. Gabet, E.J.; Burbank, D.W.; Putkonen, J.K.; Pratt-Sitaula, B.A.; Ojha, T. Rainfall thresholds for landsliding in the Himalayas of Nepal. *Geomorphology* **2004**, *63*, 131–143. [CrossRef]
30. Keunza, K.; Dorji, Y.; Wangda, D. *Landslides in Bhutan. Country Report, Department of Geology and Mines*; Royal Government of Bhutan: Thimpu, Bhutan, 2004; 8p.
31. Gansser, A. *Geology of the Bhutan Himalaya*; Birkhaüser Verlag: Basel, Switzerland, 1983; p. 181.
32. Peruccacci, S.; Brunetti, M.T.; Luciani, S.; Vennari, C.; Guzzetti, F. Lithological and seasonal control of rainfall thresholds for the possible initiation of landslides in central Italy. *Geomorphology* **2012**, *139–140*, 79–90. [CrossRef]
33. Peruccacci, S.; Brunetti, M.T.; Gariano, S.L.; Melillo, M.; Rossi, M.; Guzzetti, F. Rainfall thresholds for possible landslide occurrence in Italy. *Geomorphology* **2017**, *290*, 39–57. [CrossRef]
34. Segoni, S.; Rosi, A.; Lagomarsino, D.; Fanti, R.; Casagli, N. Brief communication: Using averaged soil moisture estimates to improve the performances of a regional-scale landslide early warning system. *Nat. Hazards Earth Syst. Sci.* **2018**, *18*, 807–812. [CrossRef]
35. Kim, S.K.; Hong, W.P.; Kim, Y.M. Prediction of rainfall-triggered landslides in Korea. In *Landslides*, 2nd ed.; Bell, D.H., Ed.; A.A. Balkema: Rotterdam, The Netherlands, 1991; pp. 989–994.
36. Heyerdahl, H.; Harbitz, C.B.; Domaas, U.; Sandersen, F.; Tronstad, K.; Nowacki, F.; Engen, A.; Kjekstad, O.; Dévoli, G.; Buezo, S.G.; et al. Rainfall-induced lahars in volcanic debris in Nicaragua and El Salvador: Practical mitigation. In Proceedings of the International Conference on Fast Slope Movements—Prediction and Prevention for risk Mitigation, IC-FSM2003, Naples, Italy, 11–13 May 2003.
37. Chleborad, A.F. *Preliminary Evaluation of a Precipitation Threshold for Anticipating the Occurrence of Landslides in the Seattle*; Open-File Report; US Geological Survey: Washington, DC, USA, 2003; Volume 03–463.
38. Polemio, M.; Sdao, F. The role of rainfall in the landslide hazard: The case of the Avigliano urban area (Southern Apennines, Italy). *Eng. Geol.* **1999**, *53*, 297–309. [CrossRef]
39. Terlien, M.T.J. The determination of statistical and deterministic hydrological landslide-triggering thresholds. *Environ. Geol.* **1998**, *35*, 124–130. [CrossRef]
40. Peruccacci, S.; Brunetti, M.T. TXT-tool 4.039-1.1: Definition and Use of Empirical Rainfall Thresholds for Possible Landslide Occurrence. In *Landslide Dynamics: ISDR-ICL Landslide Interactive Teaching Tools*; Sassa, K., Tiwari, B., Liu, K.F., McSaveney, M., Strom, A., Setiawan, H., Eds.; Springer: Berlin, Germany, 2018.
41. Pagano, L.; Picarelli, L.; Rianna, G.; Urciuoli, G. A simple numerical procedure for timely prediction of precipitation-induced landslides in unsaturated pyroclastic soils. *Landslides* **2010**, *7*, 273. [CrossRef]

42. Marques, R.; Zêzere, J.; Trigo, R.; Gaspar, J.; Trigo, I. Rainfall patterns and critical values associated with landslides in Povoação County (São Miguel Island, Azores): Relationships with the North Atlantic Oscillation. *Hydrol. Proc.* **2008**, *22*, 478–494. [CrossRef]
43. Corominas, J.; van Westen, C.; Frattini, P.; Cascini, L.; Malet, J.P.; Fotopoulou, S.; Catani, F.; Van Den Eeckhaut, M.; Mavrouli, O.; Agliardi, F.; et al. Recommendations for the quantitative analysis of landslide risk. *Bull. Eng. Geol. Environ.* **2014**, *73*, 209–263. [CrossRef]

Article

Rainfall Thresholds for Prediction of Landslides in Idukki, India: An Empirical Approach

Minu Treesa Abraham *, Deekshith Pothuraju and Neelima Satyam

Discipline of Civil Engineering, Indian Institute of Technology Indore, Bhopal 452020, Madhya Pradesh, India; deekshithpothuraju16@gmail.com (D.P.); neelima.satyam@gmail.com (N.S.)

* Correspondence: minunalloor@gmail.com

Received: 29 August 2019; Accepted: 9 October 2019; Published: 11 October 2019

Abstract: Idukki is a South Indian district in the state of Kerala, which is highly susceptible to landslides. This hilly area which is a hub of a wide variety of flora and fauna, has been suffering from slope stability issues due to heavy rainfall. A well-established landslide early warning system for the region is the need of the hour, considering the recent landslide disasters in 2018 and 2019. This study is an attempt to define a regional scale rainfall threshold for landslide occurrence in Idukki district, as the first step of establishing a landslide early warning system. Using the rainfall and landslide database from 2010 to 2018, an intensity-duration threshold was derived as $I = 0.9D^{-0.16}$ for the Idukki district. The effect of antecedent rainfall conditions in triggering landslide events was explored in detail using cumulative rainfalls of 3 days, 10 days, 20 days, 30 days, and 40 days prior to failure. As the number of days prior to landslide increases, the distribution of landslide events shifts towards antecedent rainfall conditions. The biasness increased from 72.12% to 99.56% when the number of days was increased from 3 to 40. The derived equations can be used along with a rainfall forecasting system for landslide early warning in the study region.

Keywords: rainfall thresholds; landslides; Idukki; early warning system

---

1. Introduction

The state of Kerala (India) experienced the worst disaster in its history in 2018. The disaster affected around 5.4 million people and 433 lives were lost [1]. Several landslides, particularly debris flows, were associated with the event. Among the 14 districts in the state, 13 are part of the Western Ghats and are susceptible to landslide hazards. The scarps of the Western Ghats, which are the steepest parts, are more susceptible to landslides due to heavy rainfall. Attempts have been made by researchers to study the triggering factors of landslides in the Himalayas [2–8] and the Western Ghats [9,10]. However, on a regional scale, establishing rainfall thresholds for the occurrence of landslides in the Western Ghats has not yet been attempted. This paper is an endeavor to define a regional threshold for the Idukki district (Kerala) which is a severe landslide prone zone in the Western Ghats.

A rainfall threshold can be defined using process-based or empirical methods. The process-based approach considers physical and hydrological parameters which can initiate a landslide event. This requires highly sophisticated inputs, as the spatial and temporal distribution of these parameters can only be analyzed through detailed site-specific studies [11]. Owing to the limitations of defining process-based thresholds, this study defines the rainfall conditions that when surpassed, are likely to initiate landslide events in the Idukki district in the Western Ghats. This is an empirical approach which primarily focuses on the occurrence of rainfall and landslide events. Empirical thresholds can be divided into three categories: (1) thresholds which use rainfall data for specific events, (2) thresholds which consider rainfall conditions prior to failure, and (3) others which include hydrological thresholds [12]. In the current research, thresholds in the first two categories are derived for Idukki using historical rainfall and landslide information. A rainfall event is defined by three parameters, viz., rainfall event, rainfall intensity, and rainfall duration. cumulated rainfall is the the total amount rainfall from the beginning of the rainfall event to the occurrence of failure [13]; the term duration indicates the duration of the rainfall event considered or precipitation period [14]; rainfall intensity is the amount of precipitation in a given time, i.e., the rate of precipitation over the period considered [15]. Thus the term rainfall intensity gives an idea about the average rate of rainfall during an event, not the peak intensities. Another important factor which defines the applicability of the threshold is the area considered for the study. Based on the area, thresholds are classified into local, regional, and global. The stability of the slopes depends upon the hydro-meteo-geological parameters of the region and the conditions for the triggering of landslides differ from place to place. Global thresholds give a universal minimum, below which chances of landslide occurrence is nil, without considering any physical factors. Regional thresholds deal with areas of a few to some thousands of square kilometers where climatic, physiographic, and meteorological features are similar. Local thresholds can be applied to single or a small group of landslides in regions of sizes up to the range of hundreds of square kilometers. Regional and local thresholds perform well for the area they were developed for, but they cannot be exported to other areas easily [16]. These thresholds can be used in regional/local warning systems for providing an alert level to the government and public in general.

Empirical thresholds can be classified again based on the rainfall parameters used as intensity-duration (ID) thresholds, total rainfall event-duration (ED) thresholds, and total rainfall event-intensity (EI) thresholds [12]. A general, well-accepted agreement which determines the selection of rainfall parameters is that shallow/rapid landslides are initiated by rainfalls of high intensity and short duration [16] and deep-seated landslides occur when it rains continuously over a long time [17]. This research focuses on the initiation of shallow landslides which cause maximum casualties during the monsoon time in the region and hence thresholds based on intensity-duration plane and antecedent rainfall are defined for Idukki. The objective is to start the preliminary steps towards an effective regional scale warning system for the Idukki district.

2. Study Area

In the state of Kerala, Idukki was the worst-hit district during 2018 disaster, with 143 major landslides in the state government records [1]. As shown in the slope map of Idukki, the geography of the area consists of slopes as steep as 80° (Figure 1) and the elevation ranges up to 2692 m (Figure 3). A significant share of the population of the district had houses in these unstable slopes, which were destroyed in the 2018 landslides irrespective of the building typology [1]. 97% of the major roads in the districts cut through the rugged mountains and hills, which are often blocked due to landslides in the monsoons [18]. Sprawling across an area of 4358 km^2, Idukki supplies 66% of the electric power requirements of Kerala [19]. This district with more than half of the area covered by forests is the second largest one in terms of area in the state.

The Western Ghats can be divided into two segments, north and south, separated by the Gap of Palghat. Deep-seated landslides are reported in the northern segment and the eastern flank while the southern segment mostly experiences shallow landslides [20]. Idukki belongs to the southern part, where regolith thickness ranges from 0.25–5 m [21] and is prone to shallow landslides [1]. Geomorphic classification of the terrain divides the area into four, viz., rugged hills, ridges and valleys, fringe slope, and plateau [22]. Scarps of the Western Ghats consist of frictional soil with less cohesion, thus being stable in dry conditions and losing their strength when the moisture content increases. Plateu regions have a thick layer of top soil, rich in clay content due to their morphology and tropical climate [23]. Geologically, rocks of Wayanad, Charnockite, Khondalite, and Migmatite groups contribute the formation of a part of South Indian Precambrian metamorphic shield [22]. The primary weathering process is hydrolysis in the area, which is due to the high precipitation [24].

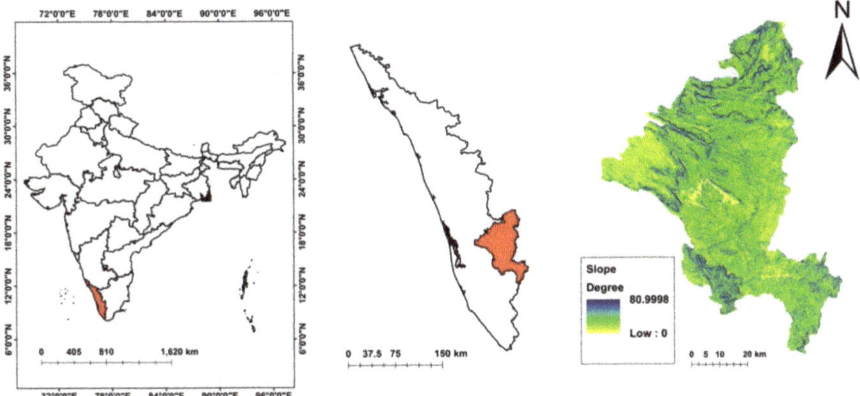

Figure 1. Location and slope map of the Idukki district.

2.1. Triggering Factors

The scarps of the Western Ghats experience an annual rainfall as high as 5000 mm as a result of the southwest monsoon, northeast monsoon, and premonsoon showers [25]. The Western flank of the Western Ghats experiences landslides during the southwest monsoon and the eastern side is affected mainly during the northeast monsoon [20]. Large amounts of high-intensity rainfalls increase the pore water pressure within the soil mass, which eventually decreases the shear strength of the soil. This is considered as the primary triggering factor of landslides in the Indian Himalayas [2,4,26] and the Western Ghats [27]. The fissures in bedrock siphons the excess rainwater to unstable zones in the slopes during the monsoon [28]. Photographs of some landslides which happened during the 2018 monsoon are shown in Figure 2. The population of this region increased rapidly after the 19th century, as the people from the midlands started migrating into the hilly region [29]. The industrially backward district was in a quest for better infrastructure due to an increase in population. As an effect, the land use has changed significantly in a short span of time, which favoured the occurrence of landslides in the region. Large scale hill-toe modifications have been done in the district in recent decades for the purpose of infrastructure development, due to which the hill slopes have become steep, without any lateral support. The terraced slopes, modified for monoculture plantations with no sufficient drainage provisions, aggravated the scenario. Due to the drain blockages, water from the intense rainfall accumulates in the top soil layers, leading to landslides.

In a detailed landslide inventory of Kerala until 2010, prepared by the Geological Survey of India (GSI), 64 major cases were reported in the Idukki district [30]. The landslide typologies vary from creep and subsidence to debris flows and avalanches. Along the major road corridors of the district, earth/debris slides have become common during monsoon period [18]. The sharp turnings and vertical cuts along the roads are highly susceptible to cut-slope failures. Incessant rainfall and the subsequent pore pressure increase adversely affects the steep slopes and results in landslides. To conclude, from the case studies conducted by GSI, a major share of the events in Idukki are of debris flow type triggered by heavy rainfall and are influenced by factors like slope, land use, overburden thickness, and disposition of streams etc. [18,23,31,32].

Figure 2. Damages that happened due to landslides in the Idukki district in 2018. (**a**) Debris slide at Anachal. (**b**) Debris flow at Kallimai. (**c**) Subsidence at Kallarkutty approach road. (**d**) Earth slide at Cheruthoni [22].

2.2. *Database for Analysis*

Building a chronology of landslides based on the historical records is the first stage of any landslide hazard study [33]. A landslide database for the research has been developed taking inputs from the Geological Survey of India [22], newspapers, state government reports [1,34], and from interactions with the people of the area. The dates of initiation of landslides were collected with a weekly accuracy, and the locations were collected with a spatial accuracy of nearest mentioned site from the reports. The database consists of the spatial (Figure 3) and temporal distribution of landslides and the typology.

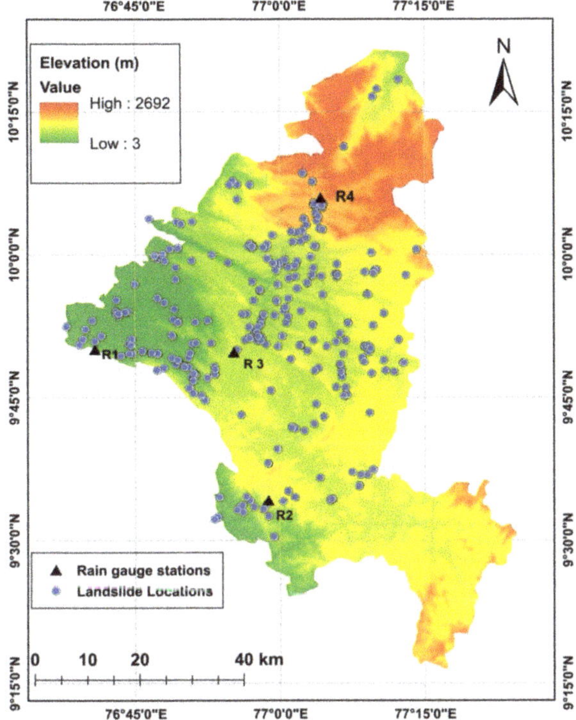

Figure 3. Digital Elevation Model [35] of the Idukki district along with the spatial distribution of landslide locations and rain gauge stations (2010–2018).

The rainfall data of daily resolution from the year 2010 was collected from four rain gauge stations in the Idukki district, maintained by the India Meteorological Department (IMD) [36], for the analysis. The locations of rain gauge stations are given in Table 1. The monthly distribution of effective rainfall in the Idukki district from 2010 to 2018 is shown in the box plot shown in Figure 4.

The distribution of rainfall is not uniform throughout the district. In a long term rainfall analysis conducted by GSI, it was found that the average annual rainfall varies from less than 1000 mm in the northeast parts of Anamudi peak to around 5000 mm near Peermedu [18]. The four rain gauges from which we collected data are located at Thodupuzha, Peermedu, Idukki, and Munnar (Figure 3). The variation of annual rainfall from the four rain gauges and the district average is plotted in Figure 5. The differences in rainfall conditions will lead to over-estimation or under-estimation of the intensity and duration values if we consider the average rainfall. Hence the rainfall event associated with each landslide was found out based on the spatial distribution of the four rain gauges [37].

Table 1. Location of rain gauge stations.

Rain Gauge Number	Place	Location
R1	Thodupuzha	9.83° N, 76.67° E
R2	Peermedu	9.57° N, 76.98° E
R3	Idukki	9.83° N, 76.92° E
R4	Munnar	10.10° N, 77.07° E

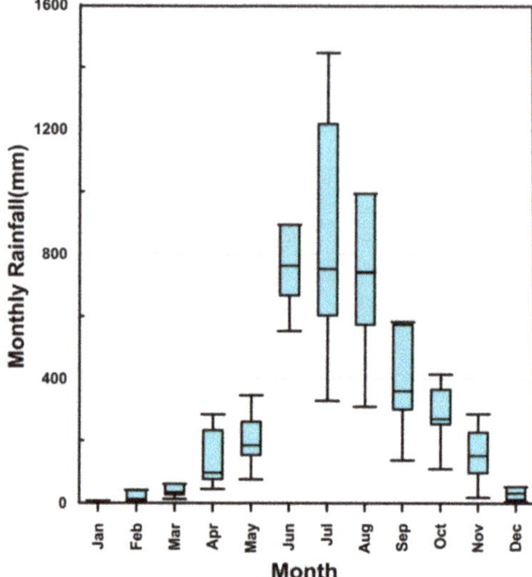

Figure 4. Box and whisker plot with monthly distributions of rainfall in the Idukki district (2010–2018). The bottom and top lines indicate minimum and maximum values respectively and the line inside the box represents the median.

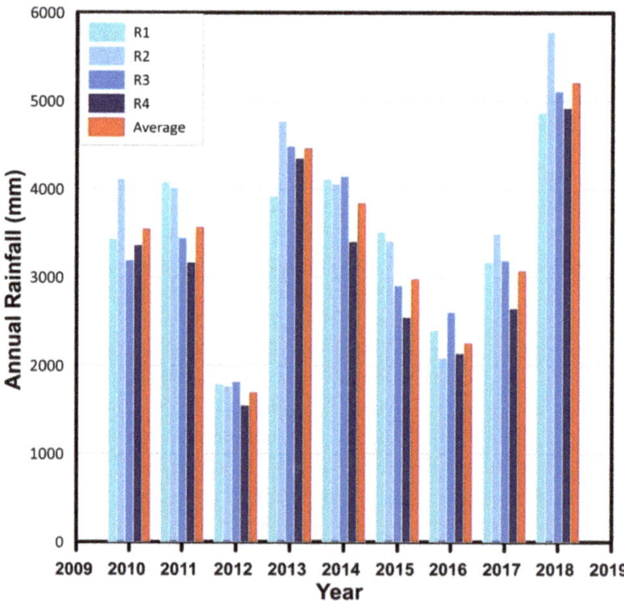

Figure 5. Variation of annual rainfall measured in four rain gauges during the study period.

Identifying a reference rain gauge is a challenging task as explained by many practitioners [14,38], especially when the number of available rain gauges is limited. One of the most common practices is to choose the rain gauge based on its proximity to the landslide location. Hence in this study, the district was divided into four Thiessen polygons, based on the location of rain gauges (Figure 6). P1 Polygon is occupied by a flat and plain territory, P2 is located in the eastern hilly sector of the study area, P3 represents the central hilly sector, and P4 contains the flanks of the mountain and the hills immediately at the foot of the mountainside, thus separating this area with peculiar physiographic characteristics from the other three. As a consequence, splitting up the area in four sectors by means of Thiessen polygons is better than operating considering the entire area as a whole.

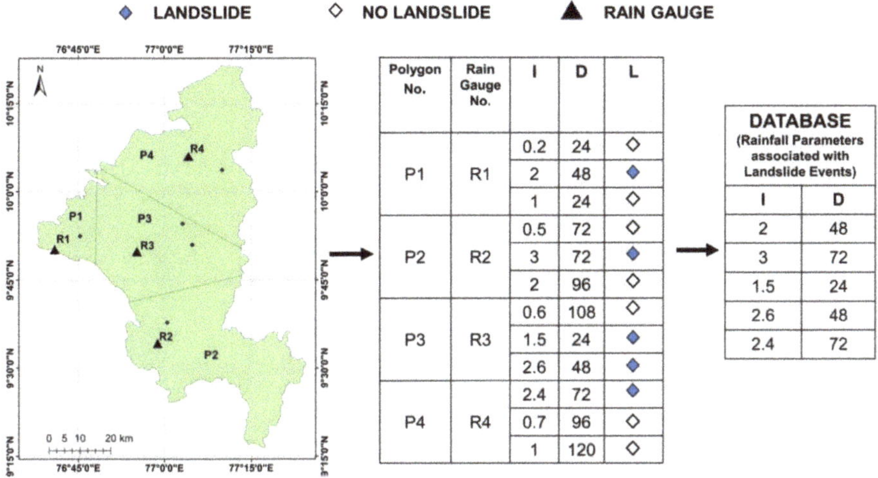

Figure 6. Conceptual sketch showing development of dataset: P1, P2, P3, and P4 represent the four Polygons and R1, R2, R3, and R4 are the reference rain gauges in each polygon. D = Duration of rainfall (hours); I = Intensity of rainfall (mmh^{-1}); L = Occurrence of landslide (Modified after [37]).

Each polygon defines a space, which is closest to the rain gauge in it (reference gauge). Each point inside a polygon is closer to the reference gauge, than the other three rain gauges. The division of polygons and the selection of reference gauge is constrained by spatial distribution only. Each polygon is assumed to be an area of similar rainfall conditions with a reference rain gauge.

The method of developing a dataset is illustrated in Figure 6 [37] using a sample dataset, i.e, the values (I,D) and the locations of landslides are not from the actual dataset, but are arbitrarily chosen for demonstrating the methodology. For all landslide events that happened in Thiessen polygon P1, the readings from R1 are considered. The procedure was same for all landslide events.

The readings corresponding to landslide events, recorded by individual reference rain gauges, were then merged to a single database. The exact number of triggered landslides and sites were not available from the reports and therefore multiple landslides on the same date within the same polygon are considered as a single landslide event. A threshold defines the possibility of occurrence of a minimum of one landslide event in the region. Thus, a total of 225 landslide events are considered in the present analysis, which happened during the time period of 2010–2018.

3. Analysis of Thresholds

The key for the development of any empirical threshold is the definition of rainfall and landslide events and the parameters related to it [39]. The necessity of developing rainfall thresholds and early warning systems for the Idukki district has been emphasized in some of the specific site studies conducted by the GSI [23]. Considering the increased number of casualties which occurred in the study area in the recent past, rainfall thresholds using intensity-duration relationships and antecedent rainfall conditions have been developed in the current research.

3.1. Intensity-Duration Thresholds

A total of 225 landslide events were recorded during the study period (2010–18), which were triggered by rainfall. The hourly intensities of all the rainfall events associated with the occurrence of landslides were calculated and plotted against the duration of events in hours in a logarithmic scale. The distribution of the events is fitted with the power-law distribution using an equation in the form

$$I = \alpha D^{\beta} \tag{1}$$

$$\text{i.e., } \log(\mathrm{I}) = \log(\alpha) + \beta \log(\mathrm{D}) \tag{2}$$

where

I is Intensity of rainfall in mmh^{-1},
D is Duration of rainfall event in hours,
and α and β are empirical parameters,
which is in the form of a straight line y = mx + c.

Use of this power-law equation has two fundamental assumptions. The first one is that with increase in the rainfall intensities, there is a nonlinear increase in the probability of occurrence of landslides. Below the threshold value, the likelihood of initiation of landslide is low, and above the threshold, the probability of occurrence of landslides increases nonlinearly. The second assumption is that the initiation of slides decreases as the duration of rainfall increases [2]. The term 'β' in Equation (1) defines this rate at which the critical intensity declines with the rise in duration. The frequentist approach of defining intensity-duration thresholds is used in this study. Empirical rainfall conditions which triggered landslides were first log-transformed and fitted using Equation (2), which is equivalent to the power-law in Equation (1). Using the Frequentist method, a best fit line for the distribution was obtained as $I = 2.54D^{-0.16}$ (Figure 7) with a coefficient of determination (R^2) of 0.04. The scattering of data results in a lower value of R^2 and hence the uncertainty associated with the fitted line is evaluated with a confidence interval of 95%. Considering the uncertainties, Equation (1) gets modified to

$$\mathrm{I} = (\alpha \pm \Delta\alpha)\ \mathrm{D}^{(\beta\ \pm\ \Delta\beta)} \tag{3}$$

The equation of the best fit line was obtained as $I = 2.54D^{-0.16}$, with a confidence interval of $I = (2.54 \pm 0.65)D^{(-0.16\ \pm\ 0.05)}$.

The approach is based on least square regression and the data is fitted using a power-law. The difference between the value on the best fit line log (I_f) and logarithm of event intensity log (I) for each event is calculated. This difference is termed as 'δI'. Kernel density estimation is used to determine the probability density function of the distribution of 'δI' and the result was fitted using a Gaussian function of the following form [40,41]:

$$f(x) = ae^{-\frac{(x-b)^2}{2c^2}} \tag{4}$$

where a and b are real constants and c is nonzero.

a,b,c, ϵ, R, and thresholds corresponding to various exceedance probabilities can be defined for the region. For a normally distributed random variable, $a = \frac{1}{\sigma\sqrt{2\pi}}$, $b = \mu$ and $c^2 = \sigma^2$ where μ and σ are the mean and standard deviation of the distribution, respectively. Hence the equation becomes

$$f(x) = \frac{1}{\sigma\sqrt{2\pi}}e^{-\frac{(x-\mu)^2}{2\sigma^2}} \tag{5}$$

This equation is used to fit the distribution of 'δI', to determine the rainfall threshold as shown in Figure 8.

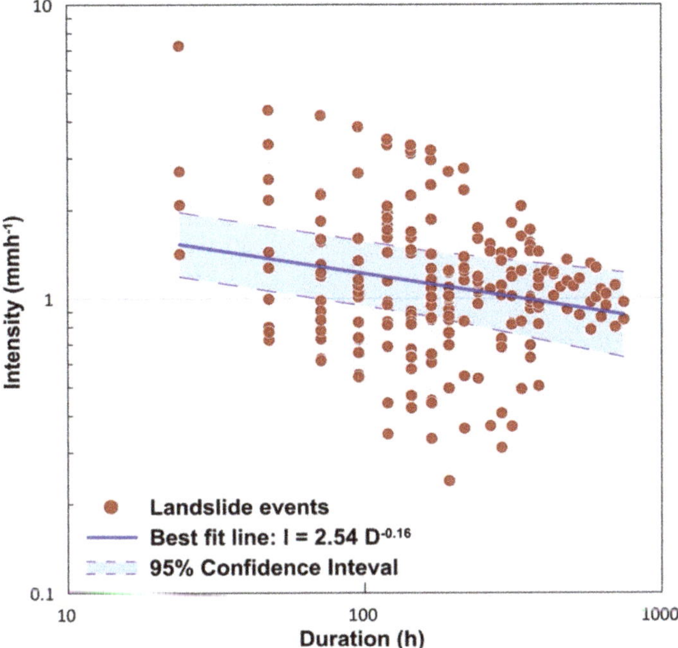

Figure 7. Rainfall Intensity vs. Duration (ID) plot on logarithmic scale for the Idukki district fitted using power-law.

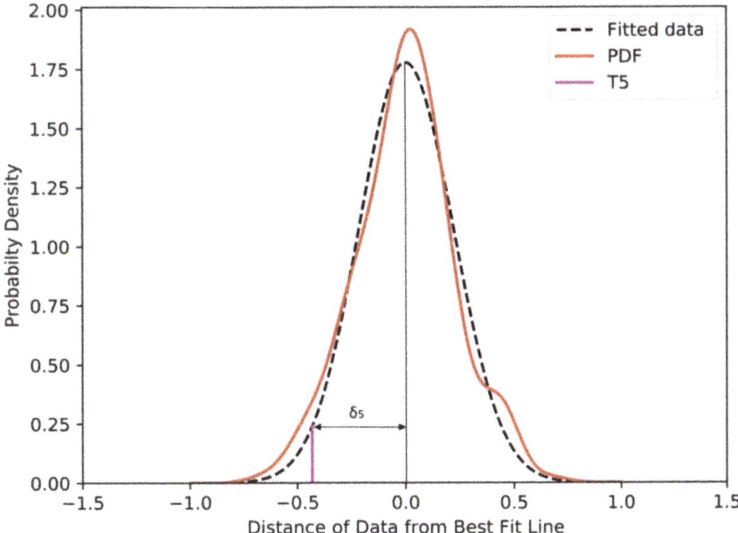

Figure 8. Probability density function of the distribution of δI, fitted using a Gaussian distribution.

The data follows a distribution similar to the standard Gaussian distribution. Hence based on standard Gaussian distribution, a T_5 line was plotted as in Figure 8, with an exceedance probability of 5%. The distance 'δ_5' indicates the deviation of threshold line from the best fit line. This deviation was used to establish the intercept of the threshold line (Figure 9).

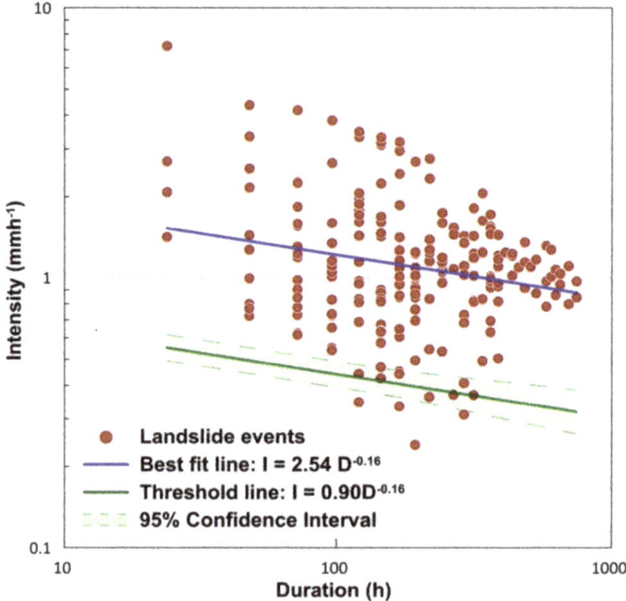

Figure 9. Intensity-duration threshold for the Idukki district on logarithmic scale.

From the threshold line, it can be inferred that for the minimum duration (24 hours), a continuous rainfall of 0.54 mmh^{-1} can trigger landslides. The maximum duration of a rainfall event observed during the study period was 31 days. The obtained results predict that an intensity of 0.3 mmh^{-1} over a period of 31 days can trigger landslides in the region. The confidence interval was obtained as $I = (0.9 \pm 0.1)D^{(-0.16 \pm 0.05)}$. The maximum number of events occurred at a duration of 7 days for which the minimum intensity to initiate a landslide event was found to be 0.4 mmh^{-1}. The lesser value of thresholds for short duration events emphasizes the need for considering antecedent rainfall conditions for defining thresholds. Hence thresholds based on antecedent rainfall conditions are also defined for the area.

3.2. Thresholds Based on Antecedent Rainfall

Intensity-duration thresholds consider only the immediate preceding rainfall event as a triggering factor of landslides. Landslides may occur as result of moisture content variation due to continuous precipitation also, which is difficult to monitor precisely. Thus a simple way is to study the effect of antecedent rainfall and define a threshold based on antecedent rainfall before the landslide event. Studies have been conducted across the globe, considering different antecedent periods ranging from 3 days to 120 days [2,4,42].

The data of 225 landslides over a period of nine years has been used for the analysis. Daily rainfall records at failure are compared with the antecedent rainfall of 3, 10, 20, 30, and 40 days before failure. The graph is plotted with antecedent rainfall (mm) and daily rainfall (mm) in x and y axes respectively. The diagonal line of the plot determines the scattering bias of the data (Figure 10).

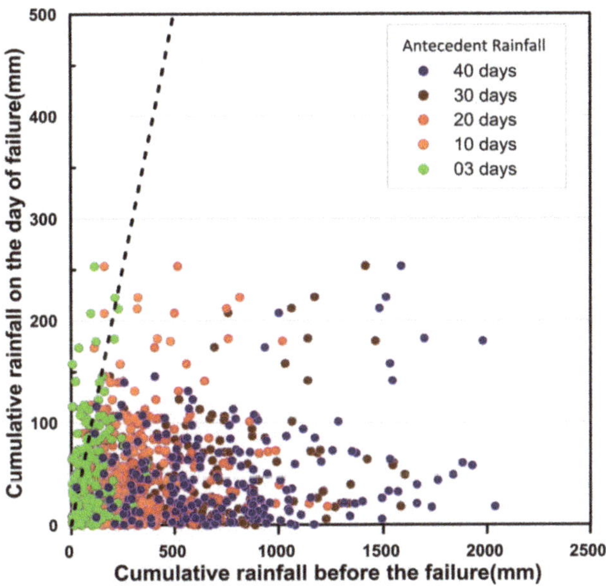

Figure 10. Plot of daily rainfall vs. antecedent rainfall (3, 10, 20, 30, and 40 days).

A significant share of landslide events is biased towards the antecedent rainfall in all cases. Hence a threshold is defined for all individual time durations of antecedent rainfall considered in the study as shown in Figure 11a–e. In the first case, three days' antecedent rainfall was considered, 28% of the total events considered are shifted towards daily rainfall, and the remaining 163 landslides are biased towards three days' antecedent rainfall. For other cases, the biasness ratio to daily rainfall and antecedent rainfall was found to be 11:214 for 10 days', 6:219 for 20 days', 3:222 for 30 days', and 1:224 for 40 days' antecedent rainfall prior to the slide event. It is evident from the analysis that the biasness towards antecedent rainfall, which was 72% in case of 3 days' antecedent rainfall increased to 99.56% when the antecedent rainfall of 40 days was considered as shown in Figure 11f. The study can be refined if the temporal resolution of the rainfall data available is improved.

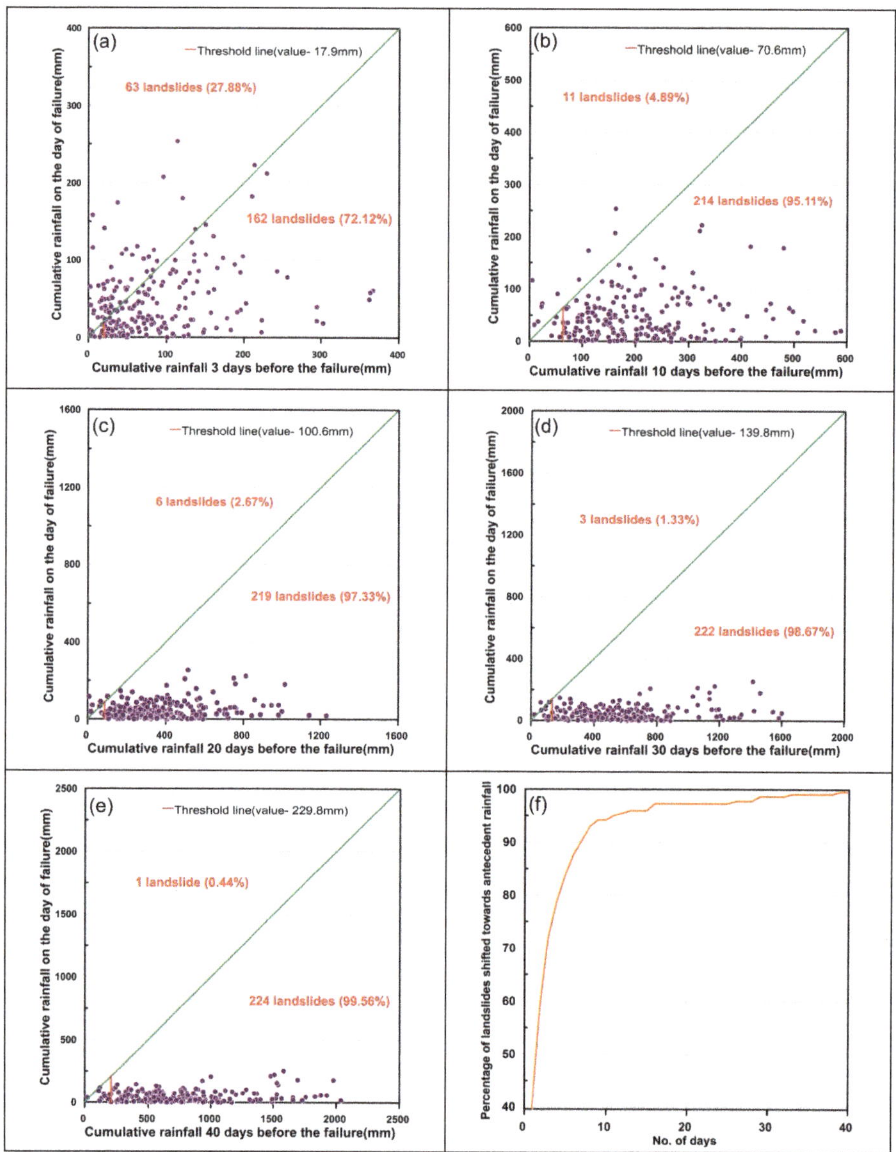

Figure 11. Plot between daily rainfall and antecedent rainfall before failure for (**a**) 3 days, (**b**) 10 days, (**c**) 20 days, (**d**) 30 days, (**e**) 40 days, and (**f**) Biasness towards antecedent rainfall.

4. Discussions

The rainfall thresholds defined in this study establish a minimum cut off below which chances of occurrence of rainfall is very low. Above these thresholds, the probability of occurrence increases exponentially, but still the chance of false alarms cannot be neglected. Even though rainfall is the major triggering factor, other physical factors also influence the stability of slopes. For a powerful Landslide Early Warning System to work effectively, parameters like soil moisture and soil movement/tilt etc. should be incorporated along with the rainfall thresholds. An integrated system with multiple sensors and rain gauges can be installed in the region for this purpose. Similar researches have been carried out for the Darjeeling Himalayas [5] using Micro Electrical Mechanical System (MEMS) tilt sensors. A network of such sensors can effectively transfer the data to the authorities in real time which can be used as an effective warning system. The frequency of available rainfall data is the key factor which determines the accuracy of thresholds. In the current scenario, the temporal resolution of rainfall data available for the region is one day, and for an area of 4358 km^2 only five rain gauge stations (as of 2019) are available. By establishing a network of sensors across the district, the spatial and temporal resolution of rainfall measurements can also be improved.

Several rainfall thresholds have been developed and periodically updated [43] for forecasting landslide events across the globe. Choosing the best method for establishing rainfall thresholds for a particular region requires detailed analysis and a quantitative comparison using statistical attributes [44]. Simple empirical models can also be modified conceptually by incorporating physical or hydrological parameters to improve the prediction power [45,46]. Further studies must be conducted for the area using existing models which are being practiced in different parts of the world [11,47–50] and the best suited rainfall threshold should be integrated with a sensor network and rainfall forecasting system. This research is a humble step towards achieving the goal of establishing an effective Landslide Early Warning System, which can minimize the casualties due to landslide hazards in the Idukki district.

5. Conclusions

This study is an effort to establish thresholds in intensity-duration plane based on antecedent rainfall data for the Idukki district in Kerala State at a regional scale. This is the first of its kind for the region and can be improved on with the availability of short term rainfall data.

The analysis was conducted using a database of 9 years from 2010 to 2018, which included 225 landslide events occurring at different parts of the district, and the principal observations can be summarized as:

- For short duration rainfall events (24 hours), a continuous rainfall intensity of 0.54 mmh^{-1} can trigger landslides. For the maximum observed duration of 31 days, a rainfall intensity as less as 0.3 mmh^{-1} can also trigger landslides. The values of thresholds are too low for a regional scale threshold, and the reason can be the biasness of occurrence of landslides to the antecedent rainfall conditions, other than the immediate preceding event.
- From the analysis of antecedent rainfall conditions, it can be stated that for the Idukki district, an antecedent rainfall of 70.6 mm over a period of 10 days and 229.8 mm over a period of 40 days can trigger a landslide event. Around 99.56% of the events are biased towards the antecedent rainfall conditions when duration of 40 days is considered.
- It is evident from the results that the occurrence of landslide events are more influenced by antecedent rainfall conditions rather than the amount of rainfall on the day of occurrence.
- It is expected that this first attempt will encourage more research for the study area, which is profoundly suffering from the increased number of landslide events in the recent hazards and this will become the first step towards establishing a regional scale warning system for the Idukki district.

Author Contributions: Data curation, M.T.A. and D.P.; Formal analysis, M.T.A. and D.P.; Supervision, N.S.; Writing—original draft, M.T.A.; Writing—review & editing, N.S.

Funding: This research received no external funding.

Acknowledgments: The authors are grateful to Geological Survey of India Kerala State Unit, Kerala State Disaster Management Authority (KSDMA) and District Soil Conservation Office, Idukki for the support they have offered for the research. The authors are also grateful to all the three reviewers for their constructive suggestions.

Conflicts of Interest: The authors declare no conflict of interest.

References

1. *Kerala Post Disaster Needs Assessment Floods and Landslides—August 2018*; Government of Kerala: Thiruvananthapuram, India, 2018; pp. 1–440.
2. Kanungo, D.P.; Sharma, S. Rainfall thresholds for prediction of shallow landslides around Chamoli-Joshimath region, Garhwal Himalayas, India. *Landslides* **2014**, *11*, 629–638. [CrossRef]
3. Dikshit, A.; Satyam, D.N. Estimation of rainfall thresholds for landslide occurrences in Kalimpong, India. *Innov. Infrastruct. Solut.* **2018**, *3*, 24. [CrossRef]
4. Dikshit, A.; Satyam, N. Rainfall Thresholds for the prediction of Landslides using Empirical Methods in Kalimpong, Darjeeling, India. In Proceedings of the JTC1 Workshop on Advances in Landslide Understanding, Barcelona, Spain, 24–26 May 2017.
5. Dikshit, A.; Satyam, N. Probabilistic rainfall thresholds in Chibo, India: Estimation and validation using monitoring system. *J. Mt. Sci.* **2019**, *16*, 870–883. [CrossRef]
6. Dikshit, A.; Sarkar, R.; Satyam, N. Probabilistic approach toward Darjeeling Himalayas landslides—A case study. *Cogent Eng.* **2018**, *5*, 1–11. [CrossRef]
7. Soja, R.; Starkel, L. Extreme rainfalls in Eastern Himalaya and southern slope of Meghalaya Plateau and their geomorphologic impacts. *Geomorphology* **2007**, *84*, 170–180. [CrossRef]
8. Prokop, P.; Walanus, A. Impact of the Darjeeling–Bhutan Himalayan front on rainfall hazard pattern. *Nat. Hazards* **2017**, *89*, 387–404. [CrossRef]
9. Kuriakose, S.L. *Effect of Vegetation on Debris Flow Initiation: Conceptualisation and Parametrisation of a Dynamic Model for Debris Flow Initiation in Tikovil River Basin, Kerala, India, using PC Raster*; International Institute of Geo-Information Science and Earth Observation and Indian Institute of Remote Sensing: Enschede, The Netherlands; Uttarakhand, India, 2006.
10. Kuriakose, S.L.; Luna, B.Q.; Portugues, S.B.; Van Westen, C.J. Modelling the runout of a debris flow of the Western Ghats, Kerala, India. *Assembly* **2009**, *11*, 4276.
11. Martelloni, G.; Segoni, S.; Fanti, R.; Catani, F. Rainfall thresholds for the forecasting of landslide occurrence at regional scale. *Landslides* **2012**, *9*, 485–495. [CrossRef]
12. Guzzetti, F.; Peruccacci, S.; Rossi, M.; Stark, C.P. The rainfall intensity-duration control of shallow landslides and debris flows: An update. *Landslides* **2008**, *5*, 3–17. [CrossRef]
13. Innes, J.L. Debris flows. *Prog. Phys. Geogr.* **1983**, *7*, 469–501. [CrossRef]
14. Aleotti, P. A warning system for rainfall-induced shallow failures. *Eng. Geol.* **2004**, *73*, 247–265. [CrossRef]
15. Caine, N. The rainfall intensity-duration control of shallow landslides and debris flows: An update. *Geogr. Ann. Ser. Phys. Geogr.* **1980**, *62*, 23–27.
16. Crosta, G. Regionalization of rainfall thresholds: An aid to landslide hazard evaluation. *Environ. Geol.* **1998**, *35*, 131–145. [CrossRef]
17. Bonnard, C.; Noverraz, F. Influence of climate change on large landslides: Assessment of long term movements and trends. In Proceedings of the International Conference on Landslides Causes Impact and Countermeasures, Davos, Switzerland, 17–21 June 2001; pp. 121–138.
18. Sajeev, R.; Praveen, K.R. *Landslide Susceptibility Mapping on Macroscale along the Major Road Corridors in Idukki District, Kerala*; Geological Survey of India: Thiruvananthapuram, India, 2014.
19. Idukki District Webpage. Available online: https://idukki.nic.in/ (accessed on 23 August 2019).
20. Kuriakose, S.L. Physically-Based Dynamic Modelling of the Effect of Land Use Changes on Shallow Landslide Initiation in the Western Ghats of Kerala, India. Ph.D. Thesis, University of Utrecht, Enschede, The Netherlands, 2010.
21. Sreekumar, S. Techniques for slope stability analysis: Site specific studies from Idukki district, Kerala. *J. Geol. Soc. India* **2009**, *73*, 813–820. [CrossRef]
22. Sulal, N.L.; Archana, K.G. *Note On Post Disaster Studies For Landslides Occurred in June 2018 At Idukki District, Kerala*; Geological Survey of India: Thiruvananthapuram, India, 2019.
23. Sajeev, R.; Sajinkumar, K.S. *Detailed Site Specific Landslide Study at Govt. College, Munnar, Idukki District, Kerala*; Geological Survey of India: Thiruvananthapuram, India, 2013.
24. Deepthy, R.; Balakrishnan, S. Climatic control on clay mineral formation: Evidence from weathering profiles developed. *J. Earth Syst. Sci.* **2005**, *114*, 545–556. [CrossRef]

25. Kuriakose, S.L.; Sankar, G.; Muraleedharan, C. History of landslide susceptibility and a chorology of landslide-prone areas in the Western Ghats of Kerala, India. *Environ. Geol.* **2009**, *57*, 1553–1568. [CrossRef]
26. Dikshit, A.; Satyam, D.N.; Towhata, I. Early warning system using tilt sensors in Chibo, Kalimpong, Darjeeling Himalayas, India. *Nat. Hazards* **2018**, *94*, 727–741. [CrossRef]
27. Jaiswal, P.; van Westen, C.J. Estimating temporal probability for landslide initiation along transportation routes based on rainfall thresholds. *Geomorphology* **2009**, *112*, 96–105. [CrossRef]
28. Muraleedharan, M.P. *Landslides in Kerala—A Drenched State Phenomena in Regolith*; Center for Earth Sciences: Thiruvananthapuram, India, 1995.
29. Jha, C.S.; Dutt, C.B.S.; Bawa, K.S. Deforestation and land use changes in Western Ghats, India. *Curr. Sci.* **2000**, *79*, 231–238.
30. Muraleedharan, C.; Sajinkumar, K.S. *Landslide Inventory of Kerala*; Geological Survey of India: Thiruvananthapuram, India, 2010.
31. Muraleedharan, C.; Praveen, M.N. *Detailed Site Specific Study Of Landslide Initiation At Kuttikanam, Peermade Taluk, Idukki District, Kerala*; Geological Survey of India: Thiruvananthapuram, India, 2011.
32. Muraleedharan, C. *Landslide Hazard Zonation on Meso-Scale for Munnar, Devikulam Taluk, Idukki District, Kerala*; Geological Survey of India: Thiruvananthapuram, India, 2010.
33. Guzzetti, F.; Reichenbach, P.; Cardinali, M.; Ardizzone, F.; Galli, M. The impact of landslides in the Umbria region, central Italy. *Nat. Hazards Earth Syst. Sci.* **2003**, *3*, 469–486. [CrossRef]
34. *Details of Landslip Damages in Agricultural Lands of Different Panchayats of Idukki District during the Monsoon 2018*; District Soil Conservation Office: Idukki, India, 2018.
35. CartoDEM. Available online: https://bhuvan-app3.nrsc.gov.in/data/download/index.php (accessed on 20 August 2019).
36. India Meteorological Department (IMD). Available online: http://dsp.imdpune.gov.in/ (accessed on 23 July 2019).
37. Berti, M.; Martina, M.L.V.; Franceschini, S.; Pignone, S.; Simoni, A.; Pizziolo, M. Probabilistic rainfall thresholds for landslide occurrence using a Bayesian approach. *J. Geophys. Res. Earth Surf.* **2012**, *117*, 1–20. [CrossRef]
38. AghaKouchak, A.; Nasrollahi, N.; Li, J.; Imam, B.; Sorooshian, S. Geometrical characterization of precipitation patterns. *J. Hydrometeorol.* **2011**, *12*, 274–285. [CrossRef]
39. Segoni, S.; Rossi, G.; Rosi, A.; Catani, F. Landslides triggered by rainfall: A semi-automated procedure to define consistent intensity-duration thresholds. *Comput. Geosci.* **2014**, *63*, 123–131. [CrossRef]
40. Brunetti, M.T.; Peruccacci, S.; Rossi, M.; Luciani, S.; Valigi, D.; Guzzetti, F. Rainfall thresholds for the possible occurrence of landslides in Italy. *Nat. Hazards Earth Syst. Sci.* **2010**, *10*, 447–458. [CrossRef]
41. Silverman, B.W. *Density Estimation for Statistics and Data Analysis*; School of Mathematics University of Bath: Bath, UK, 1986.
42. Pasuto, A.; Silvano, S. Rainfall as a trigger of shallow mass movements. A case study in the Dolomites, Italy. *Environ. Geol.* **1998**, *35*, 184–189. [CrossRef]
43. Rosi, A.; Lagomarsino, D.; Rossi, G.; Segoni, S.; Battistini, A.; Casagli, N. Updating ews rainfall thresholds for the triggering of landslides. *Nat. Hazards* **2015**, *78*, 297–308. [CrossRef]
44. Lagomarsino, D.; Segoni, S.; Rosi, A.; Rossi, G.; Battistini, A.; Catani, F.; Casagli, N. Quantitative comparison between two different methodologies to define rainfall thresholds for landslide forecasting. *Nat. Hazards Earth Syst. Sci.* **2015**, *15*, 2413–2423. [CrossRef]
45. Segoni, S.; Rosi, A.; Lagomarsino, D.; Fanti, R.; Casagli, N. Brief communication: Using averaged soil moisture estimates to improve the performances of a regional-scale landslide early warning system. *Nat. Hazards Earth Syst. Sci.* **2018**, *18*, 807–812. [CrossRef]
46. Segoni, S.; Rosi, A.; Fanti, R.; Gallucci, A.; Monni, A.; Casagli, N. A regional-scale landslide warning system based on 20 years of operational experience. *Water* **2018**, *10*, 1297. [CrossRef]
47. Peruccacci, S.; Brunetti, M.T.; Gariano, S.L.; Melillo, M.; Rossi, M.; Guzzetti, F. Rainfall thresholds for possible landslide occurrence in Italy. *Geomorphology* **2017**, *290*, 39–57. [CrossRef]
48. Melillo, M.; Brunetti, M.T.; Peruccacci, S.; Gariano, S.L.; Guzzetti, F. An Algorithm for the objective reconstruction of rainfall events responsible for landslides. *Landslide* **2015**, *12*, 311–320. [CrossRef]

49. Capparelli, G.; Tiranti, D. Application of the MoniFLaIR early warning system for rainfall-induced landslides in Piedmont region (Italy). *Landslides* **2010**, *7*, 401–410. [CrossRef]
50. Lagomarsino, D.; Segoni, S.; Fanti, R.; Catani, F. Updating and tuning a regional-scale landslide early warning system. *Landslides* **2013**, *10*, 91–97. [CrossRef]

Article

Towards a Transferable Antecedent Rainfall—Susceptibility Threshold Approach for Landsliding

Elise Monsieurs [1,2,3,*], Olivier Dewitte [1], Arthur Depicker [4] and Alain Demoulin [2,3]

1 Department of Earth Sciences, Royal Museum for Central Africa, Leuvensesteenweg 13, Tervuren 3080, Belgium; olivier.dewitte@africamuseum.be

2 Department of Geography, University of Liège, Clos Mercator 3, Liège 4000, Belgium; ademoulin@uliege.be

3 F.R.S.-FNRS, Rue d'Egmont, 5, Brussels 1000, Belgium

4 Division of Geography and Tourism, Department of Earth and Environmental Sciences, KU Leuven, Celestijnenlaan 200E, Heverlee 3001, Belgium; arthur.depicker@kuleuven.be

* Correspondence: elise.monsieurs@africamuseum.be

Received: 24 September 2019; Accepted: 21 October 2019; Published: 23 October 2019

Abstract: Determining rainfall thresholds for landsliding is crucial in landslide hazard evaluation and early warning system development, yet challenging in data-scarce regions. Using freely available satellite rainfall data in a reproducible automated procedure, the bootstrap-based frequentist threshold approach, coupling antecedent rainfall (*AR*) and landslide susceptibility data as proposed by Monsieurs et al., has proved to provide a physically meaningful regional *AR* threshold equation in the western branch of the East African Rift. However, previous studies could only rely on global- and continental-scale rainfall and susceptibility data. Here, we use newly available regional-scale susceptibility data to test the robustness of the method to different data configurations. This leads us to improve the threshold method through using stratified data selection to better exploit the data distribution over the whole range of susceptibility. In addition, we discuss the effect of outliers in small data sets on the estimation of parameter uncertainties and the interest of not using the bootstrap technique in such cases. Thus improved, the method effectiveness shows strongly reduced sensitivity to the used susceptibility data and is satisfyingly validated by new landslide occurrences in the East African Rift, therefore successfully passing first transferability tests.

Keywords: landslide hazard; antecedent rainfall threshold; landslide susceptibility; satellite-derived rainfall; TRMM Multisatellite Precipitation Analysis 3B42 (TMPA); tropical Africa

1. Introduction

Rainfall-triggered landslides pose a severe threat to societies on all continents [1,2]. Rainfall thresholds are therefore essential for characterizing landslide hazard and developing early warning systems [3–5]. Empirical approaches define thresholds on scales ranging from local [6,7] to regional and global [8,9], based on the observed relation between dated landslides and rainfall characteristics such as intensity, accumulation, duration, or antecedent rainfall (*AR*) conditions [10]. However, rainfall is only a proxy for what is regarded as the main trigger of landslides, i.e., the development of high pore-water pressure in the subsurface, constrained by water infiltration [11,12]. Interacting with retention and drainage processes [12], infiltration is a highly complex process affected by a myriad of factors such as soil physical properties (e.g., soil suction head, porosity, hydraulic conductivity) and their variations through the soil column [13–15], presence of cracks [16], hillslope morphology [17], vegetation [11,18,19], antecedent rainfall conditions [15,20–22], and rainfall intensity [23,24]. In contrast to the empirical threshold definitions, process-based approaches incorporate such hydrophysical parameters through a spatially extended infinite-slope stability model [25]. However, the large required data input for well-calibrated process-based thresholds explains their current limitation to mostly applications at the hillslope scale or through numerical simulations [4,21,25–27].

The estimation of empirical rainfall thresholds is also associated with additional sources of uncertainty. Firstly, landslide inventories are inherently biased towards high-impact landslide events and regions that are most accessible, while their accuracy is constrained by the scientific validity of the reporting sources, especially in data-scarce low-capacity environments [1,28–31]. Secondly, rainfall data comprise uncertainties related to the spatial representativeness of rain gauges or biases in satellite-derived estimates [32,33]. Thirdly, the definition of rainfall parameters, with intensity and duration forming the most frequently used parameter couple [3,5], varies strongly across studies [3]. Finally, the latter parameters' interdependence is problematic, obscuring the physical processes associated with the calculated thresholds [34].

In order to account for and characterize threshold uncertainties, a growing number of reproducible statistical techniques have been developed [3]. A weakness of such methods is, however, that they are generally tailored to a specific area and available data sets, which often prevents straightforward transferability to other regions and data sets [35]. Nevertheless, transferability is not only essential for evaluating and comparing landslide hazard over different regions of the world [10,36], but also valuable in the context of the increasing availability of ever higher-resolution data relevant for threshold analysis, such as rainfall estimates from global-scale satellite data [32].

The most influential statistical threshold techniques include the probabilistic approach through Bayesian inference [10,37], the use of receiver operating characteristics (ROC) analysis with different optimization metrics [38,39], and the frequentist approach developed by [40]. The Bayesian and ROC approaches compare conditions that resulted or not in landsliding, the former fundamentally relying on prior and marginal probabilities [37] and the latter attempting to balance the true and false positive rates derived from a confusion matrix [39]. When rainfall data are only available for conditions that triggered landslides, the frequentist method provides a quantitative way to exploit it and calculate thresholds. This method, as developed by [40] for the *(intensity, duration)* parameter couple of rainfall, calculates the least-square fit of the log-transformed data and fits a Gaussian function to the probability density function of its residuals. Next, the Gaussian curve is used to adjust the intercept of the best fit equation to the desired threshold, expressed in terms of exceedance probability [40]. Practically this means that for a threshold at, e.g., 5% exceedance probability level, there is a 0.05 probability that any landslide be triggered by rainfall conditions below the threshold. The quality of the thresholds obtained by this method depends on the size of the data set and its good covering of the whole range of the parameters used [40]. An improvement of the frequentist method lies in the adoption of a bootstrapping statistical technique to assess the parameters' uncertainty in the power-law threshold model [9]. Here, the bootstrap procedure involves many threshold calibrations (e.g., 5000 [9]), each of which based on n randomly sampled data (with replacement) from a data set of size n. The final threshold parameters and associated uncertainties are calculated as the mean and their standard deviations, respectively, of their many estimates. This approach has proved to be transferable over different regions where abundant information on landsliding and rainfall was available [9,41,42].

Recently, this frequentist approach with bootstrapping [32,41,43] has been modified by [35] through coupling a dynamic rainfall variable (*AR*) with a static indicator of the spatially varying predisposing ground conditions (landslide susceptibility, *S*) (further referred to as the *AR-S* approach). The first step in *AR-S* threshold estimation is similar to the frequentist method developed by [40] and [9], calculating the residuals of the least-square fit on the log-transformed data. Then, it proceeds to select $2x\%$ of the data with the largest negative residuals, on which a new least-square regression is applied, providing a threshold at the $x\%$ exceedance probability level. In this way, not only the intercept α but also the slope β of the threshold equation are based on the smallest *AR* data able to cause landsliding. In parallel, the following novel *AR* index, covering a period of 42 days (n) of antecedent rainfall [35], was proposed to account for the non-linear decay of the effect of rainfall on soil wetness.

$$AR_i = \sum_{k=i}^{i-n} e^{\frac{-a\times(t_i-t_k)}{r_k^b}} \times r_k, \tag{1}$$

with t referring to time (here expressed in days), and the characteristic time $\tau = r_k^b/a$ varying non-linearly with daily rainfall r_k [35].

Identifying thresholds for rainfall-triggered landsliding in data-scarce environments is challenging with respect to information on landslide occurrence and hydrophysical parameters, resulting in the quasi-absence of research on this topic in regions such as Central Africa [3] despite high hazard potential [29,44–47]. The *AR-S* approach allowed defining the first regional threshold for landsliding in the western branch of the East African Rift (WEAR) [35]. To the authors' knowledge, it has so far not been used in other regions. Moreover, the cited study relied on limited data available on landslide occurrence, global satellite-based rainfall estimates [48], and continental susceptibility data [45]. There is hence a strong need for testing the method's robustness with other data sets. A regional *S* model is now available for the WEAR [49], which outperforms the global and continental models with regard to prediction accuracy and geomorphological plausibility [49]. Moreover, the landslide event database used in [35] has now grown by about 27%. In this paper, our aim is thus to use these new data and test the transferability of the *AR-S* threshold method as designed by [35] to these new data.

2. Study Area and Data

2.1. Underreported Landslide Events in the WEAR

The WEAR covers an area of ~350,000 km^2 in tropical Africa (Figure 1). This highly populated region is characterized by high rainfall intensities, recent seismicity, deeply weathered substrates, and a complex rift topography [33,47,49–52]. These factors render the area highly susceptible to landsliding [45,49]. Indeed, recent studies incorporating observations from satellite images and fieldwork reveal high landslide activity, with hundreds of recent landslide events over the last 10 years accounting for more than 5000 individual landslides mapped in the area [44,49,50,53]. An event is defined as a single landslide or a group of landslides with a common trigger over the same area [29]. Landslide types vary greatly from dominant shallow slides and earth and debris flows to less frequent rock avalanches and deep-seated rotational slides [44,47,49,50]. Rainfall has been identified as the prevailing trigger of these recent landslides [29,35,47,49,54]. Despite this large number of landslide events, usable information about the day of their occurrence is rare [29]. Actually, for a period of observation similar to the one covered by the satellite rainfall estimates, [35] compiled an unprecedented landslide event inventory that comprises 145 events for which the location and day of occurrence is known. This clearly highlights the fact that landslides are severely underreported in this poorly accessible region [29].

In this research, the calibration data set ('CAL' in Figure 1) used for the threshold approach is the same as in [35]. The 145 dated landslide events are located with an individual accuracy better than 25 km and a mean accuracy of 7.2 km. The inventory covers the period from 2001 to 2018, with most landslides occurring from March to May, after the second rainy season [35] (Figure 2).

The event inventory used for validation ('VAL' in Figure 1) consists of newly acquired information on 39 additional events located with similar individual accuracy and a mean accuracy of 2.3 km. Information about these events was obtained through field observations and newly identified online media sources, explaining their 2002–2019 temporal coverage overlapping with that of the calibration inventory. Their seasonal distribution confirms the temporal pattern previously observed in [35] (Figure 2).

Both the calibration and validation data sets are strongly biased in space and time, due to the severe constraints inherent to the political, economic, and environmental context in the WEAR [29,35,55]. Considering the restricted accessibility of many parts of the study area, fast vegetation regrowth or land rehabilitation, and areas with poor temporal (cloud-free) satellite coverage, uncertainties are too high to claim absence of landsliding in places and periods where no event has been reported. A distinction between landslide types cannot be asserted, owing to inadequate information in media reports which constitute the principal source of both landslide event inventories. Information on the sub-daily timing of the landslide occurrence is rare. Therefore, when the day of occurrence of the event is known, there is still an uncertainty mainly related to the ambiguous reporting of landslides that occurred during the night when the date sequence is not mentioned. For this reason, we consider that a reported landslide may have occurred randomly at any time over a 36-h period centered on the reported day [35].

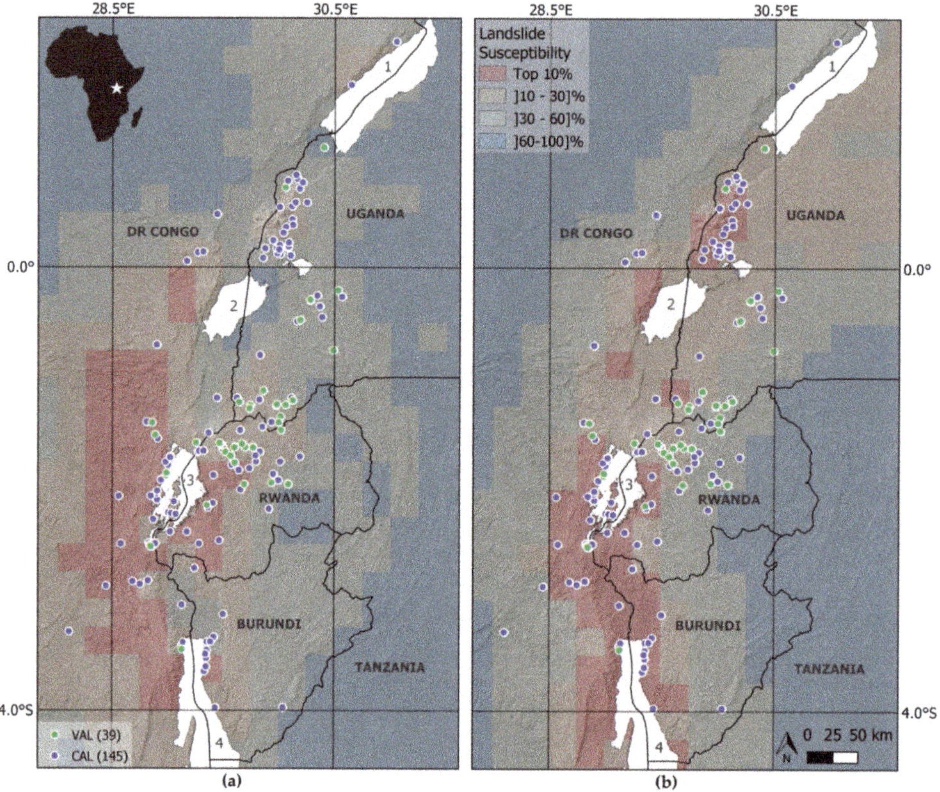

Figure 1. Landslide susceptibility at 0.25° resolution, derived from (**a**) the continental-scale model of [45] and (**b**) the regional-scale model of [49], and distribution of dated and localized landslide events in the western branch of the East African Rift. Landslide events used for calibration (CAL) are shown in blue and those for validation (VAL) in green, with their respective number between brackets. A total of 184 landslide events are distributed over 63 different pixels. 1: Lake Albert; 2: Lake Edward; 3: Lake Kivu; 4: Lake Tanganyika. Background hillshade 3 arc-second SRTM (±90 m).

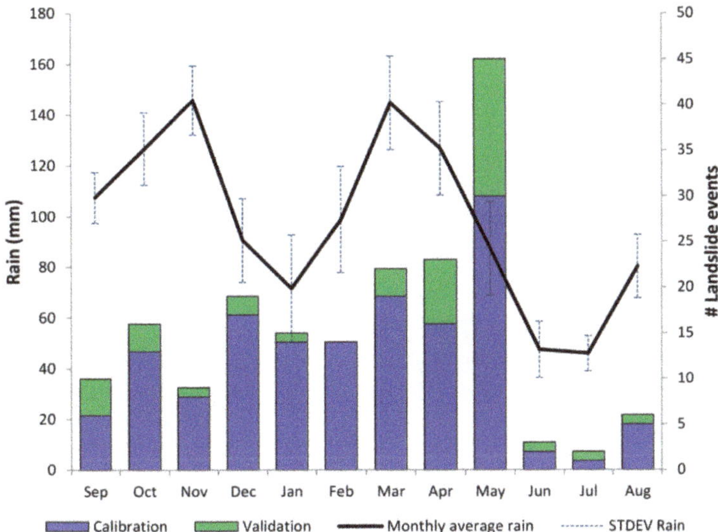

Figure 2. Monthly distribution of 145 (blue) and 39 (green) landslide events in the WEAR used for the threshold calibration and validation respectively, and mean monthly rainfall based on 20 years (2000–2019) of TRMM Multisatellite Precipitation Analysis 3B42 Real-Time, version 7 (TMPA-RT) daily data, downloaded from https://disc.gsfc.nasa.gov/ (last access: 14 April 2019).

2.2. Satellite-Based Rainfall

Due to the scarcity and poor spatial representativeness of rainfall data from ground observations [33,35], we rely on satellite-based rainfall estimates (SRE) from TRMM Multisatellite Precipitation Analysis 3B42 Real-Time, version 7 (hereafter spelled TMPA-RT). TMPA-RT data are freely available with a latency of 8 h over 50° N–50° S, at 0.25° × 0.25° and three-hourly spatiotemporal resolution, covering the period 2000 to present without gaps in space and time [48]. A recent paper on SRE over Central Africa [56] shows that TMPA has overall good skills in detecting and estimating daily rainfall as compared to ARC, CHIRPS, CMORPH, PERSIANN, TAPEER, TARCAT (see [56] for these acronyms' meanings). In their study, the Research Version of TMPA was used, a product that is outperformed by TMPA-RT over the WEAR with regard to rainfall detection skills and absolute errors [33]. Moreover, the short latency of TMPA-RT compared to the two-month latency of the Research Version, is of crucial importance in the context of early warning systems. Despite the relatively good performance of TMPA-RT in Central Africa, the WEAR is a challenging environment for SRE due to its complex topography, high rainfall variability, and presence of large lakes, with a resulting mean bias in daily rainfall estimates in the order of ~40% [33]. TMPA-RT three-hourly rainfall data have been downloaded from NASA Goddard Earth Sciences Data and Information Services Center (https://disc.gsfc.nasa.gov/, last access: 14 April 2019) for the period 2000 to 2019 and accumulated to daily rainfall to maintain consistency with the temporal resolution of the landslide inventory.

2.3. Susceptibility Models

Two *S* models are used in this study. The continental-scale *S* model of [45] is calibrated for all landslides regardless of type at a spatial resolution of 0.0033°. This model is produced through logistic regression using a ~4:1 landslide to no-landslide (L/NL) ratio and is based on four predictor variables: maximum slope (~90 m SRTM [57]), mean local relief (~90 m SRTM [57]), peak ground acceleration [58], and lithology [59]. The landslide inventory used for the model contains more than 18,000 landslides, of which 765 are located in the WEAR. The second *S* model is the regional-scale model of [49] which was calibrated for a representative part of the WEAR and extrapolated within this study for the entire WEAR. This model includes all landslide types and is trained at a 0.0003° resolution using logistic regression with a 1:1 L/NL ratio based on a local inventory and 11 global/continental predictor variables [49]: slope (~30 m SRTM [57]), peak ground acceleration [58], distance to active faults and inactive faults [52,60], lithology [59], land cover [61], distance to drainage network (~30 m SRTM [57]), planar curvature (~30 m SRTM [57]), profile curvature (~30 m SRTM [57]), aspect (~30 m SRTM [57]), and two-day 15 mm rainfall accumulation threshold exceedance [62]. Note that the rainfall predictor was of minor importance in the model and had no significant impact on the susceptibility pattern in the study area [49]. The inventory contained more than 6000 landslides and the regional model shows predictive power and geomorphological plausibility that strongly outperform the continental model [49].

In order to exploit *AR* and *S* data at the same spatial resolution, both *S* models are resampled to the coarser 0.25° resolution of TMPA-RT data while assigning the 95th percentile of the original values to the coarser pixels (Figure 1). The *S* range of the continental-scale model for pixels containing calibration (validation) landslides is 0.38–0.97 (0.31–0.97) with mean and standard deviation equal to 0.80 ± 0.15 (0.79 ± 0.16). The regional-scale *S* data range is 0.10–0.72 (0.12–0.72) with mean and standard deviation equal to 0.57 ± 0.14 (0.49 ± 0.15). The difference in the data range between the two *S* models mainly results from their different sampling strategies (L/NL). Furthermore, *S* values are scaled for different geographical extents, with the continental-scale *S* model comprising areas that are not representative for the WEAR.

3. Problem Statement

We applied the *AR-S* threshold method according to [35] at the 5% and 10% exceedance probability levels, using the same calibration landslide data set, the same TMPA-RT-based *AR* data, but the new regional-scale *S* data of [49]. We obtained the following general *AR-S* relation and threshold equations:

$$AR = (\alpha \pm \Delta\alpha) \times S^{(\beta \pm \Delta\beta)} = (38.8 \pm 1.6) \times S^{-0.06 \pm 0.06} \; \left(R^2 = 0.00\right) \tag{2}$$

$$AR\ (5\%) = (13.1 \pm 1.7) \times S^{0.24 \pm 0.16} \; \left(R^2 = 0.05\right) \tag{3}$$

$$AR\ (10\%) = (17.2 \pm 1.7) \times S^{0.22 \pm 0.16} \; \left(R^2 = 0.03\right). \tag{4}$$

Contrary to [35], the close to zero determination coefficients R^2 (averaged from 5000 bootstrap iterations) associated with the two calculated thresholds show no dependence of threshold *AR* values on *S* (Equations (3) and (4)). The meaningless character of these threshold estimates is further confirmed by the positive slope of the regression lines suggesting counter evidence that higher rainfall would be needed to trigger landslides in more susceptible areas (Figure 3). Analysis of the individual bootstrap iterations likewise uncovers a major issue lying in the estimation of parameter β, which is significant in only ~1 of 2 iterations, with relative uncertainties $\Delta\beta/\beta$ of 0.7 on average, much larger than the generally accepted 10% level [9].

Such poor thresholding cannot be ascribed to low-quality *S* data, the regional-scale data of [49] having been shown more accurate than the continental-scale *S* data of [45]. The reason for very weak and unrealistic positive correlation between *AR* and *S* has thus to be found elsewhere, most certainly in some hidden deficiency of the *AR-S* threshold method of [35]. We suggest and test hereafter that the problem arises from the way the data subset used in the threshold calibration is defined in the frequentist-based approach, based on the selection of the most negative residuals of the general fit. Indeed, in the case of the relatively small data set available in the WEAR and the unequal spread of the data across the *S* range, the frequentist method's assumption that the data set is large and well-spread [40] is not satisfied. In particular, using the regional *S* data, the distribution of the data points within the *AR-S* space is such that the 10% and 20% subsets sample (i.e., 2*x*%) comprise almost no data in the domain of low *S*, due also to the quasi horizontality of the general fit that forces the location of the most negative residuals in the high-*S* region (Figure S1). This means that a large number of the bootstrap iterations are based on data belonging exclusively to a narrow range of high *S* values, biasing the threshold *AR*-*S* relation and degrading the method's robustness. In any case, this failed test of the method highlights the need for improving it in order to overcome limitations imposed by heterogeneously distributed and relatively small-sized data sets. It also points to the possible role of the bootstrap procedure and calls for a critical evaluation of its use in such contexts. We thus propose two major methodological modifications of the *AR*-*S* approach in the next sections.

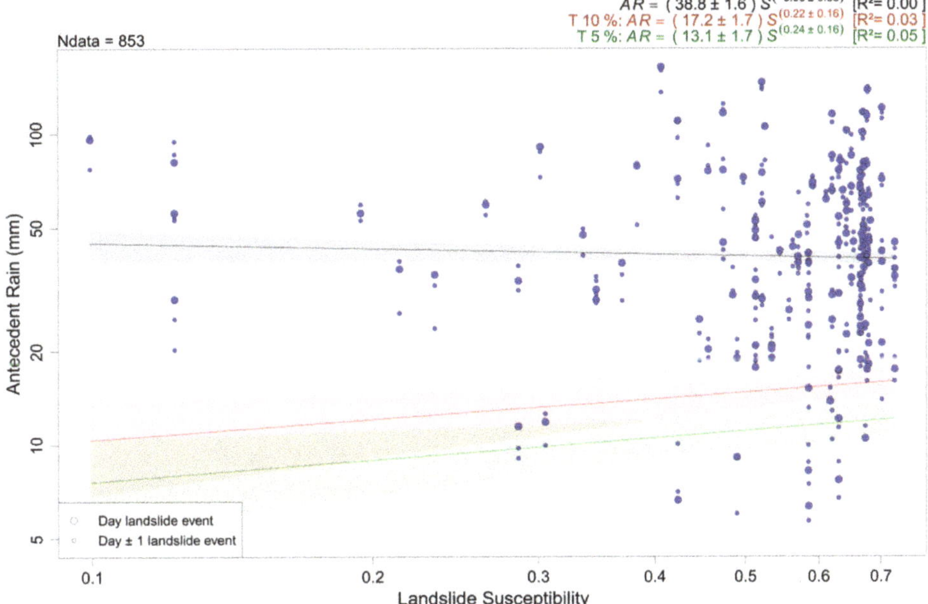

Figure 3. Log–log plot of antecedent rain (mm) vs. landslide susceptibility (regional-scale [49]) for the landslide events on the reported day and the days prior and after that date (with the point size relative to their associated weights, i.e., 0.67 and 0.17 respectively). Thresholds are obtained through the adoption of the *AR*-*S* method proposed by [35]. The black line is the regression curve obtained from the whole data set; the green and red curves are the *AR* thresholds at 5% and 10% exceedance probability levels respectively, along with their uncertainties shown as shaded areas. Ndata is the number of data in the expanded calibration set.

4. Improving the Data Distribution over the *S* Range of the Data Used for Threshold Calculation

4.1. Rationale

While sticking to the choice of [35] of defining $x\%$ thresholds from $2x\%$ subsets of data, we first propose a major modification of the *AR*-*S* method aimed at optimizing the use of the information available over the entire *S* range. Fundamentally, the data that are now considered for inclusion in the calibration subsets rely no longer on residuals of an often non-significant fit over the whole data set but rather on minimum *AR* values. The best possible distribution of the latter is obtained by stratified sampling, dividing the actual *S* range of the data set in a number of slices from which (as much as possible) equal numbers of minimum *AR* data are selected. The slices are taken of equal size in log(*S*) and their optimal number was fixed at 10 on a trial-and-error basis. As an example, suppose you want to estimate a 10% threshold based on a data set containing 150 landslide events, i.e., 450 event dates (see [35] and A.1 below). Homogeneously distributing over 10 *S* classes the 90 data of the 20% subset required for this threshold calculation implies to select the nine events with lowest *AR* in each class. Obviously, some classes may contain less than nine data, thus contributing less to the composition of the data subset, whose final size will often be slightly smaller than expected. In addition, when an *S* class does not contain enough data to fully contribute to the subset, all its data will be selected, however far their *AR* values are from minimum. However, tested through down weighting of the data proportionately to the deficit in contribution of their provenance class, this possible bias appeared to insignificantly affect the threshold estimates. The modified method is described in detail hereafter (see also Figure 4). The source code is provided in the Supplementary Material (Code S1).

A. Data preparation.

A.1. *AR* values associated with each day of a reported landslide plus the days prior and after these dates are extracted from the *AR* time series of the corresponding pixels calculated according to Equation (1) and the parameterization adopted in [35], i.e., $a = b = 1.2$, $n = 42$ days, for which the index is relevant for landslide types ranging from shallow to deep-seated landslides [35,63]. Data with $AR < 5$ mm are discarded from the data set as unlikely to have been triggered by rainfall [35]. The size of the provisional data set Q is then $q \leq 3p$, where p is the number of landslide events in the raw calibration set.

A.2. The data are weighted to account for the event date uncertainty: $w = 24/36$ for the day a landslide was reported, $w = 6/36$ for the days prior and after the landslide was reported. This weighting is implemented by expanding the data set as described in [35]. The expanded set is noted R.

B. Threshold calibration.

B.1. The number t_C of data to be selected per S class is determined as

$$t_C = \frac{2 \times \text{TPE} \times r}{10}, \tag{5}$$

where TPE refers to the desired threshold probability of exceedance, r is the number of data in R, and 10 is the number of $\log(S)$ classes.

B.2. The data of R are grouped by S class. For each S class, data with the lowest *AR* values are selected until they amount to t_C. The set of selected data points over all S classes is referred to as T and contains a number of data $t \leq (2 \times \text{TPE} \times r)$.

B.3. Thresholds are then calculated through linear least-square regressions using the log-transformed *AR* and S data from T and the bootstrap technique as in [35] to obtain threshold relations in the form of Equation (2).

C. Threshold evaluation

Threshold quality is evaluated through the correspondence between the obtained false negative rate (FNR, actual ratio of data in R below the calculated threshold) and the nominal TPE. Differences may result from t significantly smaller than $(2 \times \text{TPE} \times r)$, large outliers in T, and possibly also from bootstrap issues (see Section 5).

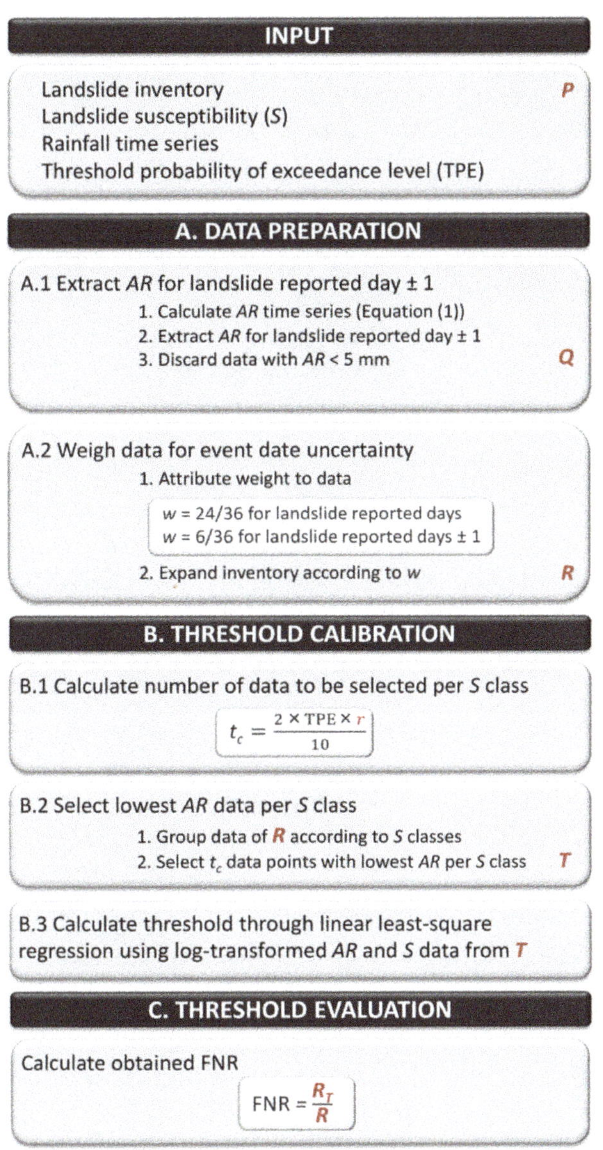

Figure 4. Workflow of the modified antecedent rainfall (*AR*)–susceptibility (*S*) threshold approach for landslides. The data sets used or derived from the respective part of the workflow are highlighted in red. R_T refers to the number of data in *R* below the threshold.

4.2. Increased Efficiency of the Method

We test the modified *AR-S* method using the regional-scale *S* data [49] and the calibration landslide data set from [35] to calculate thresholds with 0.05 and 0.10 exceedance probability. The 145 landslide events constituting the calibration set yield 435 weighted event dates, of which eight are discarded from the analysis because they do not meet the $AR \geq 5$ mm requirement, thus 427 data instances remain in the threshold analysis (constituting *Q*, Figure 4). The 5% probability level is most frequently used in landslide hazard and early warning studies [9,41,64]. We also include the 10% level because threshold estimation relies then on a larger data subset *T* (at 5% probability level, $t = 85$; at 10% level, $t = 171$). Significance measures mentioned throughout the paper are associated to the significance level p = 0.05. The R open-source software, release 3.4.3 (http://www.r-project.org, last access: 14 April 2019) was used for all analyses. *AR* thresholds at the 5% and 10% exceedance probability levels were estimated as (Figure 5)

$$AR\ (5\%) = (4.8 \pm 0.6) \times S^{(-1.16 \pm 0.08)}\ \left(R^2 = 0.69\right) \quad (6)$$

$$AR\ (10\%) = (6.4 \pm 0.6) \times S^{(-1.08 \pm 0.07)}\ \left(R^2 = 0.62\right) \quad (7)$$

Contrary to the unrealistic results obtained from the original *AR-S* approach (Equations (3) and (4)), we get here plausible marked inverse relations between *S* and *AR* [65,66]. Moreover, the threshold equations are now associated with meaningful average R^2 coefficients of 0.69 and 0.62. All bootstrap iterations provide significant α and β parameters for both thresholds. We remind that here the bootstrap procedure consists in repeating the threshold calibration phase 5000 times, each iteration being based on a random sampling (with replacement) out of the *R* data set until the number of sampled data equals that of the *r* points of the data set. The subset of lowest-*AR* data is then selected from the random sample before threshold estimation. The mean and standard deviations of the 5000 estimates of α and β define the parameter values and uncertainties ($\Delta\alpha$ and $\Delta\beta$). The results indicate an excellent performance of the modified *AR-S* threshold approach where the spread of the data subset for threshold calibration is forced over the entire *S* range. Obviously, strongly negative slopes result in decreased values of intercept α in Equations (6) and (7) as compared to Equations (3) and (4), respectively.

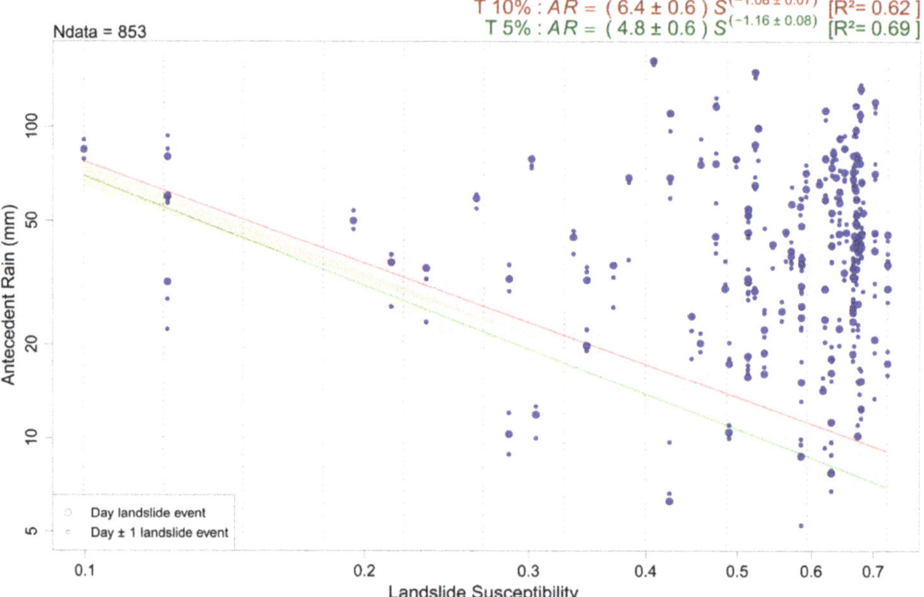

Figure 5. Log–log plot of antecedent rain (mm) vs. landslide susceptibility (regional-scale [49]) for the landslide events on the reported day and the days prior and after that date (with the point size relative to their associated weights, i.e., 0.67 and 0.17 respectively). The green and red curves are the *AR* thresholds at 5% and 10% exceedance probability levels respectively, along with their uncertainties shown as shaded areas and have been obtained with the modified *AR-S* method (Figure 4). Ndata is the number of data in the expanded calibration set. The dashed lines delimit the log(*S*) classes.

Though performing satisfyingly well, the modified *AR-S* threshold method leaves two minor issues open. The first one is related to the very close parameterization of the 5% and 10% thresholds and finds its cause in the similar actual FNRs of 0.05 and 0.07 obtained from 5% and 10% thresholding, respectively. In particular, the too low actual FNR associated with the 10% threshold equation betrays the real nature of the problem, which lies in the insufficient number of data in the low-*S* classes preventing the constitution of a complete data subset *T* to estimate the desired threshold. This issue is independent of the size of the original data set because, however large the number of recorded events might be, their distribution across the *S* range will remain similarly unequal, with low-*S* classes relatively deficient in data, especially for thresholds with higher exceedance probability demanding larger calibration subsets. Owing to the specific distribution of the data in the *AR-S* space, the *AR-S* approach inevitably implies to make a trade-off between high exceedance probability levels and degraded distribution of the data from which the threshold is estimated. Fortunately, more conservative low-exceedance probability thresholds (typically 5%) are the least affected by this issue.

High relative uncertainties (in the order of 10%) on parameter α might be another source of concern. However, beyond being subjective, the criterion chosen by [9] to qualify the threshold quality, namely a > 10% relative uncertainty, is barely usable in the *AR-S* space, where the many outliers in data distribution alter the efficiency of the bootstrap technique of uncertainty estimation (see Section 5). Moreover, in addition to the fit uncertainty, the bootstrap-based errors on the parameters obtained here from our weighted approach include the event date uncertainty and are also affected by the effect of the partly erratic character of the data distribution, inherent to the combination of ground (*S*) and meteorological (*AR*) variables on which the method relies. We thus conclude that the benefits of a method yielding thresholds directly modulated by the environmental conditions greatly outweigh the shortcomings of slightly higher uncertainty mainly on the threshold line intercept.

5. Bootstrapping Called into Question

The non-parametric bootstrap statistical techniques, including that introduced by [9] in the frequentist approach of threshold estimation, were designed to estimate the sampling distribution of a variable based on an empirical data set and assign measures of accuracy to statistical estimates [67]. While [9] acknowledge that, owing to the use of the same data for calculating the regression and estimating its parameters' uncertainties, the bootstrap may yield optimistic estimates of the latter, other possible drawbacks are not discussed in studies having incorporated the bootstrap technique in threshold estimations [35,41,64]. However, the bootstrap may fail when the data set is incomplete, resulting in overestimation of the uncertainty, or when there are outliers in the data set, to which least-square regression estimates are highly sensitive [67]. Therefore, in the light of the observed uncertainty level and hints of variability in the bootstrap results, we decided to evaluate the pros and cons of applying this technique by performing a run of threshold estimation without using it.

Performing a single threshold calculation (no bootstrap), we obtained the following *AR* thresholds (Figure 6a):

$$AR\ (5\%) = 4.6 \times S^{-1.18} \quad \left(R^2 = 0.70\right) \tag{8}$$

$$AR\ (10\%) = 6.2 \times S^{-1.10} \quad \left(R^2 = 0.65\right). \tag{9}$$

Parameters α and β are significant for both threshold levels, with α barely smaller and β barely larger compared to the thresholds obtained using the bootstrap method (Equations (6) and (7)), thus well within the bootstrap-defined uncertainty boundaries. Opposed changes in α and β might be anticipated from the inverse correlation that links coefficient and exponent of power law fits to a given data set. Therefore, the two parameter changes damp each other, thus inducing almost no difference in thresholds calculated with or without bootstrap (Table 1). Using no bootstrap, only the information about fit uncertainty is lost because date uncertainty is still accounted for through data weighting. Moreover, in the case of the *AR-S* approach, the inherent poor *S*-spread of the data and the presence of large outliers in the data subset used for threshold estimation imply that the bootstrap procedure, which, sampling with replacement n data from a set of size n, is nothing more than a kind of random data weighting, includes a number of iterations with oversampled outliers. These iterations yield erratic results and may alter the final mean threshold estimate and exaggerate the fit uncertainty to an unknown extent. This is highlighted here by the better coefficients of determination of the *AR-S* thresholds obtained from the no-bootstrap approach. Furthermore, with or without bootstrap, the *AR-S* method does not account for crucial uncertainties affecting *AR* and *S* data themselves, so that providing bootstrap-derived uncertainties is actually misleading. We thus conclude that the *AR-S* threshold procedure is more meaningful when no bootstrap is applied. The corresponding source code of the *AR-S* threshold method is provided in the Supplementary Material (Code S2). As for the other issue affecting the modified *AR-S* approach mentioned in the previous section, namely the bias in higher exceedance probability threshold estimates (FNR < TPE), it is essentially linked to the lack of data in the low-*S* classes. It is thus independent of the use of a bootstrap technique and cannot be solved by discarding the latter.

Table 1. *AR* threshold values (in mm) at 5% and 10% exceedance probability with (Equations (6) and (7)) and without (Equations (8) and (9)) bootstrap for the extreme susceptibility values *S* observed in the data set.

Threshold	with Bootstrap	without Bootstrap
5% threshold, $S = 0.10$	69.4	69.6
5% threshold, $S = 0.72$	7.0	6.8
10% threshold, $S = 0.10$	76.9	78.1
10% threshold, $S = 0.72$	9.1	8.9

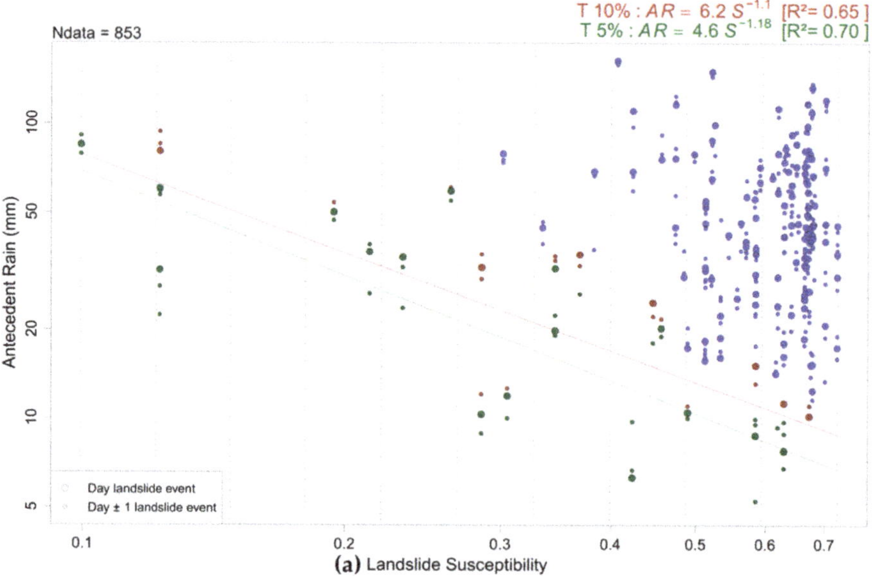

Figure 6. *Cont.*

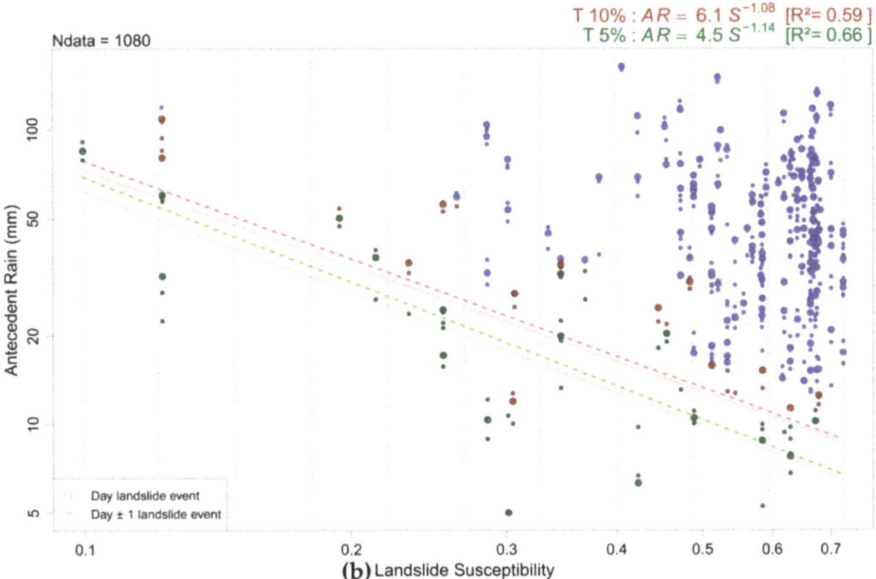

Figure 6. Log–log plots of antecedent rain (mm) vs. landslide susceptibility (regional-scale [49]) for the landslide events on the reported day and the days prior and after that date (with the point size relative to their attributed weights, i.e., 0.67 and 0.17 respectively). Thresholds are based on the calibration inventory (**a**), and the complete (calibration + validation) inventory (**b**). The threshold method applied is outlined in Figure 4 without adopting the bootstrapping statistical technique. Data subsets used for the calibration of thresholds at the 5% (green dots) and 10% (green and red dots) exceedance probability are highlighted (*T* in Figure 4). Dashed green and red lines in (**b**) present the thresholds based on the calibration data set only, as shown in (**a**). Ndata is the number of data in the respective expanded data set. The dashed lines delimit the log(*S*) classes.

6. Robustness of the Modified *AR-S* Threshold Method

6.1. First Test: Sensitivity to New Data on Landslide Occurrence

The modified *AR-S* method with no bootstrap is tested firstly by using the recent addition to the WEAR data set of dated landslide events. Taking into account their date uncertainty, the 39 landslide events constituting this validation set yield 117 new weighted event dates, of which four are discarded from the analysis because they do not meet the $AR \geq 5$ mm requirement. The 113 remaining data instances (constituting *Q*, Figure 4) are distributed in the log(*AR*)–log(*S*) space in such a manner that 6% of them are located below the 5% threshold line derived from the calibration (Equation (8)) and 8% below the 10% threshold line (Equation (9)), indicating a good performance of the calculated thresholds (Figure S2) considering the small sample size.

Another test using the validation set, which in the same time should improve the accuracy of the calibrated thresholds, has consisted in combining the data of the calibration and validation sets into a larger data set of 540 event dates in order to recalculate the thresholds. The new thresholds read as

$$AR\ (5\%) = 4.5 \times S^{-1.14} \ \left(R^2 = 0.66\right) \quad (10)$$

$$AR\ (10\%) = 6.1 \times S^{-1.08} \ \left(R^2 = 0.59\right) \quad (11)$$

and do not much differ from those derived from the calibration set only (Equations (8) and (9)) (Figure 6b), confirming the relevance of the modified *AR-S* method. Though slightly decreased by additional noise brought in the middle- to low-S classes by the new data (Figure S2), their coefficients of determination remain highly significant. Likewise, their FNRs (0.04 and 0.06 for the 5% and 10% thresholds, respectively) are slightly degraded mainly as a result of an increased deficit in data in these S classes. Owing to the larger size of the data set, we nevertheless consider these thresholds (Equations (10) and (11)) more reliable than those based only on the calibration set, especially the 5% threshold, for which FNR ≈ TPE.

6.2. Second Test: Robustness to Different S Data Sets

We test the modified AR-S approach for the adoption of a different data set for S, using the continental-scale S data [45] and the complete (calibration+validation) data set of landslide events, obtaining the following AR thresholds (Figure 7)

$$AR\ (5\%) = 5.7 \times S^{-2.10}\ \left(R^2 = 0.73\right) \quad (12)$$

$$AR\ (10\%) = 7.6 \times S^{-2.08}\ \left(R^2 = 0.61\right) \quad (13)$$

with significant and meaningful values for R^2 and threshold parameters α, and β. Moreover, these thresholds show a stronger relation between threshold AR values and S with increased values for parameters (α,) β and R^2, explained by the increased dispersion of the data over the S range (Figure 7) relative to when the regional-scale S data was applied (Figure 6b). Where the AR-S approach developed by [35] posed problems for adopting a different S model than that used for its development (Figure 3), these results show that the modified AR-S approach proved to solve this matter. The threshold at the higher exceedance probability remains affected by a bias similar to that in Equation (11) with the actual FNR lower than the TPE (FNRs equal 0.05 and 0.07 for the 5% and 10% thresholds, respectively).

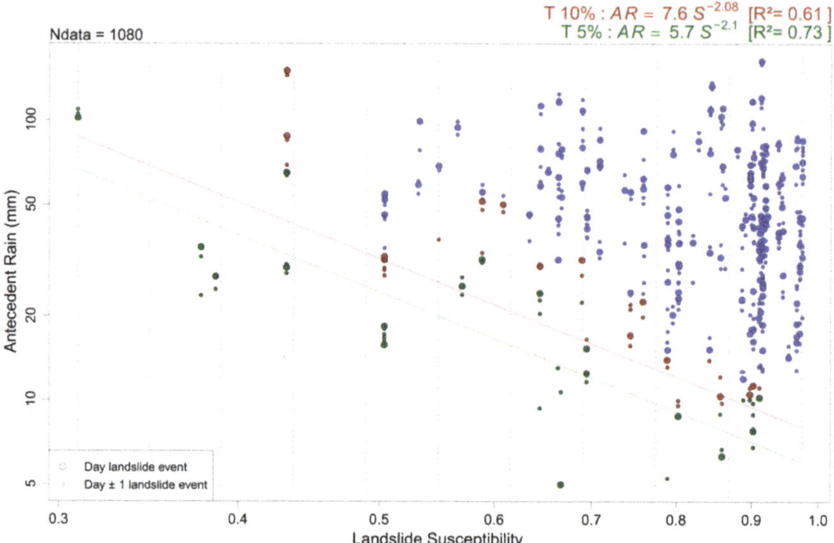

Figure 7. Log–log plot of antecedent rain (mm) vs. landslide susceptibility (continental-scale [45]) for the landslide events on the reported day and the days prior and after that date (with the point size relative to their attributed weights, i.e., 0.67 and 0.17 respectively). The green and red curves are the AR thresholds at 5% and 10% exceedance probability levels respectively, obtained with the modified AR-S method (Figure 4) without the bootstrapping statistical technique. Ndata is the number of data in the expanded (calibration+validation) data set. The dashed lines delimit the log(S) classes.

Another indicator for the robustness of the modified *AR-S* method is the generally satisfying correspondence between *AR* threshold results for the continental- and regional-scale *S* models (Table 2). The single main discrepancy is observed for the 10% threshold in the low *S* data range, related to the actual FNRs of these thresholds being significantly smaller than their TPE. The latter is explained by the sensitivity of the threshold slope to the deficient number and exact location of data in the low-*S* classes, causing the largest threshold difference to appear for the 10% threshold (implying a greater lack of data in low-*S* classes) at the low end of the *S* range. By contrast, the intercept of the threshold equations, being located in the *AR-S* space with the highest density of data, remains quasi stable for different *S* models (Figure 6b, Figure 7).

Table 2. *AR* threshold values (in mm) calculated using continental- (Equations (12) and (13)) vs. regional-scale (Equations (10) and (11)) *S* data, provided at 5% and 10% exceedance probability for the extreme susceptibility values *S* observed in the data sets. The estimations are based on the complete calibration+validation data set of landslide events.

Threshold	Continental *S* Data	Regional *S* Data
5% threshold, min *S*	66.7	62.1
5% threshold, max *S*	6.1	6.5
10% threshold, min *S*	86.9	73.3
10% threshold, max *S*	8.1	8.7

Because of the enhanced relation between threshold *AR* values and *S* in Equations (12) and (13), it is tempting to suggest that thresholds based on the continental-scale *S* data would be more efficient when adopted in a landslide early warning system. However, the spatial pattern of the *AR* thresholds based on the regional-scale *S* data are closer to the reality, given that this regional-scale *S* model has a higher predictive power and geomorphological plausibility as compared to the continental-scale model [49]. The respective *AR* threshold maps are presented at the 5% probability of exceedance level in Figure 8. In general, we observe lower *AR* thresholds within the Rift. Nevertheless, there are some major differences between the two threshold maps, caused by differences in the quality of the susceptibility models. First, the threshold model using the continental susceptibility map of [45] assigns low *AR* thresholds to the rainforest in DR Congo south of the equator, despite the fact that the area is characterized by high amounts of rain and few landslides [29,35]. Second, the regional susceptibility data of [49] overall shows a much lower threshold in Uganda. In conclusion, we confirm the earlier observation that the *AR* thresholds based on the regional-scale *S* data and the currently most extensive landslide event inventory are to date the most accurate available thresholds for landsliding in the WEAR (Equations (10) and (11)).

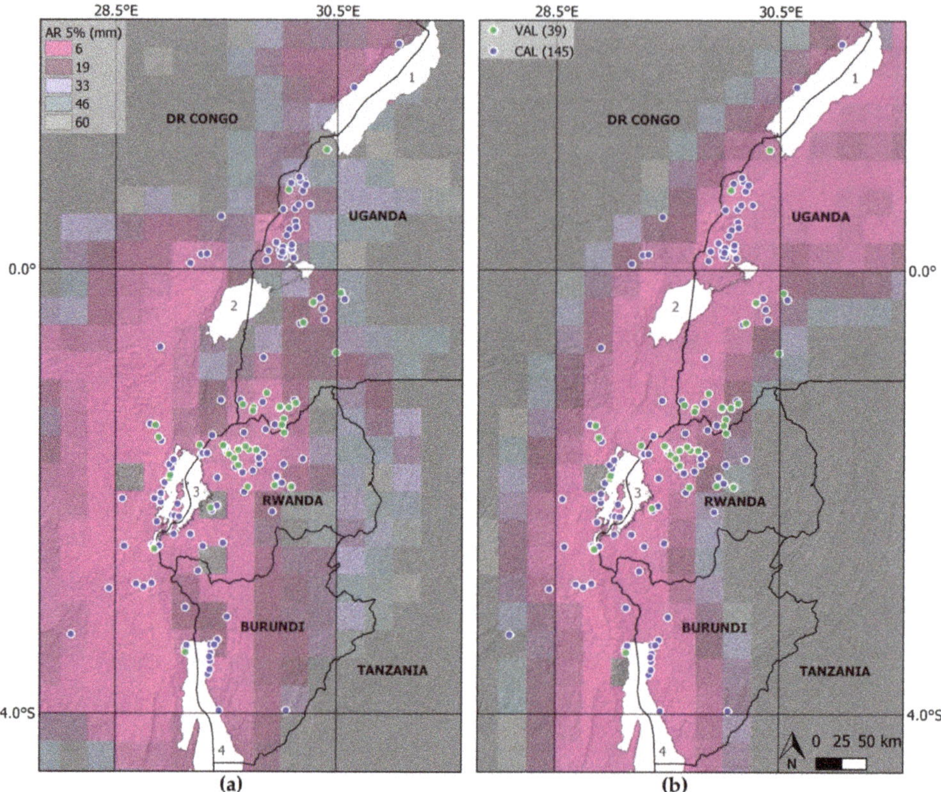

Figure 8. Antecedent rainfall (*AR*) threshold maps (0.25° resolution) at 5% exceedance probability, based on the complete (calibration+validation) landslide inventory, and the (**a**) continental-scale *S* model [45] (Figure 7, Equation (12)) and (**b**) regional-scale *S* model [49] (Figure 6b, Equation (10)). *AR* threshold values are only shown for the *S* range covered by the 184 landslide events used for the threshold estimations (i.e., (**a**): *S* 0.31–0.97; (**b**): *S* 0.10–0.72). 1: Lake Albert; 2: Lake Edward; 3: Lake Kivu; 4: Lake Tanganyika. Background hillshade 3 arc-second SRTM (±90 m).

7. Relevance to Landslide Hazard and Early Warning Studies

The modified *AR-S* approach is relevant for the increased accuracy of the resulting *AR* thresholds, which is partly also on account of the regional-scale *S* data [49] and the enlarged landslide inventory. In particular, the conservative low-exceedance probability thresholds are most reliable, being least affected by a degraded distribution of data used for the threshold calibration (FNR ≈ TPE). Depending on the local susceptibility, thresholds at the 5% exceedance probability range from *AR* = 62 mm in the least susceptible areas to *AR* = 7 mm in the highest susceptibility pixels, respectively, for which landslide have been reported (Equation (10)). These triggering *AR* conditions might seem low at first sight when compared to values obtained in other studies that look into antecedent rainfall conditions based on gauge measurements, e.g., a required minimum of 139 mm cumulated over 20 days to trigger landslides in the NE Himalaya [68]; a mean triggering rainfall accumulation of 376 mm for periods ranging between 15 and 40 days in NW Spain [69]; a critical rainfall amount of 450 mm over a two-week period in the greater Durban region in South Africa [70]. However, the triggering values obtained in our study are conceivable given the following main factors contributing to their relative lower values: (1) the exponential decay function applied in our *AR* calculation (Equation (1)) in contrast to the values obtained in the above cited studies through mere accumulation; (2) the high weathering conditions in the tropical context of the WEAR that may increase the sensitivity to landsliding [50]; and (3) the underestimation of the area-averaged SRE [33,56] used in the calculation of *AR* (Equation (1)). The latter is not necessarily an issue when thresholds are evaluated with the same SRE used for their calibration [32]. To date, SRE-based studies form only a small fraction in landslide threshold research [3,32]. A TMPA-RT-based threshold was established for Italy at the 20% exceedance probability, obtaining a critical rainfall accumulation of 189 mm for an extrapolated duration of 42 days [32]. However, the extrapolation is doubtful, given the limited threshold calibration range of ~11 days [32]. In addition, no decay function is applied to this extrapolated value for accumulated antecedent rainfall and the higher exceedance probability level obviously renders an elevated threshold compared to the 5% and 10% levels deployed in our study. On the other hand, the 5% thresholds calibrated for central Italy by [71] based on TMPA (Research Version) data, estimate critical accumulated rainfall in the order of 30 mm over an extrapolated duration of 42 days, which falls in our estimated triggering range at the 5% exceedance probability.

Furthermore, the modified *AR-S* threshold method is relevant in the context of landslide hazard analysis when rainfall data are only available for conditions that triggered landslides, for it proved to be a robust alternative for frequentist-based threshold approaches [9,40] when the method's assumptions (i.e., large and well-spread data set [40]) are not met. To the authors' knowledge, it is the first time that a stratified data selection technique is adopted in the threshold calibration approach, which effectively showed to enhance the data distribution over the whole range of the causative threshold variable (*S*). This allows the method to be transferable not only to other data sets for *S* (and hence other study areas), but to any parameter that might be considered as a possible cause for landsliding [34] without the requirement of the data to be homogeneously distributed, to which further research should be carried out. This is significant in the context of the 'trigger-cause' conceptual framework of threshold definition as proposed by [34], in response to the shortcomings of the rainfall-only thresholds with regard to their limited physical meaning [34]. The framework was designed to introduce hydrological information on different timescales, with the choice of the parameters and timescales depending on their expected significance for slope failure given the physiographic context and considered landslide types [34]. In the *AR-S* approach, the causative hydrological status of the slope is substituted by information on spatially varying predisposing ground conditions, while *AR* presents the progressive build-up of the landslide trigger [35]. We could anticipate, however, that improvements in satellite-based soil moisture data, with regard to their spatial resolution and performance over dense vegetated areas or complex topography [72–74], would allow in the future to replace the static *S* variable by a dynamic causative hydrological factor over data-scarce regions.

In addition, the enhanced relation between *AR* threshold values and *S* renders a potential landslide early warning system more efficient. In this respect, the integrated spatial component of the *AR-S* approach (Figure 8) has a clear advantage over traditional thresholds, such as rainfall intensity-duration thresholds, the latter only informing 'when' the probability of a landslide occurrence increases but not 'where' [65]. The spatial component furthermore avoids data to be partitioned according to homogeneous physiographic units to enhance the accuracy of thresholds [9,75], which is of particular relevance in data-scarce contexts. Finally, the method was designed to use area-averaged SRE, allowing its adoption in regions where a dense rain gauge network is absent and evaluate hazard in near-real time.

However, the efficient use of the *AR-S* approach in hazard or early warning studies is hampered by a number of factors that might contribute to an obscured relation between *AR* and *S*, including: (1) the level of accuracy and completeness of the landslide inventory [28,42]; (2) a lack of differentiation in landslide processes whose triggering conditions are different [63]; (3) the accuracy of *S* and SRE data [33,49]; (4) the resampling of *S* data to the coarser SRE data resulting in inaccurate *S*-classifications of landslide data; (5) the anthropogenic influence on the environment, such as in the context of exponential demographic pressure in the WEAR [76,77]; and (6) the empirically defined parameters in the *AR* equation (Equation (1)) by [35]. The highlighted obstacles and limitations serve as pathways for further investigation and improvements in the *AR-S* threshold approach.

8. Conclusions

We propose a modified antecedent rainfall–susceptibility (*AR-S*) threshold approach that improves on the initial *AR-S* method of [35], being transferable to other data sets for landsliding and *S*. For its development and evaluation we exploit the most current and extensive landslide inventory for the western branch of the East African Rift comprising 184 dated landslide events from 2001 to 2019, satellite-based rainfall estimates from TMPA 3B42 RT, and two *S* models, i.e., the continental-scale model of [45] and the regional-scale *S* model of [49]. The main novelty in the modified *AR-S* approach is the stratified selection of data associated with the lowest *AR* values able to cause landsliding, allowing to deploy data sets that are not necessarily homogenously distributed over the *S* range. Furthermore, we highlight that the threshold procedure is more meaningful when no bootstrapping statistical technique is applied, as the uncertainties in the parameters that define the power-law threshold model are mainly introduced by the bootstrapping related random sampling in combination with the presence of outliers in the data set. We obtain improved *AR* thresholds with an increased susceptibility-dependent gradient, and *AR* threshold maps with a higher accuracy through the use of the regional-scale *S* model in the modified *AR-S* approach. The improved *AR* threshold values at the 5% exceedance probability range from 7 mm in the most susceptible areas ($S = 0.72$) to 62 mm in the lowest susceptible areas ($S = 0.10$) where landslides have been recorded (uncorrected for underestimation by TMPA). Our approach is foremost relevant in data-scarce regions, where the lack of abundant data from rain gauges and in particular on landslide occurrence hampers the use of homogenously distributed data sets. Moreover, we suggest that this modified method is transferable not only to other data sets for *S*, but to any parameter that might be considered as a possible cause for landsliding.

Supplementary Materials: The following are available online at http://www.mdpi.com/2073-4441/11/11/2202/s1, Figure S1: Distribution of the data (white bars) in the calibration data set over 10 logarithmic equidistant *S* classes for the (**a**) continental-scale [45] and (**b**) regional-scale [49] susceptibility models. "10%" and "20%" refer to the ratio of the data with the lowest *AR* values that are selected from the data set (presented here without random sampling) for the calibration of the 5% and 10% thresholds respectively; Figure S2: Log–log plot of antecedent rain (mm) vs. landslide susceptibility (regional-scale [49]) for the landslide events on the reported day and the days prior and after that date (with the point size relative to their attributed weights, i.e., 0.67 and 0.17 respectively). The green and red curves are the *AR* thresholds at 5% and 10% exceedance probability levels respectively, obtained with the modified *AR-S* method (Figure 4) without adopting the bootstrapping statistical technique, using the calibration (CAL) data set only. Ndata is the number of data in the expanded calibration and validation (VAL) data sets; Code S1: R code for the modified *AR-S* threshold approach (Figure 4) with the bootstrapping statistical technique; Code S2: R code for the modified *AR-S* threshold approach (Figure 4) without the bootstrapping statistical technique.

Author Contributions: Conceptualization, E.M., O.D. and A.D. (Alain Demoulin); Data curation, E.M.; Formal analysis, E.M.; Funding acquisition, E.M. and O.D.; Investigation, E.M.; Methodology, E.M.; Project administration, E.M. and O.D.; Resources, E.M., O.D. and A.D. (Arthur Depicker); Software, E.M.; Supervision, O.D. and A.D. (Alain Demoulin); Validation, E.M.; Visualization, E.M.; Writing—original draft, E.M.; Writing—review & editing, E.M., O.D., A.D. (Arthur Depicker) and A.D. (Alain Demoulin)

Funding: This study was supported by the Belgium Science Policy (BELSPO) through (1) the PAStECA project (BR/165/A3/PASTECA) entitled 'Historical Aerial Photographs and Archives to Assess Environmental Changes in Central Africa' (http://pasteca.africamuseum.be/), (2) the RESIST project (SR/00/305) entitled 'Remote Sensing and In Situ Detection and Tracking of Geohazards' (http://resist.africamuseum.be/), (3) the GeoRisCA project (SD/RI/02A), entitled 'Geo-Risk in Central Africa: integrating multi-hazards and vulnerability to support risk management' (http://georisca.africamuseum.be), and (4) the AfReSlide project (BR/121/A2/AfReSlide) entitled 'Landslides in Equatorial Africa: Identifying culturally, technically and economically feasible resilience strategies' (http://afreslide.africamuseum.be/). E.M. was funded by F.R.S.–FNRS.

Acknowledgments: The authors thank Jente Broeckx for providing the continental landslide susceptibility model. Special thanks go to our partners at Université Officielle de Bukavu (DR Congo) and Centre de Recherche en Sciences Naturelles de Lwiro (DR Congo), who facilitated fieldwork in the study area and provided information on the timing of the landslides. We acknowledge the NASA Goddard Earth Sciences Data and Information Services Center for providing full access to the precipitation data sets exploited in this study. We also thank the reviewers for their constructive feedback.

Conflicts of Interest: The authors declare no conflict of interest.

References

1. Froude, M.J.; Petley, D.N. Global fatal landslides 2004 to 2016. *Nat. Hazards Earth Syst. Sci.* **2018**, *18*, 2161–2181. [CrossRef]
2. Haque, U.; da Silva, P.F.; Devoli, G.; Pilz, J.; Zhao, B.; Khaloua, A.; Wilopo, W.; Andersen, P.; Lu, P.; Lee, J.; et al. The human cost of global warming: Deadly landslides and their triggers (1995–2014). *Sci. Total Environ.* **2019**, *682*, 673–684. [CrossRef] [PubMed]
3. Segoni, S.; Piciullo, L.; Gariano, S.L. A review of the recent literature on rainfall thresholds for landslide occurrence. *Landslides* **2018**, *15*, 1483–1501. [CrossRef]
4. Intrieri, E.; Carlà, T.; Gigli, G. Forecasting the time of failure of landslides at slope-scale: A literature review. *Earth Sci. Rev.* **2019**, *193*, 333–349. [CrossRef]
5. Piciullo, L.; Calvello, M.; Cepeda, J.M. Territorial early warning systems for rainfall-induced landslides. *Earth Sci. Rev.* **2018**, *179*, 228–247. [CrossRef]
6. Giannecchini, R.; Galanti, Y.; D'Amato Avanzi, G. Critical rainfall thresholds for triggering shallow landslides in the Serchio River Valley (Tuscany, Italy). *Nat. Hazards Earth Syst. Sci.* **2012**, *12*, 829–842. [CrossRef]
7. Giannecchini, R. Relationship between rainfall and shallow landslides in the southern Apuan Alps (Italy). *Nat. Hazards Earth Syst. Sci.* **2006**, *6*, 357–364. [CrossRef]
8. Guzzetti, F.; Peruccacci, S.; Rossi, M.; Stark, C.P. The rainfall intensity-duration control of shallow landslides and debris flows: An update. *Landslides* **2008**, *5*, 3–17. [CrossRef]
9. Peruccacci, S.; Brunetti, M.T.; Luciani, S.; Vennari, C.; Guzzetti, F. Lithological and seasonal control on rainfall thresholds for the possible initiation of landslides in central Italy. *Geomorphology* **2012**, *139–140*, 79–90. [CrossRef]
10. Guzzetti, F.; Peruccacci, S.; Rossi, M.; Stark, C.P. Rainfall thresholds for the initiation of landslides in central and southern Europe. *Meteorol. Atmos. Phys.* **2007**, *98*, 239–267. [CrossRef]
11. Bittelli, M.; Valentino, R.; Salvatorelli, F.; Rossi Pisa, P. Monitoring soil-water and displacement conditions leading to landslide occurrence in partially saturated clays. *Geomorphology* **2012**, *173–174*, 161–173. [CrossRef]
12. Bogaard, T.A.; Greco, R. Landslide hydrology: From hydrology to pore pressure. *Wiley Interdiscip. Rev. Water* **2016**, *3*, 439–459. [CrossRef]
13. Green, W.H.; Ampt, G. The flow of air and water through soils. *J. Agric. Sci.* **1911**, *4*, 1–24.
14. Richards, L.A. Capillary conduction of liquids through porous mediums. *J. Appl. Phys.* **1931**, *1*, 318–333. [CrossRef]
15. Bierman, P.R.; Montgomery, D. *Key Concepts in Geomorphology*; W. H. Freeman and Company Publishers: New York, NY, USA, 2013; ISBN 9781429238601.
16. McGlynn, A.L.; McDonnel, J.J.; Brammer, D.D.; Mcglynn, B.L.; Mcdonnel, J.J.; Brammer, D.D. A Review of the Evolving Perceptual Model of Hillslope Flowpaths at the Maimai Catchment, New Zealand. *J. Hydrol.* **2002**, *257*, 1–26. [CrossRef]
17. Haitjema, H.M.; Mitchell-Bruker, S. Are water tables a subdued replica of the topography? *Groundwater* **2005**, *43*, 781–786. [CrossRef]
18. Thompson, S.E.; Harman, C.J.; Heine, P.; Katul, G.G. Vegetation-infiltration relationships across climatic and soil type gradients. *J. Geophys. Res. Biogeosci.* **2010**, *115*, 1–12. [CrossRef]
19. Wilson, G.V.; Luxmoore, R.J. Infiltration, Macroporosity, and Mesoporosity Distributions on Two Forested Watersheds. *Soil Sci. Soc. Am. J.* **1988**, *52*, 329–335. [CrossRef]
20. Johnson, K.A.; Sitar, N. Hydrologic conditions leading to debris-flow initiation. *Can. Geotech. J.* **1990**, *27*, 789–801. [CrossRef]
21. Rahardjo, H.; Leong, E.C.; Rezaur, R.B. Effect of antecedent rainfall on pore-water pressure distribution characteristics in residual soil slopes under. *Hydrol. Process.* **2008**, *22*, 506–5023. [CrossRef]
22. Wilson, R.C.; Wieczorek, G. Rainfall thresholds for the initiation of debris flows at La Honda, California. *Environ. Eng. Geosci.* **1995**, *1*, 11–27. [CrossRef]
23. Watakabe, T.; Matsushi, Y. Lithological controls on hydrological processes that trigger shallow landslides: Observations from granite and hornfels hillslopes in Hiroshima, Japan. *Catena* **2019**, *180*, 55–68. [CrossRef]

24. Xue, J.; Gavin, K. Effect of rainfall intensity on infiltration into partly saturated slopes. *Geotech. Geol. Eng.* **2008**, *26*, 199–209. [CrossRef]
25. Iverson, M.R. Landslide triggering by rain infiltration. *Water Resour. Res.* **2000**, *36*, 1897–1910. [CrossRef]
26. Thomas, M.A.; Mirus, B.B.; Collins, B.D. Identifying Physics-Based Thresholds for Rainfall-Induced Landsliding. *Geophys. Res. Lett.* **2018**, *45*, 9651–9661. [CrossRef]
27. Kim, D.; Im, S.; Lee, S.H.; Hong, Y.; Cha, K.S. Predicting the rainfall-triggered landslides in a forested mountain region using TRIGRS model. *J. Mt. Sci.* **2010**, *7*, 83–91. [CrossRef]
28. Peres, D.J.; Cancelliere, A.; Greco, R.; Bogaard, T.A. Influence of uncertain identification of triggering rainfall on the assessment of landslide early warning thresholds. *Nat. Hazards Earth Syst. Sci.* **2018**, *18*, 633–646. [CrossRef]
29. Monsieurs, E.; Jacobs, L.; Michellier, C.; Basimike Tchangaboba, J.; Ganza, G.B.; Kervyn, F.; Maki Mateso, J.C.; Mugaruka Bibentyo, T.; Kalikone Buzera, C.; Nahimana, L.; et al. Landslide inventory for hazard assessment in a data-poor context: A regional-scale approach in a tropical African environment. *Landslides* **2018**, *15*, 2195–2209. [CrossRef]
30. Taylor, F.E.; Malamud, B.D.; Freeborough, K.; Demeritt, D. Enriching Great Britain's National Landslide Database by searching newspaper archives. *Geomorphology* **2015**, *249*, 52–68. [CrossRef]
31. Kirschbaum, D.B.; Stanley, T.; Zhou, Y. Spatial and temporal analysis of a global landslide catalog. *Geomorphology* **2015**, *249*, 4–15. [CrossRef]
32. Brunetti, M.T.; Melillo, M.; Peruccacci, S.; Ciabatta, L.; Brocca, L. How far are we from the use of satellite rainfall products in landslide forecasting? *Remote Sens. Environ.* **2018**, *210*, 65–75. [CrossRef]
33. Monsieurs, E.; Kirschbaum, D.B.; Tan, J.; Maki Mateso, J.C.; Jacobs, L.; Plisnier, P.D.; Thiery, W.; Umutoni, A.; Musoni, D.; Bibentyo, T.M.; et al. Evaluating TMPA Rainfall over the Sparsely Gauged East African Rift. *J. Hydrometeorol.* **2018**, *19*, 1507–1528. [CrossRef]
34. Bogaard, T.; Greco, R. Invited perspectives: Hydrological perspectives on precipitation intensity-duration thresholds for landslide initiation: Proposing hydro-meteorological thresholds. *Nat. Hazards Earth Syst. Sci.* **2018**, *18*, 31–39. [CrossRef]
35. Monsieurs, E.; Dewitte, O.; Demoulin, A. A susceptibility-based rainfall threshold approach for landslide occurrence. *Nat. Hazards Earth Syst. Sci.* **2019**, *19*, 775–789. [CrossRef]
36. Kirschbaum, D.B.; Adler, R.; Hong, Y.; Kumar, S.; Peters-Lidard, C.; Lerner-Lam, A. Advances in landslide nowcasting: Evaluation of a global and regional modeling approach. *Environ. Earth Sci.* **2012**, *66*, 1683–1696. [CrossRef]
37. Berti, M.; Martina, M.L.V.; Franceschini, S.; Pignone, S.; Simoni, A.; Pizziolo, M. Probabilistic rainfall thresholds for landslide occurrence using a Bayesian approach. *J. Geophys. Res. Earth Surf.* **2012**, *117*, 1–20. [CrossRef]
38. Postance, B.; Hillier, J.; Dijkstra, T.; Dixon, N. Comparing threshold definition techniques for rainfall-induced landslides: A national assessment using radar rainfall. *Earth Surf. Process. Landf.* **2018**, *43*, 553–560. [CrossRef]
39. Mirus, B.; Morphew, M.; Smith, J. Developing Hydro-Meteorological Thresholds for Shallow Landslide Initiation and Early Warning. *Water* **2018**, *10*, 1274. [CrossRef]
40. Brunetti, M.T.; Peruccacci, S.; Rossi, M.; Lucian, S.; Valigi, D.; Guzzetti, F. Rainfall thresholds for the possible occurrence of landslides in Italy. *Nat. Hazards Earth Syst. Sci.* **2010**, *10*, 447–458. [CrossRef]
41. Melillo, M.; Brunetti, M.T.; Peruccacci, S.; Gariano, S.L.; Roccati, A.; Guzzetti, F. A tool for the automatic calculation of rainfall thresholds for landslide occurrence. *Environ. Model. Softw.* **2018**, *105*, 230–243. [CrossRef]
42. Gariano, S.L.; Brunetti, M.T.; Iovine, G.; Melillo, M.; Peruccacci, S.; Terranova, O.; Vennari, C.; Guzzetti, F. Calibration and validation of rainfall thresholds for shallow landslide forecasting in Sicily, southern Italy. *Geomorphology* **2015**, *228*, 653–665. [CrossRef]
43. Marra, F.; Destro, E.; Nikolopoulos, E.I.; Zoccatelli, D.; Creutin, J.D.; Guzzetti, F.; Borga, M. Impact of uncertainty in rainfall estimation on the identification of rainfall thresholds for debris flow occurrence. *Geomorphology.* **2017**, *21*, 4525–4532. [CrossRef]
44. Nobile, A.; Dille, A.; Monsieurs, E.; Basimike, J.; Bibentyo, T.M.; d'Oreye, N.; Kervyn, F.; Dewitte, O. Multi-temporal DInSAR to characterise landslide ground deformations in a tropical urban environment: Focus on Bukavu (DR Congo). *Remote Sens.* **2018**, *10*, 626. [CrossRef]
45. Broeckx, J.; Vanmaercke, M.; Duchateau, R.; Poesen, J. A data-based landslide susceptibility map of Africa. *Earth Sci. Rev.* **2018**, *185*, 102–121. [CrossRef]

46. Stanley, T.; Kirschbaum, D.B. A heuristic approach to global landslide susceptibility mapping. *Nat. Hazards* **2017**, *87*, 145–164. [CrossRef]
47. Jacobs, L.; Dewitte, O.; Poesen, J.; Sekajugo, J.; Nobile, A.; Rossi, M.; Thiery, W.; Kervyn, M. Field-based landslide susceptibility assessment in a data-scarce environment: The populated areas of the Rwenzori Mountains. *Nat. Hazards Earth Syst. Sci.* **2018**, *18*, 105–124. [CrossRef]
48. Huffman, G.J.; Bolvin, D.T.; Nelkin, E.J.; Wolff, D.B.; Adler, R.F.; Gu, G.; Hong, Y.; Bowman, K.P.; Stocker, E.F. The TRMM Multisatellite Precipitation Analysis (TMPA): Quasi-Global, Multiyear, Combined-Sensor Precipitation Estimates at Fine Scales. *J. Hydrometeorol.* **2007**, *8*, 38–55. [CrossRef]
49. Depicker, A.J.S.P.; Jacobs, L.; Delvaux, D.; Havenith, H.B.; Maki Mateso, J.C.; Govers, G.; Dewitte, O. The added value of a regional landslide susceptibility assessment: The western branch of the East African Rift. *Geomorphology* **2019**, in press.
50. Dille, A.; Kervyn, F.; Mugaruka Bibentyo, T.; Delvaux, D.; Bamulezi Ganza, G.; Mawe Ilombe, G.; Buzera Kalikone, C.; Safari Makito, E.; Moeyersons, J.; Monsieurs, E.; et al. Causes and triggers of deep-seated hillslope instability in the tropics—Insights from a 60-year record of Ikoma landslide (DR Congo). *Geomorphology* **2019**. [CrossRef]
51. Moeyersons, J.; Tréfois, P.; Lavreau, J.; Alimasi, D.; Badriyo, I.; Mitima, B.; Mundala, M.; Munganga, D.O.; Nahimana, L. A geomorphological assessment of landslide origin at Bukavu, Democratic Republic of the Congo. *Eng. Geol.* **2004**, *72*, 73–87. [CrossRef]
52. Delvaux, D.; Mulumba, J.L.; Sebagenzi, M.N.S.; Bondo, S.F.; Kervyn, F.; Havenith, H.B. Seismic hazard assessment of the Kivu rift segment based on a new seismotectonic zonation model (western branch, East African Rift system). *J. Afr. Earth Sci.* **2017**, *134*, 831–855. [CrossRef]
53. Maki Mateso, J.C.; Dewitte, O. Towards an inventory of landslide processes and the elements at risk on the Rift flanks west of Lake Kivu (DRC). *Geo-Eco-Trop* **2014**, *38*, 137–154.
54. Jacobs, L.; Dewitte, O.; Poesen, J.; Maes, J.; Mertens, K.; Sekajugo, J.; Kervyn, M. Landslide characteristics and spatial distribution in the Rwenzori Mountains, Uganda. *J. Afr. Earth Sci.* **2017**, *134*, 917–930. [CrossRef]
55. Monsieurs, E.; Kirschbaum, D.B.; Thiery, W.; van Lipzig, N.; Kervyn, M.; Demoulin, A.; Jacobs, L.; Kervyn, F.; Dewitte, O. Constraints on Landslide-Climate Research Imposed by the Reality of Fieldwork in Central Africa. In Proceedings of the 3rd North American Symposium on Landslides: Landslides: Putting Experience, Knowledge, and Emerging Technologies into Practice, Roanoke, VA, USA, 4–8 June 2017; pp. 158–168.
56. Camberlin, P.; Barraud, G.; Bigot, S.; Dewitte, O.; Makanzu Imwangana, F.; Maki Mateso, J.; Martiny, N.; Monsieurs, E.; Moron, V.; Pellarin, T.; et al. Evaluation of remotely sensed rainfall products over Central Africa. *Q. J. R. Meteorol. Soc.* **2019**, *145*, 2115–2138. [CrossRef]
57. USGS. *Shuttle Radar Topography Mission, Global Land Cover Facility*; University of Maryland: College Park, MD, USA, 2006.
58. Giardini, D.; Grunthal, G.; Shedlock, K.M.; Zhang, P. The GSHAP Global Seismic Hazard Map. *Ann. Geophys.* **1999**, *42*, 1225–1228. [CrossRef]
59. Hartmann, J.; Moosdorf, N. The new global lithological map database GLiM: A representation of rock properties at the Earth surface. *Geochem. Geophys. Geosystems* **2012**, *13*, 1–37. [CrossRef]
60. Smets, B.; Delvaux, D.; Ross, K.A.; Poppe, S.; Kervyn, M.; d'Oreye, N.; Kervyn, F. The role of inherited crustal structures and magmatism in the development of rift segments: Insights from the Kivu basin, western branch of the East African Rift. *Tectonophysics* **2016**, *683*, 62–76. [CrossRef]
61. European Space Agency (ESA). *Climate Change Initiative—Land Cover Project 2017. 20 m Resolution*; European Space Agency: Paris, France, 2016.
62. Maidment, R.I.; Grimes, D.; Black, E.; Tarnavsky, E.; Young, M.; Greatrex, H.; Allan, R.P.; Stein, T.; Nkonde, E.; Senkunda, S.; et al. A new, long-term daily satellite-based rainfall dataset for operational monitoring in Africa. *Sci. Data* **2017**, *4*, 170082. [CrossRef]
63. Sidle, R.C.; Bogaard, T.A. Dynamic earth system and ecological controls of rainfall-initiated landslides. *Earth Sci. Rev.* **2016**, *159*, 275–291. [CrossRef]
64. Peruccacci, S.; Brunetti, M.T.; Gariano, S.L.; Melillo, M.; Rossi, M.; Guzzetti, F. Rainfall thresholds for possible landslide occurrence in Italy. *Geomorphology* **2017**, *290*, 39–57. [CrossRef]
65. Pradhan, A.M.S.; Lee, S.R.; Kim, Y.T. A shallow slide prediction model combining rainfall threshold warnings and shallow slide susceptibility in Busan, Korea. *Landslides* **2019**, *16*, 647–659. [CrossRef]

66. Segoni, S.; Lagomarsino, D.; Fanti, R.; Moretti, S.; Casagli, N. Integration of rainfall thresholds and susceptibility maps in the Emilia Romagna (Italy) regional-scale landslide warning system. *Landslides* **2015**, *12*, 773–785. [CrossRef]
67. Davison, A.C.; Hinkley, D.V. *Bootstrap Methods and Their Application*; Cambridge University Press: Cambridge, UK, 1997; ISBN 0521573912.
68. Koley, B.; Nath, A.; Saraswati, S.; Bandyopadhyay, K.; Ray, B.C. Assessment of Rainfall Thresholds for Rain-Induced Landslide Activity in North Sikkim Road Corridor in Sikkim Himalaya, India. *J. Geogr. Environ. Earth Sci. Int.* **2019**, *19*, 1–14. [CrossRef]
69. Valenzuela, P.; Zêzere, J.L.; Domínguez-Cuesta, M.J.; Mora García, M.A. Empirical rainfall thresholds for the triggering of landslides in Asturias (NW Spain). *Landslides* **2019**. [CrossRef]
70. Bell, F.G.; Maud, R.R. Landslides associated with the colluvial soils overlaying the Natal group in the greater Durban region of Natal, South Africa. *Environ. Geol.* **2000**, *39*, 1029–1038. [CrossRef]
71. Rossi, M.; Luciani, S.; Valigi, D.; Kirschbaum, D.B.; Brunetti, M.T.; Peruccacci, S.; Guzzetti, F. Statistical approaches for the definition of landslide rainfall thresholds and their uncertainty using rain gauge and satellite data. *Geomorphology* **2017**, *285*, 16–27. [CrossRef]
72. Liu, Y.Y.; Dorigo, W.A.; Parinussa, R.M.; De Jeu, R.A.M.; Wagner, W.; McCabe, M.F.; Evans, J.P.; Van Dijk, A.I.J.M. Trend-preserving blending of passive and active microwave soil moisture retrievals. *Remote Sens. Environ.* **2012**, *123*, 280–297. [CrossRef]
73. Dorigo, W.A.; Scipal, K.; Parinussa, R.M.; Liu, Y.Y.; Wagner, W.; De Jeu, R.A.M.; Naeimi, V. Error characterisation of global active and passive microwave soil moisture datasets. *Hydrol. Earth Syst. Sci.* **2010**, *14*, 2605–2616. [CrossRef]
74. Liu, Y.; Weerts, A.H.; Clark, M.; Hendricks Franssen, H.J.; Kumar, S.; Moradkhani, H.; Seo, D.J.; Schwanenberg, D.; Smith, P.; Van Dijk, A.I.J.M.; et al. Advancing data assimilation in operational hydrologic forecasting: Progresses, challenges, and emerging opportunities. *Hydrol. Earth Syst. Sci.* **2012**, *16*, 3863–3887. [CrossRef]
75. Crosta, G. Regionalization of rainfall thresholds: An aid to landslide hazard evaluation. *Environ. Geol.* **1998**, *35*, 131–145. [CrossRef]
76. Michellier, C.; Pigeon, P.; Kervyn, F.; Wolff, E. Contextualizing vulnerability assessment: A support to geo-risk management in central Africa. *Nat. Hazards* **2016**, *82*, 27–42. [CrossRef]
77. López-Carr, D.; Pricope, N.G.; Aukema, J.E.; Jankowska, M.M.; Funk, C.; Husak, G.; Michaelsen, J. A spatial analysis of population dynamics and climate change in Africa: Potential vulnerability hot spots emerge where precipitation declines and demographic pressures coincide. *Popul. Environ.* **2014**, *35*, 323–339. [CrossRef]

Article

Empirical and Physically Based Thresholds for the Occurrence of Shallow Landslides in a Prone Area of Northern Italian Apennines

Massimiliano Bordoni [1,*], Beatrice Corradini [1], Luca Lucchelli [1], Roberto Valentino [2], Marco Bittelli [3], Valerio Vivaldi [1] and Claudia Meisina [1]

1 Department of Earth and Environmental Sciences, University of Pavia, Via Ferrata 1, 27100 Pavia, Italy; beatrice.corradini01@universitadipavia.it (B.C.); luca.lucchelli01@universitadipavia.it (L.L.); valerio.vivaldi@unipv.it (V.V.); claudia.meisina@unipv.it (C.M.)

2 Department of Chemistry, Life Sciences and Environmental Sustainability, University of Parma, Parco Area delle Scienze, 157/A, 43124 Parma, Italy; roberto.valentino@unipr.it

3 Department of Agricultural Sciences, University of Bologna, Viale Fanin 44, 40127 Bologna, Italy; marco.bittelli@unibo.it

* Correspondence: massimiliano.bordoni@unipv.it or massimiliano.bordoni01@universitadipavia.it; Tel.: +39-0382985830

Received: 20 November 2019; Accepted: 13 December 2019; Published: 16 December 2019

Abstract: Rainfall thresholds define the conditions leading to the triggering of shallow landslides over wide areas. They can be empirical, which exploit past rainfall data and landslide inventories, or physicallybased, which integrate slope physical–hydrological modeling and stability analyses. In this work, a comparison between these two types of thresholds was performed, using data acquired in Oltrepò Pavese (Northern Italian Apennines), to evaluate their reliability. Empirical thresholds were reconstructed based on rainfalls and landslides triggering events collected from 2000 to 2018. The same rainfall events were implemented in a physicallybased model of a representative testsite, considering different antecedent pore-water pressures, chosen according to the analysis of hydrological monitoring data. Thresholds validation was performed, using an external dataset (August 1992–August 1997). Soil hydrological conditions have a primary role on predisposing or preventing slope failures. In Oltrepò Pavese area, cold and wet months are the most susceptible periods, due to the permanence of saturated or close-to-saturation soil conditions. The lower the pore-water pressure is at the beginning of an event, the higher the amount of rain required to trigger shallow failures is. physicallybased thresholds provide a better reliability in discriminating the events which could or could not trigger slope failures than empirical thresholds. The latter provide a significant number of false positives, due to neglecting the antecedent soil hydrological conditions. These results represent a fundamental basis for the choice of the best thresholds to be implemented in a reliable earlywarning system.

Keywords: shallow landslides; rainfall; thresholds; physicallybased model; hydrological monitoring

1. Introduction

Shallow landslides are slope instabilities of a mass of soil and/or debris, which could involve the most superficial colluvial layers till around 2.0 m from ground level. Although they involve small volumes (10^1–10^5 m^3) of soil, they can be densely distributed across small catchments [1] and can affect slopes close to urbanized areas, provoking significant damages to cultivations and infrastructures, and sometimes causethe loss of human lives [2].

Rainfall is generally the main triggering factor [3]. Rainfall features leading to shallow landslides and the consequent temporal probability of occurrence at regional scale are generally estimated by means of rainfall thresholds, defined for different geological, geomorphological, and environmental settings [4]. These thresholds represent the main tool to estimate the daily or hourly level of hazard across a territory prone to shallow landslides or to implement earlywarning systems [5], representing the lower bound of rainfall conditions that caused the triggering of shallow landslides [3,6]. These thresholds are expressed as curves which separate the rainfall conditions leading to shallow slope failures from the ones where stability is maintained, sometimes with associated different probabilities of occurrence with uncertainties related to the possible incompleteness of the input data required to define the same thresholds [5,7].

The most widespread type of rainfall thresholds is the empirical one. These thresholds are reconstructed through the statistical analysis of empirical distributions of rainfall conditions that presumably resulted in the triggering of shallow landslides in a particular testsite [8]. The comparison between a multi-temporal inventory of shallow-landslides events and rainfall parameters measured in several points of the study area (e.g., in correspondence of raingauges) during the same analyzed time span is required in order to estimate these types of thresholds. Several authors proposed different methods for the estimation of empirical rainfall thresholds in different contexts all over the world [4,8–20]. In all cases, two different rainfall parameters were considered to build up boundary thresholds, namely cumulated event rainfall vs. rainfall duration or mean rainfall intensity vs. rainfall duration.

The use of only easily measurable rainfall data and the reconstruction based on the analysis of real past events, whether or not they triggered shallow landslides, makes empirical thresholds a reliable tool to estimate temporal probability of occurrence of shallow landslides at a large scale (catchment, regional, and national) [4,5]. Instead, these are sometimes limited in their effectiveness for different reasons. First, the shape of the thresholds is affected by the following: (i) the availability and quality of rainfall and of landslide information across the analyzed study area [21,22]; and (ii) the correct definition of the real rainfall features responsible for slope failures during a particular triggering event, generally linked to leakage of precise information about the moment of shallow landslides occurrence during a particular event [4,5]. Moreover, these types of thresholds do not take into account the unsaturated/saturated flow processes and the hydromechanical conditions of soils at the beginning of a particular rainfall event. The mechanical processes, which lead to shallow-slope failures, are in fact related to rainwater flows and water accumulation in the subsurface that provoke the increase in pore-water pressure and the consequent decrease of soil shear strength [22–26].

To overcome these limitations, rainfall thresholds can be estimated by means of a physicallybased model that can provide the assessment of the link between the rainfall features, the soil hydromechanical conditions before a rainfall event, and the shear strength response of the soils during the rainwater infiltration. In this case, the deterministic model estimates the response of the typical geological–geomorphological frame prone to shallow landsliding toward a particular rainfall event, defined by those parameters that are generally involved also for the reconstruction of an empirical threshold (cumulated event rainfall vs. rainfall duration; mean rainfall intensity vs. rainfall duration). This response is represented by the trend in time of the slope safety factor (Fs), during the modeled event. Triggering conditions are then represented by the rainfall patterns, which provoke the decrease of Fs below 1 (unstable conditions). Instead, if Fs stay higher than 1, shallow failures are not modeled (stable conditions). Some attempts were proposed to build up reliable physicallybased thresholds in some areas prone to shallow landslides, such as in Italian alpine catchments [26], catchments of the Central Italian Apennines [27], hilly catchments of Southern Italy [20,28–31], western hilly and mountainous settings of United States [32], and Chinese areas susceptible to shallow landsliding [33,34].

The main limitations of physicallybased thresholds are related to the most important disadvantages of the deterministic methods [35]: (i) requiring a significant amount of geotechnical, mechanical, and hydrological parameters for model simulation; and (ii) reconstructing the boundary conditions which represent, in the best way, the real soil and slope behaviors. Integration of meteorological measurements (e.g., rainfall) and hydrological soil parameters (e.g., pore-water pressure and water content) could help in obtaining a better insight into the quantitative effects of antecedent soil conditions on the triggering mechanism of shallow landslides. Thus, field monitoring allows us to improve the calibration of the physicallybased models used to reconstruct rainfall thresholds [23,32–34].

This paper aims to reconstruct and compare empirically and physicallybased rainfall thresholds for the occurrence of shallow landslides in a susceptible area of the Northern Italian Apennines (Figure 1). The main objectives of this work can be summarized as follows: (i) assessing empirical thresholds through the analysis of time series of rainfall data and of shallow-landslide inventories for the identification of the triggering and non-triggering events; (ii) calibrating a physicallybased model by the comparison between monitored and simulated soil hydrological parameters in correspondence of a test-site slope, which can be assumed to be representative of the typical geological, geomorphological, and environmental settings prone to shallow landsliding in the study area; (iii) assessing physicallybased thresholds through the application of the calibrated deterministic model in correspondence with the representative testsite for different rainfall events; (iv) comparing the two typologies of estimated thresholds and verifying their predictive capabilities through different inventories of occurred shallow landslides not used for the threshold reconstruction. Considered rainfall events corresponded to the ones that occurred in the 2000–2018 period and to other synthetic rainfalls characterized by strong average intensities and limited durations, which are not typical of the current climate of Oltrepò Pavese. Instead, their probability of occurrence may increase in the future due to the effects of climate change, which could cause an increase in very intense and short-duration rainfalls in Northern Italy, where the study area is located [36,37].

2. Materials and Methods

2.1. The Study Area

The study area is the hilly sector of Oltrepò Pavese (265 km^2 wide, Figure 1) that corresponds to the northern termination of the Italian Apennines. It is characterized by a complex geological and geomorphological setting [38–40] (Figure 1c). The northern part of the area presents a bedrock geology composed by sandstones and conglomerates overlying marls and evaporitic deposits. In this sector, superficial soils, derived from bedrock weathering, are mostly clayey or clayey–sandy silts. Their thickness, measured in micro-boreholes and in trenches, have a thickness ranging between a few tens of centimeters and 2 m. Hillslopes are steep, with an average slope angle between 15° and 20° and maximum values up to 35°. Instead, the central and southern parts of the study area are characterized by calcareous and marlyflyshes, alternated with sandstones, marls, and mélanges with a peculiar block-in-matrix at the outcrop scale. Due to the different lithology of the bedrock, superficial soils have a clayey or a silty clayey texture. Their thickness is generally in the order of more than 1 m, mostly ranging between 1.5 and 2 m from ground level, as measured in micro-boreholes and in trenches. Hillslopes have a medium steepness, with a typical slope angle of 8°–15°.

The slope elevation ranges between 60 and 500 m a.s.l. According to Koppen's classification, the climatic regime of the Oltrepò Pavese area is temperate/mesothermal, with a mean yearly temperature of 12 °C and an average yearly rainfall amount between 700 and 1000 mm, increasing from western to eastern sectors and from northern to southern sectors.

The area is significantly prone to shallow landsliding [24,39]. Several triggering events have occurred in Oltrepò Pavese since 1970s [38–40]. In the last 10 years, more than 2500 shallow landslides (Figure 1b) occurred in this area as a consequence of several rainfall triggering events during the winter and spring months. Most of the shallow landslides are classified as complex phenomena, starting as roto-translational slides and evolving into flows [41]. They are generally 10–70 m wide and 10–500 m long. Sliding surfaces are generally located at 1 m in depth [24]. Rainfall-induced shallow landslides affect medium–steep and steep slopes, with a slope angle of at least 8°–10°.

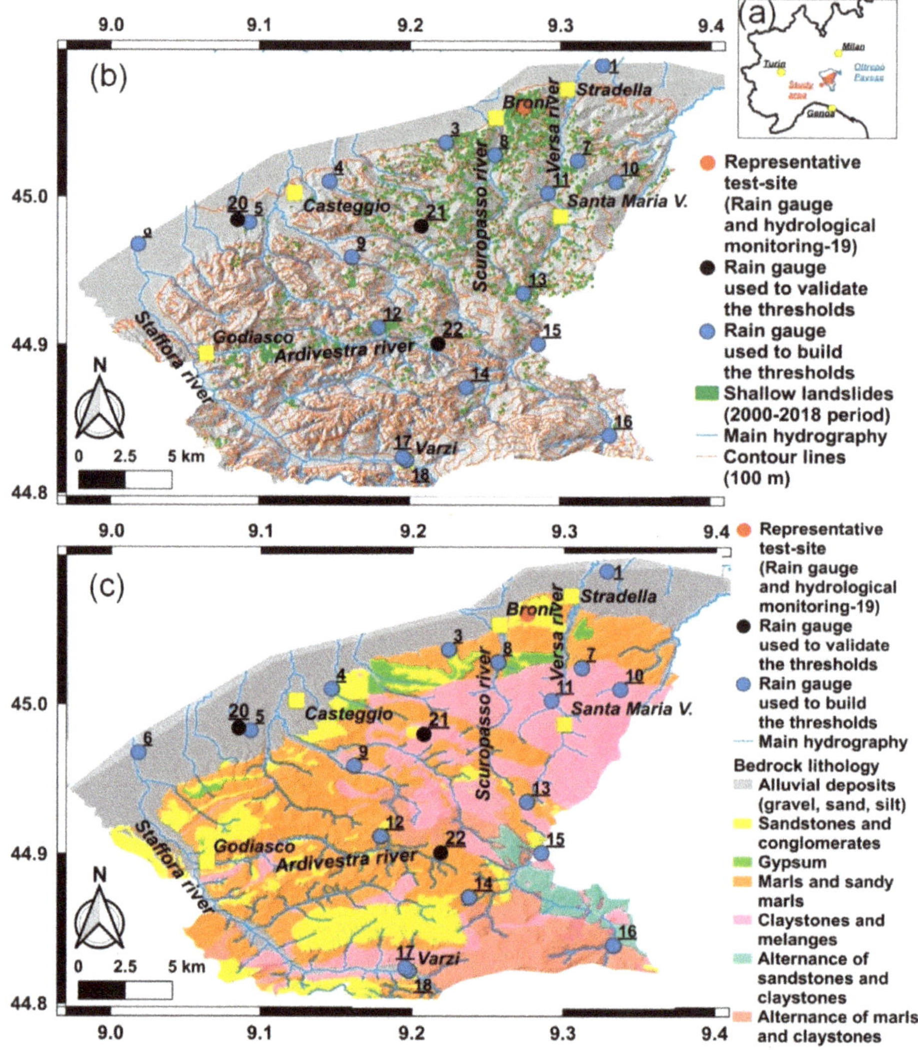

Figure 1. The study area, Oltrepò Pavese hilly zone: (**a**) location of the study area; (**b**) main geomorphological features and distribution of shallow landslides occurred in a 2000–2018 time span; (**c**) bedrock lithological features [36–38].

An integrated hydrometeorological monitoring station was installed on 27 March 2012 in a test-site slope located near the village of Montuè (red circle in Figure 1), in the northeastern part of Oltrepò Pavese. This testsite (Figure 2) is representative of the typical geological and geomorphological settings of Oltrepò Pavese areas most prone to shallow landslides for the following reasons: (i) the presence of triggering zones of past shallow landslides; (ii) its position in areas with medium–high susceptibility to shallow landslides according to previous studies [1,24]; and (iii) the typical geomorphological (hillslopes with medium–high thickness) and lithological features (clayey and silty soils) of the sectors most prone to shallow landsliding in the study area. In this station, rainfall amounts are measured through a rain gauge with an accuracy of 0.1 mm. Soil water content is measured by means of Time Domain Reflectometer (TDR) probes, with an accuracy of 0.01–0.02 m^3/m^3. Soil pore-water pressure is measured through a combination of tensiometers, with an accuracy of 1.5–2.0 kPa and measuring values higher than −100 kPa, and heat dissipation (HD) sensors, with an accuracy of 1.5–2.0 kPa and range of measure till -10^5 kPa.

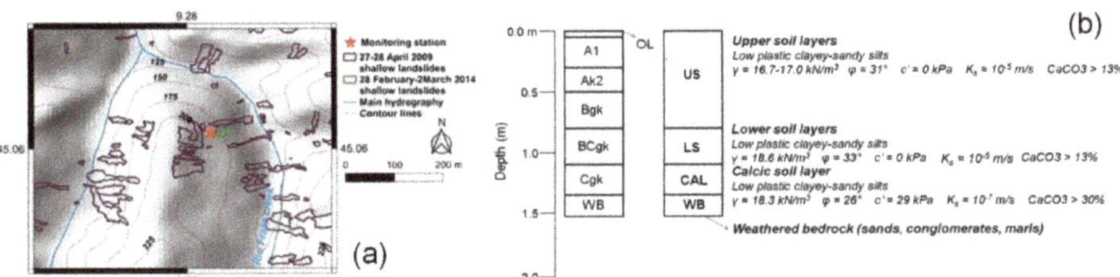

Figure 2. Representative test-site slope of Montuè: (**a**) morphology and shallow landslides distribution; (**b**) typical soil profile.

2.2. Reconstruction of the Empirical Thresholds

Empirical rainfall thresholds were reconstructed by implementing CTRL-T tool, written in R open-source software and freely available at: http://geomorphology.irpi.cnr.it/tools/rainfall-events-and-landslides-threshold. A detailed description of the algorithm is reported by Melillo et al. [8].

Figure 3 illustrates the logical framework of this method to assess the empirical rainfall threshold for a set of rain gauges and a multi-temporal shallow landslide inventory.

1. Identification of distinct rainfall events, along the hourly time series of each rain gauge. Different lengths of dry periods were considered, i.e., considering different lengths of dry periods, meaning significant time spans without rain, which depend on the climatic feature of the area. The length of a dry period separating two distinct events depends on the time required for the soil to dry out and on the season, namely a cold period with low temperatures and limited amount of evapotranspiration (C_c) and a warm one with high temperatures and significant evapotranspiration (C_W). The length in months of C_c and C_W was calculated for the study area, following the procedure described in Melillo et al. [8], based on the application of the monthly soil water balance (MSWB) model [42]. The average monthly potential evapotranspiration PET (Figure 4a) of the study area was estimated, since the data acquired from 2000 to 2018 for the meteorological stations of the study area (Figure 1b). The average monthly real evapotranspiration RET (Figure 4a) was then estimated, considering a maximum field capacity of 208 mm/m, which is typical of the soil types (clayey and silty soils) and of the land uses (shrubs, woods, and grapevines) of the study area. For each month, RET was divided by PET, obtaining the parameter of the monthly aridity index AI [43]. C_W was, then, the period when the soil exhibits a water deficit (RET<PET, AI<1) and was from May to September (Figure 4b). Conversely, C_c was the period when RET>PET and AI>1, from October to April (Figure 4b). The total amount of RET for C_W period was then divided by the total amount of RET for C_c period, obtaining an R index equal to 2.1. R is defined as the factor of difference between the length of the dry periods (i.e., time span between two different rainfall events) in C_W and C_c periods. The dry intervals used for the definition of the rainfall events was, then, the following (Table 1): (i) for the definition of isolated rainfall events, the dry period P_1 was of 3 and 6 h in C_W and C_c, respectively; (ii) for the definition of the sub-events, the dry period P_2 was of 6 and 12 h in C_W and C_c, respectively; (iii) for the definition of a rainfall event, the dry period P_4 was of 24 and 48 h in C_W and C_c, respectively. According to Melillo et al. [8,44], irrelevant rainfall sub-events (P_3) with a cumulated amount less than or equal to 1 mm had to be excluded in the calculation of the final events.
2. Linking rainfall data to shallow landslide events. For each shallow failure, related rain gauge was located in a circular buffer with a radius Rad of 10 km centered on the landslide location.

This radius was chosen according to the morphology of the study area (no significant variations on slope height, which could influence rainfall amount) and to the density of rain gauges in the study area (an average of one gauge per 13 km^2).

3. Estimation of rainfall conditions leading to shallow landslide triggering. For each event in the inventory, the algorithm estimates possible rainfall conditions (in terms of duration and cumulated rainfall amount) leading to slope failure. This allows us to consider a possible inaccuracy in the estimation of the rain features triggering a landslide due to the distance between the slope failure and the related rain gauge. A weight, W, was assigned according to the inverse square distance between the rain gauge and the landslide (d^{-2}), the cumulated rainfall amount (E), and the rainfall mean intensity (I) (Equation (1)):

$$W = d^{-2}E^{2}I^{-1} \quad (1)$$

Furthermore, a parameter, k, assumed equal to 0.84, allowed us to take into account the antecedent soil moisture condition depending on the amount of rain fallen in the previous days.

4. Reconstruction of rainfall threshold, based only on events triggering shallow landslides. Moreover, for each event, only the rainfall condition with the highest W value was selected. The threshold is defined as a power law curve which relates the cumulated rainfall amount (E) and the duration (D) of the events (Equation (2)):

$$E = (\alpha \pm \Delta\alpha)\ D^{(\omega \pm \Delta\omega)} \quad (2)$$

where α is the intercept of the curve;ω is the slope of the power law curve; and $\Delta\alpha$ and $\Delta\omega$ are the uncertainties of α and ω, respectively.

The threshold was defined by means of a frequentist method for reconstructing a 5% exceedance probability threshold, according to Brunetti et al. [13]. The fitting parameters of the curve and the related uncertainties were estimated through the calculation of thresholds of 5000 synthetic series of rainfall events. These series contained the same number of rainfall events that triggered landslides, but selected randomly with replacement, according to a bootstrap technique [45]. Analysis of these series allowed us to estimate the final threshold, that had α and ω corresponding to the mean values of the different bootstrap thresholds with their respective uncertainties ($\Delta\alpha$ and $\Delta\omega$).

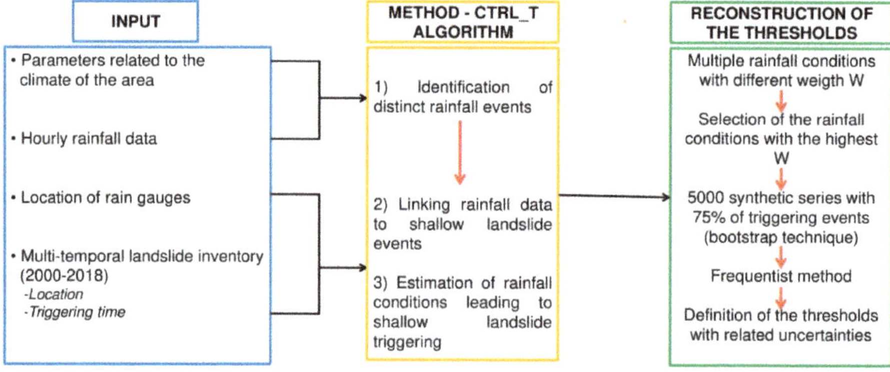

Figure 3. Flowchart of the methodology adopted for the reconstruction of the empirical thresholds.

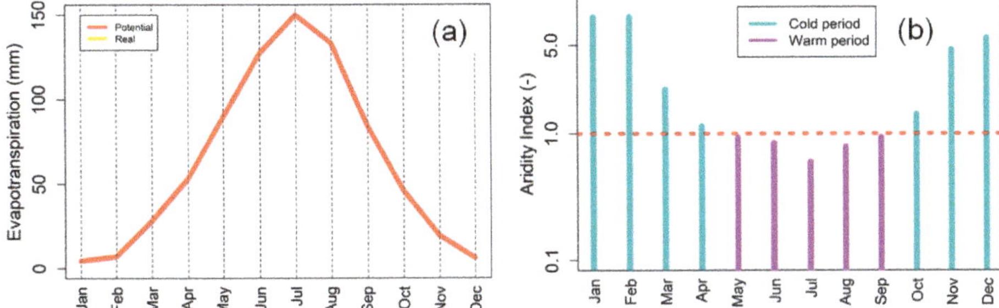

Figure 4. (**a**) Average monthly potential and real evapotranspiration in the study area; (**b**) warm and cold periods identified through the trend of the aridity index.

Table 1. Parameters used in CTRL-T tool for reconstructing rainfall events and defining the empirical threshold. (C_W) warm period in a year (May–September); (C_C) cold period in a year (October–April); (G_s) resolution of the rain gauge; (P_1, P_2, and P_4) time periods used to remove irrelevant amount of rain and to reconstruct rainfall events; (P3) irrelevant rainfall sub-events that had to be excluded in the calculation of the final events; (Rad) radius of the buffer to assign each landslide to the closest rain gauge.

Parameter	Value		Unit
	C_w	C_c	
G_s	0.1	0.1	mm
P_1	3	6	h
P_2	6	12	h
P_3	1	1	mm
P_4	24	48	h
Rad	10	10	km

The empirical threshold for the study area was assessed through this procedure, using hourly rainfall measurements collected in the period from January 2000 to December 2018, by a network of 19 rain gauges (blue circles in Figure 1), with a resolution (G_s) of 0.1 mm. Shallow landslides inventory of the same time span grouped the spatial and the temporal information of 143 triggering events. The spatial resolution of these events was about 1 km^2. For 44% of the events, the exact triggering hour was known, while for the remaining 56%, only the part of the day (generally, each 6 h in a day), when slope failure occurred, was identified. Among the landslide inventories, 30 events (11% of the inventory) were located by using information related to field surveys, 155 events (55% of the inventory)by means of aerial or satellite images [1,24], and 96 events (34% of the inventory) from newspapers and online chronicles.

2.3. Reconstruction of the Physicallybased Thresholds

The adopted procedure for the reconstruction of the physicallybased thresholds is composed of a series of consequent steps (Figure 5):

1. Identification of the representative testsite. Montuè was chosen as testsite exhibiting the typical geological and geomorphological settings prone to shallow landsliding in the study area (Figure 2a). The typical soil profile is shown in Figure 2b and described in detail in Bordoni et al. [24]. Test-site soils are low plastic clayey–sandy silts with a thickness mostly between 0.5 and 1.5 m. From the ground surface till 0.7 m, the upper soil layer (US) is characterized by a high content in carbonates (15%), as soft concretions, and unit weight in the order of 16.7–17.0 kN/m^3. Below this level, the lower soil layer (LS), from 0.7 to 1.1 m from the ground level, is characterized by similar carbonate content with respect to the US, but it presents a higher unit weight, ranging around 18.6 kN/m^3.

Between 1.1 and 1.3 m from ground level, there is a layer (CAL) characterized by a significant increase in carbonate content, till 35.3%. The weathered bedrock (WB), which is composed of sand and poorly cemented conglomerates, is at 1.3 m from the ground surface. US and LS are characterized by similar values of friction angle, equal to 31° and 33°, respectively, and by a nil effective cohesion. Instead, the CAL horizon has a lower value of friction angle (26°), but a higher effective cohesion (29 kPa). Saturated hydraulic conductivity (K_s) is in the order of 10^{-5} m/s, except for CAL level that is characterized by a saturated hydraulic conductivity of 10^{-7} m/s. Soil water characteristic curves (SWCCs) of the soil layers, fitted through Van Genuchten's [46] model, are quite similar [47], with wetting paths having saturated (θ_s) and residual (θ_r) water contents of 0.37–0.40 and 0.01 m^3/m^3, respectively. λ and n fitted parameters of the model range between 0.011 and 0.017 kPa^{-1} and between 1.20 and 1.40, respectively.

From the geomorphological point of view, the testsite has steep slopes (steepness higher than 15° and mostly between 26° and 35°) and is east-facing. The slope elevation ranges from 170 to 210 m a.s.l. The land use is mainly constituted by grass and shrubs passing to a woodland of black robust trees at the bottom of the slope. Shallow landslides occurred on this slope in response to two events: (i) on 27 and 28 April 2009, as a consequence of an extreme rainfall event characterized by 160 mm of cumulated rain in 62 h; (ii) between 28 February and 2 March 2014, as a consequence of an event of 68.9 mm in 42 h. The source areas of these shallow landslides have a slope angle higher than 25°, especially between 30° and 35°. The failure surfaces are located at a depth of around 1.0–1.2 m from ground level.

In the test-site slope, an integrated meteorological and hydrological monitoring station has been installed since 27 March 2012, and is still functioning [24]. The station allows meteorological parameters (rainfall depth, air temperature and humidity, air pressure, net solar radiation, wind speed and direction) to be measured, together with soil pore-water pressure, at depths of 0.2, 0.6, and 1.2 m from ground level, and soil water content, at depths of 0.2, 0.4, 0.6, 1.0, 1.2, and 1.4 m from ground level. Further details on this monitoring station are reported in Bordoni et al. [24,47].

2. Physicallybased model, to model the hydrological and the mechanical responses of the slope to different rainfall events. The well-established physicallybased methodology named TRIGRS [48] was used. It has been already applied successfully for modeling the occurrence of these phenomena [1,49–54]. This physicallybased model considers the method outlined by Srivastava and Yeh [55] and Iverson [56] to explain shallow landslide triggering in relation to rainwater infiltration both in saturated or unsaturated soil conditions, assuming an impermeable basal boundary at a finite depth and a simple runoff-routing scheme. The model assesses the pore-water pressure and the slope safety factor (Fs) during different moments of a rainfall event, due to rainwater infiltration.

A 5 × 5 m digital elevation model (DEM), realized through LiDAR data acquired in 2008 and 2010 by the Italian Ministry for Environment, Land, and Sea, provided the topographic basis for the study area and was used to derive the slope angle and the flow direction maps required by the model. Hydrological parameters required for the hydrological model of TRIGRS were K_s, θ_s, θ_r, and the ξ parameter, which represents the fitting parameter of Gardner's [57] SWCC equation. ξ was estimated based on the λ and n fitting parameters of Van Genuchten's model through the method proposed by Ghezzehei et al. [58] (Equation (3)):

$$\xi = \lambda(1.3 \times n) \tag{3}$$

Hydrological boundary condition of the model corresponded to the presence of a low permeable soil level, which limits the infiltration of the rainwater and causes the uprising of a perched-water table in correspondence of the most intense rainfall events. As demonstrated by Bordoni et al. [24], this can be assumed as the main triggering mechanism of shallow failures in the study area. This water table developed in correspondence of the CAL layer, due to its lower permeability than the overlying soil levels, at about 1.1–1.2 m from ground level. The perched water table could rise up to 0.8–1.0 m from ground level, in LS layer, during the most intense rainfall events. TRIGRS modeled water table depth in the soil profile and the corresponding pore-water pressure profiles during a rainfall event, considering also the water table depth at the beginning of a modeled event, which was estimated through the most superficial measured pore-water pressure (in US soil layer; ρ_{US}) [59] (Equation (4)):

$$d_W = \rho_{US}/(\gamma_W \cos^2\beta), \quad (4)$$

where γ_W is the water unit weight (9.8 kN/m^3), and β is the slope angle.

In TRIGRS model, an infinite slope stability analysis is coupled with the hydrological model to compute the Fs at different time instants at different points and depths in the analyzed area (Fs(z, t)), considering its change over time during a rainfall event, due to change in pore-water pressure ρ(z, t) (Equation (5)):

$$Fs(z, t) = (\tan\varphi'/\tan\beta) + [(c' - \rho(z, t)\gamma_W \tan\varphi')/(\gamma z \sin\beta\cos\beta)], \quad (5)$$

where φ' is the soil shear strength angle, c′ is the effective cohesion, γ is the soil unit weight, and z is the soil depth.

3. Table 2 lists the soil hydrological and geotechnical values of TRIGRS input parameters. Since US and LS layers had similar values of the different required parameters (K_s, θ_s, θ_r, ξ, φ', c', and γ), a uniform soil profile was assumed, and then a unique set of input values was inserted in TRIGRS. In regard to the parameters of SWCCs (θ_s, θ_r, ξ), the values of the wetting path of SWCCs were taken into account, according to the modeling of the process of rainwater infiltration [22,24]. The sliding surface depth (z) was assumed equal to 1.0 m, according to the typical depths observed in past shallow failures.
4. Corroboration of the physicallybased model. The modeled pore-water pressures at 1.2 m from ground level, which corresponds to a level very near to the typical observed sliding surface depth, were then compared with the measured values for different rainfall events. The reliability of the model fit was evaluated with the root mean square error (RMSE) statistical index (Equation (6)):

$$RMSE = \sqrt{\frac{\sum_{i=1}^{n_{tot}} (\rho_{o,i} - \rho_{m,i})^2}{n_{tot}}} \quad (6)$$

where $\rho_{o,i}$ is the observed pore-water pressure at the ith hour of the considered rainfall event, $\rho_{m,i}$ is the pore-water pressure estimated by the model at the same ith hour of the same event, and n_{tot} is the number of observations, which corresponds to the duration of the rainfall event (in hours).
5. Modeling slope safety factor (Fs) for different rainfall events. Once both the implementation and validation had been completed, the physicallybased model was used to estimate Fs of the testsite for rainfall scenarios corresponding to the ones identified by CTRL-T algorithm during the phase of reconstruction of rainfall events. Furthermore, synthetic rainfalls characterized by average intensities of 50, 75, and 100 mm/h, for a duration ranging between 1 and 12 h, were also modeled. A modeled rainfall event represented a triggering moment for shallow landslides if Fs dropped below 1 (unstable conditions) in correspondence of the sectors of the testsite where typically shallow landslides source areas developed in the past, namely the sectors with a slope angle higher than 25°. Instead, if Fs stayed higher than 1 (stable conditions) in all the testsite, the rainfall event was not considered to be a triggering event. Each event was modeled by considering different initial pore-water pressures representative of the typical antecedent conditions before

landslide triggering, particularly in correspondence of the depth where typically sliding surfaces developed in the testsite (1.0 m).

6. Reconstruction of the rainfall thresholds. The method used for the reconstruction of the physicallybased thresholds was the same applied for the assessment of the empirical ones. In this case, only rainfall scenarios leading to shallow-landslide triggering based on the model application were considered. As for empirical thresholds, the physicallybased ones had fitting parameters corresponding to the mean values of the different bootstrap thresholds, with their respective uncertainties. Different rainfallthresholds could be reconstructed, according to the different initial pore-water pressure conditions used in modeling the rainfall events.

Table 2. Soil input parameters of TRIGRS model. (K_s) saturated hydraulic conductivity; (θ_s) saturated water content; (θ_r) residual water content; (ξ) Gardner's model fitting parameter; (φ') soil friction angle; (c') soil cohesion; and (γ) soil unit weight; (z) soil depth.

Parameter	Value	Unit
	C_w	
K_s	$1.0 \cdot 10^{-5}$	m/s
θ_s	0.39	m^3/m^3
θ_r	0.01	m^3/m^3
ξ	0.014	kPa^{-1}
φ'	32	°
c'	0.0	kPa
γ	17.8	kN/m^3
z	1.0	m

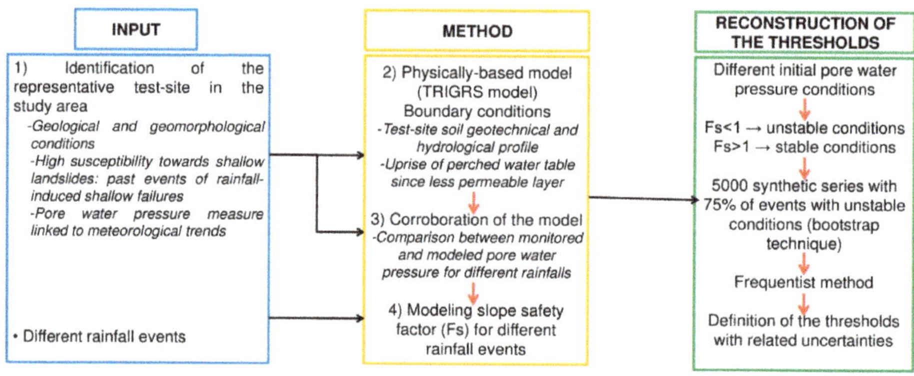

Figure 5. Flowchart of the methodology adopted for the reconstruction of the physicallybased thresholds.

2.4. Corroboration of the Reconstructed Threshold

A comparison between empirical and physicallybased thresholds estimated for Oltrepò Pavese area was performed in order to evaluate the predictive capability of both these models, as well as their advantages and limitations. Their validation was carried out through a dataset of events that took place during the period of August 1992–August 1997. For this time span, rainfall data were collected in correspondence of 3 rain gauges (black circles in Figure 1), while location and triggering moment of shallow landslides were recorded from local newspapers and information of fire brigades.

CTRL-T tool was used to reconstruct the different rainfall events also for this dataset, using the same required input parameters listed in Table 1. For the empirical thresholds, the final position on the graph below or above the defined thresholds was evaluated. In regard to the physicallybased thresholds, similarly to what done for their definition, the identified rainfall conditions were used as input data to apply the TRIGRS model and to assess slope Fs at the representative testsite. Thus, each event was estimated as responsible to trigger or not to trigger shallow landslides based on the assessed values of Fs in correspondence of the sectors typically affected by shallow landslides. Then, the position on the graph below or above the defined physicallybased thresholds was evaluated.

In the case of the estimated physicallybased thresholds validation, it is required to assess the pore-water pressure condition at the same depth of the observed sliding surfaces (1.0 m). For these reasons, time series of pore-water pressures at 1.0 m were modeled through HYDRUS-1D [60] software, considering the same physical and hydrological boundary conditions used for TRIGRS implementation. This model was chosen because it can assess long time series of soil hydrological parameters influenced by intermittent dry and rainy periods in a reliable way [60]. HYDRUS-1D was implemented for each of the 3 meteorological stations included in the validation dataset. Soil hydrological properties (Table 2) and boundary conditions corresponded to those used for the application of the TRIGRS model. Meteorological conditions required by the model were the rainfall amounts and air temperatures that were used to model the evapotranspiration effects through Hargreaves et al.'s [61] equation. Modeled time spans started from a significant dry period of a year, corresponding to 1 August 1992. In Oltrepò Pavese, early August is characterized every year by dry conditions of soils, which keep steady along depth, due to low rainfalls and high evapotranspiration rates in the previous summer months (June–July). In particular, a pore-water pressure equal to −993 kPa was assumed, according to the field measurements reported in Bordoni et al. [24,47]. This modeling approach was already implemented for the estimation of time series of soil hydrological parameters in other Italian regions prone to shallow landslides [53,62], obtaining a good estimation of the initial pore-water pressure conditions of a triggering event.

Statistical indexes were then calculated to evaluate the predictive capabilities of both types of thresholds for the validation dataset. Considering a rainfall threshold as a binary classifier of rainfall conditions leading to shallow landslides, its performance can be assessed by computing a 2 × 2 a posteriori contingency table [15]. Thus, each rainfall event can correspond to occurrence (true) or nonoccurrence (false) of shallow landslides, while the model can be considered as positive (successful prediction) or negative (wrong prediction). Accordingly, the following indexes can be classified [17,63]: true positives (TP), i.e., rainfall conditions exceeding the threshold causing shallow landslides; false positives (FP), i.e., rainfall conditions exceeding the threshold but without real triggering of shallow landslides; true negatives (TN), i.e., rainfall conditions below the threshold and without shallow landslides occurrence; false negatives (FN), i.e., rainfall conditions below the threshold but causing shallow landslides.

3. Results

3.1. Reconstructed Rainfall Events

A total of 5231 rainfall events were identified by exploiting the database for the period of 2000–2018. Seventy-two percent of the events were detected in the cold season, while only 28% of the events were identified in warm periods, confirming the higher aridity index of warm months calculated for the study area (Figure 4b). In cold periods, the duration of the events ranged between 1 and 280 h, while cumulated amounts ranged between 1.0 and 285.0 mm. About 80% of the events were characterized by a duration lower than 50 h, and cumulated amounts were lower than 50 mm (Figure 6a,b). Instead, in warm periods, duration and cumulated amounts of the events ranged between 1 and 67 h and between 1.0 and 155.2 mm, respectively. Ninety-five percent of the events had a duration lower than 30 h and cumulated amounts lower than 50 mm (Figure 6a,b).

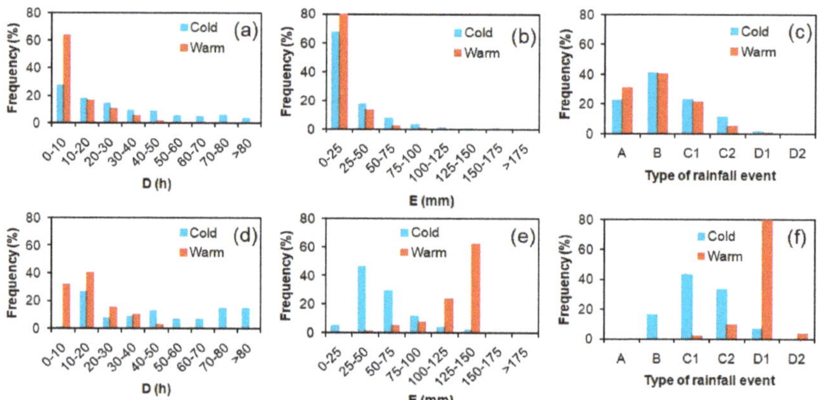

Figure 6. Frequency distribution of duration, cumulated amount, and typology of the rainfall events identified in the period 2000–2018: (**a**) frequency distribution of duration (D) of the rainfall events; (**b**) frequency distribution of cumulated amount (E) of the rainfall events; (**c**) frequency distribution of typology of rainfall events; (**d**) frequency distribution of duration (D) of the rainfall events triggering shallow landslides; (**e**) frequency distribution of cumulated amount (E) of the rainfall events triggering shallow landslides; and (**f**) frequency distribution of typology of rainfall events triggering shallow landslides.

The intensity of the events was classified according to Alpert et al.'s [64] classification. According to the duration and cumulated amount of rainfalls, 87% of the events were classified as light (A), light–moderate (B) or moderate–heavy (C1) (Figure 6c). These events represent the typical rainfalls due to regional frontal systems, which characterize cold months in Northern Italy. Only 13% of the events were heavy (C2) or heavy–torrential (D1) (Figure 6c). Similar results were obtained for warm months, when light (A) and light–moderate (B) were more abundant (Figure 6c). These rainfalls are caused by local convective storms, which are typical of the warm months in Northern Italy. In warm periods, more intense rainfalls were less probable (6% of the total events). Triggering events of shallow landslides occurred more frequently during cold months (Figure 7). Eighty-four percent of the events occurred between January and April and between October and December, with the highest amount in March (20%). Only 16% of landslides triggering occurred in warm months, between May and September, with a higher percentage in August (10%). In cold months, 93% of the total triggering events were light–moderate (B), moderate–heavy (C1), or heavy (C2) rainfalls, with duration between 4 and 105 h and cumulated amount between 30.4 and 133.7 mm (Figure 6d–f). Instead, in warm months, triggering rainfalls were mostly (87%) heavy–torrential (D1) or torrential (D2) rainfalls, characterized by duration between 5 and 15 h and cumulated amounts between 106.8 and 155.2 mm (Figure 6d–f).

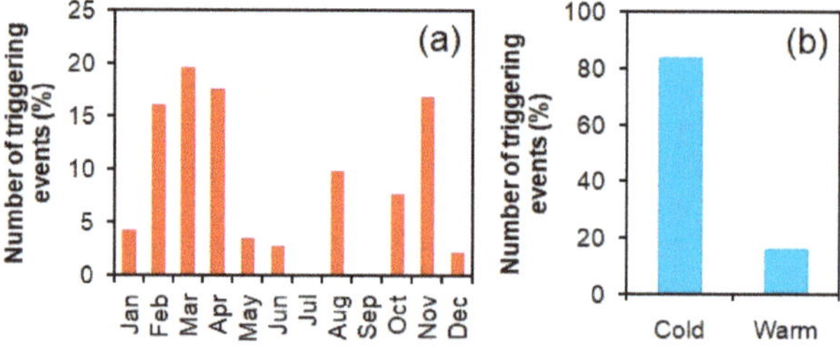

Figure 7. Distribution of shallow-landslide-triggering events in the period 2000–2018 along the different months of the year (**a**) and cold and warm periods in a year (**b**).

3.2. Pore-Water Pressure Distribution at Sliding Surface Depth

The characterization of pore-water pressure distribution during different seasons in correspondence of the typical depth of sliding surface was performed through the monitoring of data acquired at 1.2 m from ground level, in the test-site slope, in the period March 2012–December 2018. Averagely, the soil horizon kept in unsaturated conditions during the year, reaching minimum values during summer warm months (till −993 kPa), when strong evapotranspiration was not compensated due to limited rainfall amounts. During cold months, soil horizon re-wetted due to a more significant infiltration of rainwater and a more limited evapotranspiration. In these timespans, pore-water pressure grew to values close to 0 kPa, testifying conditions close to saturation in this soil level. In several periods during the cold time span of a year, pore-water pressure could reach positive values (till 12 kPa), especially when several rainfall events were spaced out by limited dry periods.

The distribution of the measured values of pore-water pressure (Table 3 and Figure 8) presents a certain degree of Gaussian trend, as confirmed by the values of the skewness very close to 0 and by the results of Shapiro–Wilk test, whose $W_{S\text{-}W}$ statistic did not allow us to reject the null hypothesis of gaussianity at 95% confidence level ($W_{S\text{-}W}$ = 0.92; *p*-value = 0.06). The first quartile of this distribution was equal to −846 kPa, while the third one was of −20 kPa.

Table 3. Main statistics of the distribution of the pore-water pressure values at typical depth of shallow-landslide sliding surface (1.0–1.2 m from the ground level) during the monitored time span (March 2012–December 2018) at the test-site slope: (Sd) standard deviation; (Median) median; (Min) minimum value; (Max) maximum value; (I quart) first quartile; (III quart) third quartile; (Skew) skewness; ($W_{S\text{-}W}$) statistic of the Shapiro–Wilk test, applied to test the gaussianity of the distribution; (*p*-value) confidence level of the statistic of the Shapiro–Wilk test.

Mean (kPa)	Sd (kPa)	Median (kPa)	Min (kPa)	Max (kPa)	I quart. (kPa)	III quart. (kPa)	Skew (kPa)	$W_{S\text{-}W}$ (–)	*p*-Value (–)
−464	387	−470	−993	12	−846	−20	0.007	0.92	0.006

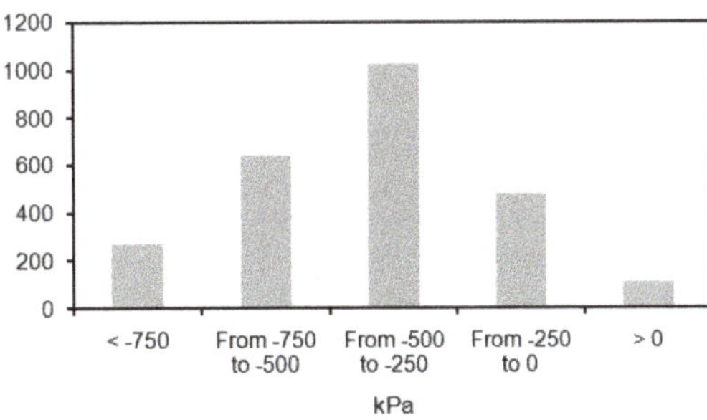

Figure 8. Histogram of distribution of the pore-water pressure values at typical depth of shallow-landslides sliding surface (1.0–1.2 m from the ground level) during the monitored time span (March 2012–December 2018) at the test-site slope.

Monitoring data allowed to exploit information on triggering events occurred in cold periods. Bordoni et al. [24] showed that during the observed event of 28 February–2 March 2014 at the test-site slope, pore-water pressure was about 0 kPa, at the beginning of the rainfall event which caused the shallow landslide triggering. Only this information is not enough to characterize exhaustively the antecedent hydrological conditions immediately before a rainfall able to provoke landslides in the study area. To analyze a higher range of soil hydrological conditions causing shallow landslides triggering and to test the effect of initial pore-water pressure on the definition of a threshold, physicallybased thresholds were then estimated by modeling the response of the soil to different rainfall events, starting from an initial pore-water pressure condition of −20, −10, or 0 kPa. For simplicity, they are named TRIGRS/-20, TRIGRS/-10, and TRIGRS/0, respectively. In this way, a significant amount of the typical pore-water pressure values at depths of 1.0–1.2 m was considered in the definition of physicallybased threshold, as the third quartile of the measured values was in fact equal to −20 kPa.

3.3. Comparison between Measured and Estimated Pore-Water Pressure at Sliding Surface Depth

Figure 9 shows the comparison between measured and modelled pore-water pressure trends at the typical sliding surface depth (1.0–1.2 m from ground level) in the representative test-site slope for different rainfall events reported in Table 4. The selected events represented typical rainfall scenarios occurring in the study area during the analyzed time span and were characterized by initial pore-water pressure conditions similar to the ones chosen for the reconstruction of the physicallybased thresholds.

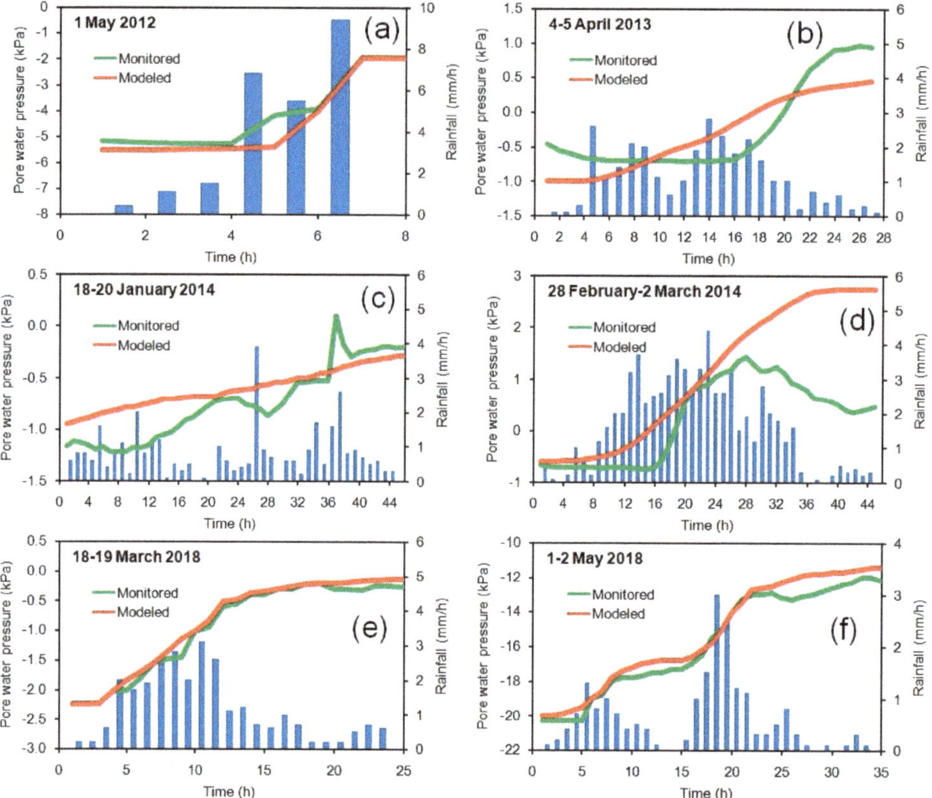

Figure 9. Comparison of measured and estimated by TRIGRS pore-water pressure trends at the typical depth of shallow-landslide sliding surface (1.0–1.2m from ground) for the selected rainfall events at the monitoring station in the representative testsite: (**a**) 1 May 2012; (**b**) 4–5 April 2013; (**c**) 18–20 January 2014; (**d**) 28 February–2 March 2014; (**e**) 18–19 March 2018; (**f**) and 1–2 May 2018.

Table 4. Measured initial and final pore-water pressure values versus those computed by TRIGRS at the typical depth of shallow-landslide sliding surface for the selected rainfall events at the monitoring station in the representative testsite. The number related to each rainfall event corresponds to the following: (1) event of 1 May 2012; (2) event of 4–5 April 2013; (3) event of 18–20 January 2014; (4) event of 28 February–2 March 2014; (5) event of 18–19 March 2018; and (6) event of 1–2 May 2018.

Rainfall Event	Duration (h)	Cumulated Amount (mm)	Initial Pore-Water Pressure at Potential Sliding Surface (1.0–1.2 m from Ground) (kPa)		Final Pore-Water Pressure at Potential Sliding Surface (1.0–1.2 m from Ground) (kPa)		RMSE (kPa)
			Measured	Modeled by TRIGRS	Measured	Modeled by TRIGRS	
1	6	24.7	−5.2	−5.5	−1.9	−2.0	0.5
2	26	29.5	−0.6	−1.0	0.9	0.5	0.4
3	44	34.6	−1.1	−1.0	−0.2	−0.8	0.3
4	43	68.9	−0.7	−0.6	0.4	2.8	1.2
5	23	27.4	−2.2	−2.2	−0.2	−0.1	0.1
6	33	20.0	−20.3	−20.0	−12	−11.4	0.6

Despite the different features of the tested events, the trend of the pore-water pressure modeled through the physicallybased method (TRIGRS model) seems to simulate in a reliable way the field measurements during each analyzed rainfall event. Differences between measured and estimated values are always lower than 2 kPa at the analyzed soil depth. RMSE values of 0.1–1.2 kPa confirmed the reliability of these simulations. The highest pore-water pressure value at the end of each rainfall event was generally attained through the physicallybased method, unless for the event occurred on 28 February–2 March 2014. Although the model results were in very good agreement with the real measured values, modeling errors in pore-water pressure trends could be linked to the simplification provided by the TRIGRS model with regard to soil hydrological features. In particular, TRIGRS model does not consider a layered soil profile, thus forcing to assume average values of the required soil parameters across the analyzed soil profile.

3.4. Reconstruction of Empirical and Physicallybased Thresholds

Rainfall thresholds reconstructed with different methodologies are shown in Figure 10. All of these functions were characterized by a low uncertainty of the two fitting parameters (0.2–1.9 for α, 0.01–0.04 for ω). Instead, equations of the reconstructed thresholds were very different from each other. Average values of the α parameter ranged between 11.2 and 225.0, while mean values of the ω parameter ranged between 0.08 and 0.30. Empirical threshold and physicallybased threshold considering initial pore-water pressure of −20 kPa (TRIGRS/-20) were steeper than the other two functions, as testified by significantly higher values of the ω parameter (0.25–0.30 against 0.08–0.12, respectively). The empirical threshold had the lowest value of intercept α (11.2 ± 0.2). Within physicallybased thresholds, the lower was the value of α the higher is the initial pore-water pressure used to reconstruct the threshold. The α parameter of TRIGRS/0 was about 5 times and 10 times lower than the values for the thresholds TRIGRS/–10 and TRIGRS/–20, respectively.

The practical effects of these differences are clearer when the cumulated amount of rain able to trigger shallow landslides is calculated for different rainfall durations (between 10 and 50 h), based on the defined thresholds (Table 5). For the same duration, the amount of rainfall able to trigger shallow landslides was lower by considering the empirical threshold than physicallybased threshold.

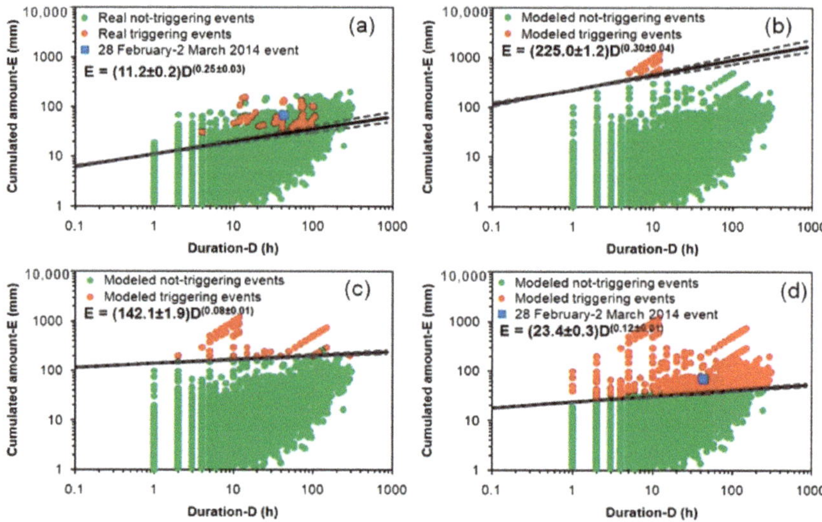

Figure 10. Rainfall thresholds (black line for the average function, gray dot lines for the average functions plus or minus the uncertainties) for the occurrence of shallow landslides in the study area: (**a**) threshold reconstructed through the empirical method; (**b**) threshold reconstructed through the physicallybased method considering an initial pore-water pressure of −20 kPa at the depth of the sliding surface (TRIGRS/–20); (**c**) threshold reconstructed through the physicallybased method considering an initial pore-water pressure of −10 kPa at the depth of the sliding surface (TRIGRS/–10); (**d**) threshold reconstructed through the physicallybased method considering an initial pore-water pressure of 0 kPa at the depth of the sliding surface (TRIGRS/0).

Table 5. Ranges of different rainfall cumulated amount enough to trigger shallow landslides for different rainfall duration, calculated using the different reconstructed thresholds.

Duration (h)	Cumulated Amount (mm)			
	Empirical Thresholds	PhysicallyBased Thresholds (−20 kPa) (TRIGRS/–20)	PhysicallyBased Thresholds (−10 kPa) (TRIGRS/–10)	PhysicallyBased Thresholds (0 kPa) (TRIGRS/0)
10	18.3–21.7	407.2–494.8	162.5–174.8	29.8–32.0
20	21.3–26.4	487.7–626.4	170.6–186.1	32.1–35.0
30	23.2–29.5	541.9–719.0	175.5–193.0	33.6–36.9
40	24.8–32.0	584.0–792.9	179.0–198.1	34.7–38.3
50	26.0–34.1	618.9–855.3	181.9–202.1	35.5–39.4

Using the TRIGRS/0 threshold, the amount of critical cumulated rainfall increases of 5.3–10.5 mm, for the same rainfall duration. For the other physicallybased thresholds, the increase of the critical cumulated amount was more significant. Considering the TRIGRS/-20 threshold, the critical cumulated rain was about 22–25 times higher than that defined by using empirical threshold, for the same duration. Instead, considering the TRIGRS/-10 threshold, the required rainfall able to trigger shallow landslides was about 6–9 times higher than that defined using empirical threshold, for the same duration.

For the empirical thresholds, it is important to highlight that 26.2% of the rainfall events which did not cause the real triggering of shallow landslides (green circles in Figure 10a) was located above the defined thresholds (false positives). Instead, the percentage of rainfall events modeled as not able to trigger landslides but located above the thresholds was lower than 0.5% for all types of physicallybased thresholds. Considering the only triggering event when also the initial pore-water pressure at the depth of the sliding surface was known (28 February–2 March 2014 event at the testsite, 68.9 mm of rain fallen in 42 h), the empirical threshold and the TRIGRS/0 threshold correctly identified this rainfall scenario as a triggering event, since it was located above the defined thresholds (blue square in Figure 10a,d).

To verify the reliability of a rainfall threshold, it is required to quantify its effectiveness in distinguishing rainfall events able to or not able to trigger shallow landslides. This procedure could not be performed for both empirically and physicallybased thresholds by using only the database of the events already utilized to build these models. In fact, a direct comparison between the reliability of different types of thresholds could not be performed, due to the intrinsic outputs of the methods used to reconstruct each threshold. In particular, in the definition of each physicallybased threshold, all the modeled events whose Fs was lower than 1.0 potentially represented a triggering event. Instead, in the database of the triggering events that occurred between 2000 and 2018 and were used as input to build the different thresholds, the initial pore-water pressure at the depth of the sliding surface was measured only for the event of 28 February–2 March 2014 monitored at the testsite. Thus, it is not possible to link an initial pore-water pressure to all the events, neglecting the possibility to quantify the predictive capability of the different thresholds in identifying triggering or non-triggering events.

For these reasons, the validation and the evaluation of the predictive capability of the thresholds were performed by using an external database of rainfall and shallow-landslide events available for another period.

3.5. Validation of the Reconstructed Thresholds

For the validation period of August 1992–August 1997, 488 rainfall events were identified (Figure 11a). Twenty of these events represented conditions able to trigger shallow landslides in the study area. The triggering events occurred in the cold period of the year, especially in November and in February–March and were classified as light–moderate (B), moderate–heavy (C1), or heavy (C2) rainfalls according to Alpert et al.'s [64] classification, with a duration between 9 and 218 h and cumulated amount between 38.0 and 129.4 mm.

During the modelled time-span, evapotranspiration rates (Figure 11b) ranged between 0 and 12 mm/day. Values close to 0 mm/day occurred in cold winter months, while summer dry months were characterized by a warmer condition that allowed evapotranspiration.

Pore-water pressure trend at typical depths of the sliding surfaces (Figure 11c) was characterized by the typical hydrological behaviors of the soil layers at the same depth, as inferred by field data at the test-site slope during the monitoring period since March 2012 [24,47]. These soil layers reached the driest condition during warm months of the year, especially between June and October, when few thunderstorms were interspersed by prolonged periods without rain and with significant evapotranspiration rates. The first significant rainfall events of October–November, characterized by at least 30 mm of rain fallen in 24 h, caused a slight increase in pore-water pressure. A more evident pore-water pressure increase was observed in the following wet period, between November and January, when rainfall events of at least 20–30 mm/day were rather close to each other, and evapotranspiration rates were limited (<1 mm/day) (Figure 11c). During both cold and wet months, pore-water pressure generally remained lower than –20 kPa, reaching saturated conditions in correspondence of other important rainfall events of at least 20 mm/day. Saturated conditions and the development of positive pressures (corresponding to the formation of a perched water table) were most probable till the end of March. In April, pore-water pressure began to decrease down to values lower than –20 kPa, due to an increase in evapotranspiration rates (about 4–5 mm/day) and to an increase in dry days between two different rainfalls. Instead, till the end of June, after very intense events of at least 50 mm of rain fallen in 12 h, a transient increase of pore-water-pressure till values of about −10 kPa was observed. Pore-water pressure tended to decrease very fast, till the driest soil conditions, since the end of June-beginning of July.

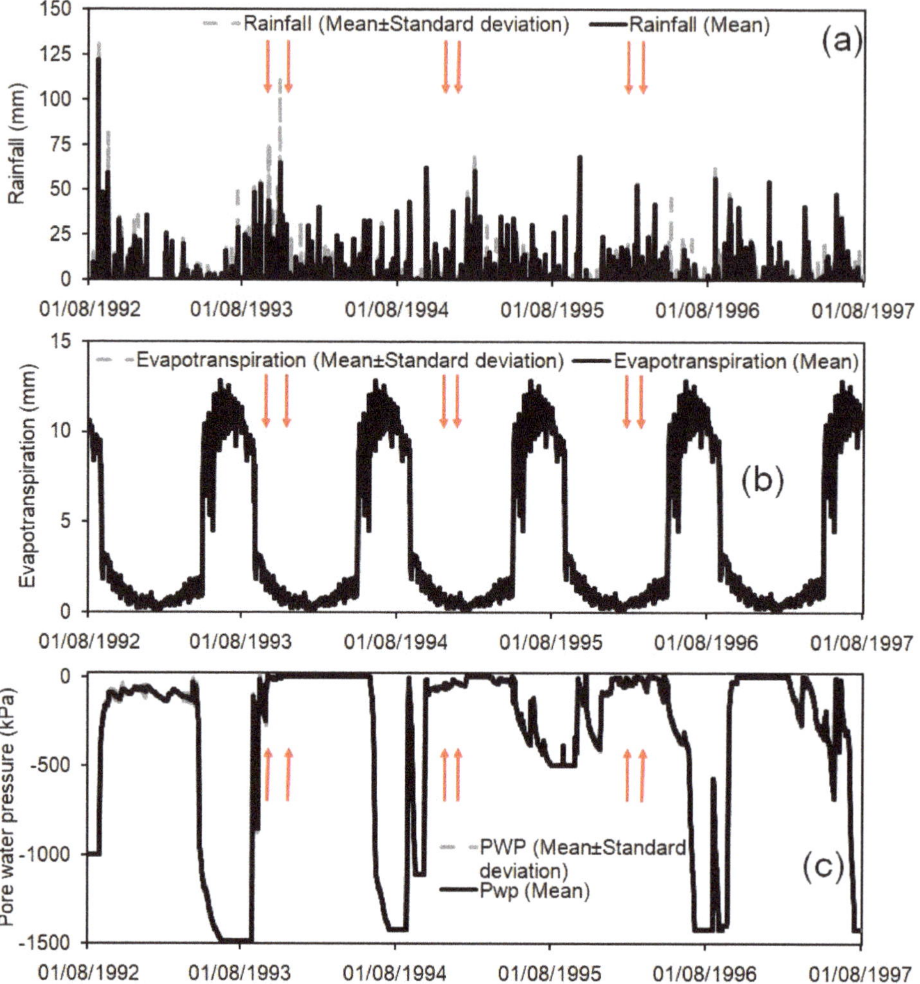

Figure 11. Rainfall amounts (**a**), evapotranspiration rates (**b**), and modeled pore-water pressure at the typical depth (1.0–1.2 m form ground) of the shallow-landslide sliding surface (**c**) for the time span August 1992–August 1997.

The distribution of the modeled values of pore-water pressure (Table 6 and Figure 12) was similar to that observed in the field since March 2012, confirming the reliability of the model in representing the real soil hydrological conditions. Main significant differences between monitored and modeled distributions regarded the lowest value of pore-water pressure (−993 and −1483 kPa for monitored and modeled trends, respectively) and the first quartile (−846 and −484 for monitored and modeled trends, respectively), together with the degree of gaussianity, which was not shown in the distribution of the modeled values (skewness of −1.18; $W_{S\text{-}W}$ statistic of Shapiro–Wilk test of 0.71, p-value < 0.01; Table 6 and Figure 12). Instead, the third quartile of the distribution of the modeled pore-water pressure was equal to −22 kPa, which is very similar to that one of the monitored values (−20 kPa). These results confirm the reasonable choice of considering initial conditions of pore-water pressure higher than −20 kPa for the reconstruction of the physicallybased thresholds. The modeled value of pore-water pressure at the sliding surface depth at the beginning of a triggering event in the time span of the validation phase was around 0 kPa every time, which is also in agreement with the initial conditions in correspondence of the monitored triggering event of 28 February–2 March 2014 at the test-site slope [24].

Table 6. Main statistics of the distribution of the pore-water pressure values at typical depth of shallow landslides sliding surface (1.0–1.2 m from the ground level), modeled for the period August 1992–August 1997 at the test-site slope: (Sd) standard deviation; (Min) minimum value; (Max) maximum value; (I quart) first quartile; (III quart) third quartile; (Skew) skewness; ($W_{S\text{-}W}$) statistic of the Shapiro–Wilk test, applied to test the gaussianity of the distribution; (*p*-value) confidence level of the statistic of the Shapiro–Wilk test.

Mean (kPa)	Sd (kPa)	Median (kPa)	Min (kPa)	Max (kPa)	I quart. (kPa)	III quart. (kPa)	Skew (kPa)	$W_{S\text{-}W}$ (–)	*p*-Value (–)
−368	406	−215	−1483	10	−494	−22	−1.18	0.71	<0.001

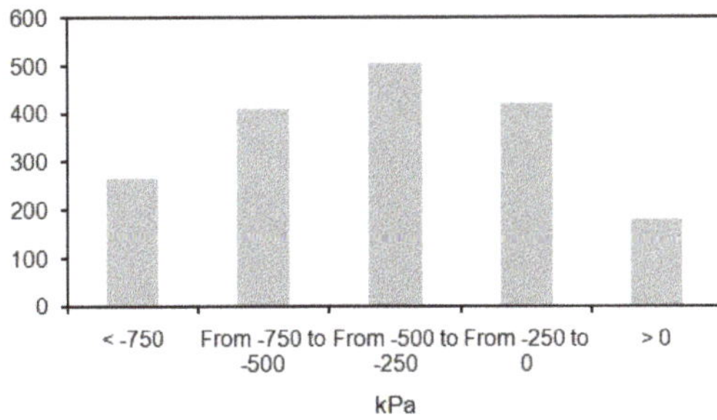

Figure 12. Histogram of distribution of the pore-water pressure values at typical depth of shallow-landslide sliding surface (1.0–1.2 m from the ground level) for the modeled time span (August 1992–August 1997) at the test-site slope.

Pore-water pressure at the beginning of each identified rainfall was linked to each reconstructed rainfall scenario. This was done to relate rainfall events with a certain initial pore-water pressure to the correct physicallybased threshold. For the validation of the empirical thresholds, all the rainfall events considered for the validation of each physicallybased threshold (TRIGRS/−20, TRIGRS/−10, and TRIGRS/0) were used, in order to make homogeneous the comparison between the validation phases of all the defined thresholds (Figure 13).

Table 7 Listingof the results of the validation phase. All the thresholds correctly identified the rainfall events able to trigger shallow landslides (true positives), as testified by TP values of 95 ± 2% and 100 ± 0% and by FN values of 5 ± 2% and 0 ± 0% for empirical and TRIGRS/0 thresholds, respectively.TP and FN indexes were not calculated for both TRIGRS/−10 and TRIGRS/−20, because no events triggered shallow landslides, starting from initial conditions of pore-water pressure equal to either −10 or −20 kPa. Instead, the reliability of these thresholds in identifying non-triggering rainfall events was assessed by means of TN and FP values. For events with initial pore-water pressure conditions of −20 or −10 kPa, the respective thresholds are characterized by TN of 100 ± 0% and by FP of 0 ± 0%, confirming the capability of these models in distinguishing events able to trigger or not trigger shallow landslides. Moreover,the TRIGRS/0 threshold assessed the conditions which could not trigger slope instabilities well, as testified by TN of 93 ± 1% and by FP of 7 ± 1%. Instead, the empirical threshold was characterized by a lower ability in distinguishing triggering or non-triggering events. Its TN was of 76 ± 3%, while its FP was of 24 ± 3%. In these terms, these thresholds overestimated the conditions able to trigger shallow landslides, classifying 24 ± 3% of real non-triggering events as able to cause shallow landsliding (false positives).

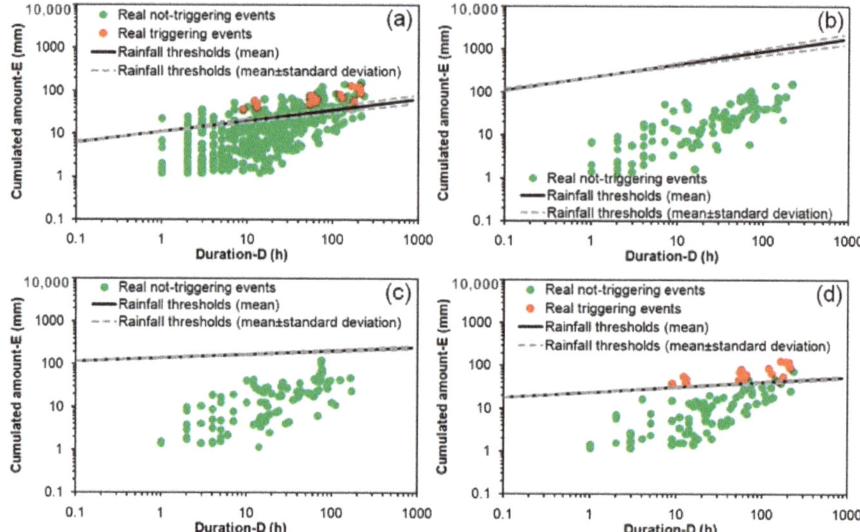

Figure 13. Duration(D) and Cumulated amount (E) conditions for the rainfall events recorded in the period August 1992–August 1997 and used for the validation phase, and the corresponding thresholds: (**a**) threshold reconstructed through the empirical method; (**b**) threshold reconstructed through the physicallybased method, considering an initial pore-water pressure of −20 kPa at the depth of the sliding surface (TRIGRS/–20); (**c**) threshold reconstructed through the physicallybased method considering an initial pore-water pressure of −10 kPa at the depth of the sliding surface (TRIGRS/–10); (**d**) threshold reconstructed through the physicallybased method, considering an initial pore-water pressure of 0 kPa at the depth of the sliding surface (TRIGRS/0).

Table 7. Mean ± standard deviation of the statistical indexes used in the validation phase of the reconstructed thresholds. (TP) true positives; (TN) true negatives; (FP) false positives; and (FN) false negatives.

Threshold	TP (%)	TN (%)	FP (%)	FN (%)
Empirical thresholds	95 ± 2	76 ± 3	24 ± 3	5 ± 2
Physicallybased thresholds (−20 kPa) (TRIGRS/–20)	-	100 ± 0	0 ± 0	-
Physicallybased thresholds (−10 kPa) (TRIGRS/–10)	-	100 ± 0	0 ± 0	-
Physicallybased thresholds (0 kPa) (TRIGRS/0)	100 ± 0	93 ± 1	7 ± 1	0 ± 0

4. Discussion

Rainfall thresholds can be considered a fundamental tool for assessing hazard toward slope instabilities and for defining reliable early warning system tools for their prediction [65].

One of the major challenges in establishing effective thresholds is obtaining a threshold able to correctly distinguish the triggering scenarios (true positives) from the events which cannot cause the development of slope failures (true negatives), also avoiding numerous erroneous alerts, corresponding to rainfall conditions that could not cause real instabilities (false positives).

Rainfall thresholds answering these issues are mostly reconstructed through an empirical/statistical approach, exploiting past inventories formed by the events able to or not able to trigger shallow landslides [4,5]. Uncertainties and limitations of these thresholds (i.e., availability and quality of rainfall data and landslides information, correct definition of the triggering times, and neglecting the antecedent soil hydrological conditions) induce researchers in order to verify the possibility of using physicallybased procedures that can provide the assessment of the link between rainfall features, soil hydromechanical conditions before a rainfall event, and shear strength response of soils during the rainwater infiltration.

Mirus et al. [32] and Fusco et al. [30] aimed to perform a robust comparison between these two types of approach, in steep slopes covered by colluvial soils derived from glacial and till deposits in the coastal area of the Northwestern United States [32], and in steep slopes covered by thick pyroclastic deposits in Southern Italy [30], respectively. The present paper compares empirical and physicallybased rainfall thresholds estimated for a wide area of Northern Italy (Oltrepò Pavese), significantly prone to shallow landslides and representative of the typical geological, geomorphological, and environmental features of Apennine area [66].

Empirical thresholds of the study area were reconstructed by means of the typical exploitation of a long multi-temporal inventory (2000–2018) of rainfall events able to trigger or not shallow landslides. Instead, the second type of thresholds were estimated through a physicallybased slope model (TRIGRS), representative of the real geological and geomorphological conditions where shallow landslides develop in the study area, allowing to couple the monitoring of soil hydrological responses toward atmosphere-soil interface, the modeling of slope hydrological responses, and the slope stability analysis. In this way, the estimation of rainfall thresholds was performed by considering not only rainfall attributes, but also the typical antecedent soil hydrological conditions.

Monitoring data acquired during a significant time span, covering more than seven years (March 2012–October 2019) [24,45], and the modeled ones for a five-year period (August 1992–August 1997) demonstrate variations of soil pore-water pressure trends in deep soil horizons, where sliding surface could form. Monitoring and modeling data confirm that the soil pore-water pressure regime is linked to the seasonal and interannual meteorological variability, showing similar trends during warm/dry and cold/wet months across different years. Unsaturated conditions are typical of warm months in the year, especially between May and September. A certain increase of pore-water pressure till values close to 0 kPa was observed only after the most intense events, of at least 50 mm of rain fallen in at least 12 h. After re-wetting events in the first weeks of Autumn months (at least 30 mm of rain fallen in at least 24 h), the coldest time of the year, which lasts from October to April, is characterized by pore-water pressure closer to saturated conditions, generally in the order of −20 and −10 kPa. Development of nil or positive values in correspondence of other important rainfall events, corresponding to the formation of a perched water table, occurs when further strong rainfall events, of at least 20 mm/day, or prolonged rainy periods, affect the study slope.

Such a seasonal hydrological behavior explains why shallow-landslide-triggering events occurred mostly during cold months between October and April. Antecedent pore-water pressure close to 0 kPa in soil horizons where shallow landslides develop, combined with further heavy rainfall events or a prolonged rainy period (duration between 4 and 105 h, with cumulated amount between 30.4 and 133.7 mm), cause the typical scenario which induces widespread slope instabilities in the study area. This scenario also confirms the monitored conditions of triggering during 28 February–2 March 2014 event that is shown in Bordoni et al. [23].

Conversely, in warm months between May and September, only heavy–torrential (D1) or torrential (D2) rainfalls, according to Alpert et al.'s [64] classification (duration between 5 and 15 h and cumulated amounts between 106.8 and 155.2 mm), have the potential to trigger shallow failures, only when they are preceded by other rainfalls which cause the increase in pore-water pressure to go up to around −20 kPa.

Triggering conditions in cold months of the study area are similar to those identified in different contexts all over the world, which are characterized by a cold and wet season in a year like in the study area [32,67–71]. Instead, triggering events of warm months have features similar to those commonly occurring in the coastal zones of the Mediterranean region [16,28,30,51,63,72,73], when strong convective thunderstorms affect those areas especially at the end of summer (September) or in the first weeks of autumn (October–November).

The differences in triggering conditions and the significant effect of soil hydrological conditions at the beginning of a rainfall event influence the reconstructed thresholds for shallow landslides' occurrence. By comparing the different physicallybased thresholds, it is clear that the drier the soil is, the bigger the amount of rain required to trigger a landslide is, considering the same duration of the event. For a certain temporal length of the rainfall, the cumulated amount able to trigger shallow landslides for an initial pore-water pressure condition of −20 kPa is about 20–25 times higher than that required if the initial pore-water pressure is of 0 kPa. This amount decreases if the initial pore-water pressure is of −10 kPa, even if it is still 6–8 times higher than that obtained considering an initial pore-water pressure of 0 kPa. This estimation matches with the datasets of triggering events analyzed for the study area, where rainfalls able to trigger shallow landslides were more severe, in terms of cumulated amount (higher than 100 mm), when they occurred in periods with soil in unsaturated conditions. Instead, the amount of rain able to trigger shallow landslides decreased significantly, till more than 3 times, when the soil was saturated.

The empirical threshold is very close to the physicallybased one estimated based on an initial pore-water pressure condition of 0 kPa. For the same duration, the amount of triggering cumulated rainfall for an initial pore-water pressure of 0 kPa is 5.3–10.5 mm higher than that estimated by the empirical threshold. This is in agreement with comparisons between physicallybased and empirical thresholds performed in other areas prone to shallow landsliding worldwide [28,30,32].

In the dataset used to validate the reconstructed thresholds, triggering events occurred only in conditions of pore-water pressure equal to 0 kPa. Both empirical threshold and TRIGRS/0 threshold correctly identified rainfall events able to trigger shallow landslides (TP higher than 95%, FN lower than 5%), although only the TRIGRS/0 threshold recognized all the triggering events. However, the empirical threshold significantly overestimated the rainfall conditions able to trigger shallow landslides, as testified by FP = 24±3%. Instead, TRIGRS/0 threshold worked well for assessing the conditions which could not trigger slope instabilities, strongly limiting the false positives (FP = 7±1%).

These results confirm the fundamental role played by the soil hydrological conditions present at the beginning of a rainfall event on the development of shallow slope failures. All the false positives identified by the empirical threshold correspond to rainfall event occurred when the soil was not completely saturated, especially (90%) when pore-water pressure was lower than −10 kPa (Figure 14). These results are confirmed also by an event that occurred on 21 October 2019, when a strong thunderstorm hit the northern portion of the study area, in particular close to rain gauges 3 and 6 (Figure 1). In total, 118 mm of rain fell in 24 h, with a peak of 97 mm of cumulated rain in 6 h, between 5:00 p.m. and 11:00 p.m. local time. These rainfall conditions are located above the empirical thresholds, but they did not cause any triggering of shallow failures due to pore-water pressure conditions, at the beginning of the rainfall, of −800 kPa, as measured by the monitoring station in the study area.

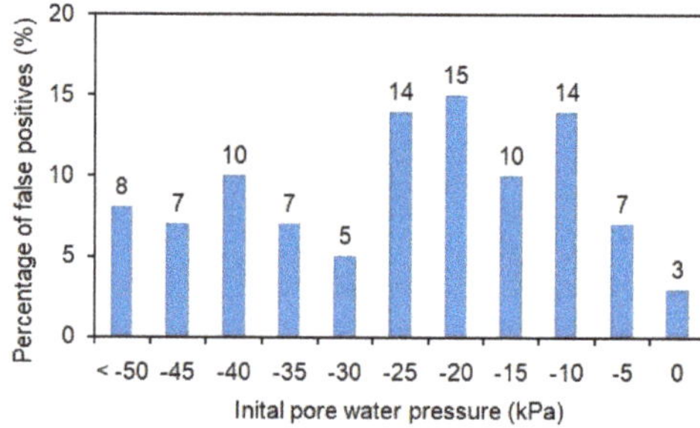

Figure 14. Modeled values of initial pore-water pressure conditions in correspondence of the false positives of empirical thresholds for the validation dataset.

Empirical threshold causes an overestimation of triggering events, determining false positives in correspondence of rainfall conditions similar to the ones that provoked observed shallow failures, but with an initial soil condition drier than that corresponding to the real triggering events. Thus, physicallybased thresholds which also take into account the antecedent soil conditions in terms of pore-water pressure can represent an improvement, both in terms of objectively predicting shallow-landslide occurrence and also limiting false positives.

Reconstructed rainfall thresholds for the Oltrepò Pavese area were then compared with other duration(D) and cumulated amount (E) thresholds of other Italian areas (Figure 15). Regional and national thresholds in Italy [8,74–76] were derived by using an empirical approach similar to that adopted for the empirical thresholds of the Oltrepò Pavese area. Thresholds for Oltrepò Pavese were also compared to a world threshold defined by Innes [77] for the occurrence of debris flows.

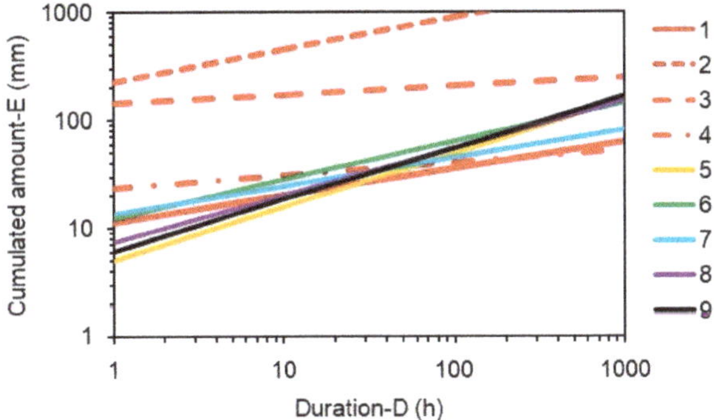

Figure 15. Comparison between the reconstructed thresholds for Oltrepò Pavese area, with some regional, national, and world thresholds. Source: (1) empirical threshold (mean fitting parameters) of Oltrepò Pavese area; (2) physicallybased threshold for initial pore-water pressure of −20 kPa (mean fitting parameters) of Oltrepò Pavese area (TRIGRS/−20); (3) physicallybased threshold for initial pore-water pressure of −10 kPa (mean fitting parameters) of Oltrepò Pavese area (TRIGRS/−10); (4) physicallybased threshold for initial pore-water pressure of 0 kPa (mean fitting parameters) of Oltrepò Pavese area (TRIGRS/0); (5) world [77]; (6) Italy [74]; (7) Liguria [8]; Sicily [75]; and (9) Italian Alps [76].

Physicallybased thresholds obtained on the basis of antecedent pore-water pressure equal to −20 kPa (TRIGRS/−20) or −10 kPa (TRIGRS/−10) are located above all the other thresholds, in agreement with the need of a higher amount of rain to trigger shallow landslides in unsaturated soil conditions. TRIGRS/0 threshold and the empirical thresholds are located close to each other, with the former slightly above the empirical curves. Physicallybased thresholds reconstructed for completely saturated soils (TRIGRS/0) intercept all other considered thresholds (at world, Italian, and regional scale) for an event duration of 40 h,whereas the empirical thresholds intercept world and some regional (Sicily, Italian Alps) thresholds at the same duration. Moreover, both these thresholds show a lower steepness, which implies that the rainfall amount required to trigger shallow landslides for event with duration less than 40 h is higher than the one of the compared world, Italian, and regional thresholds. Instead, for event longer than 40 h, TRIGRS/0 thresholds and the empirical thresholds are below the other thresholds. For these rainfall features, the cumulated amount able to trigger shallow failures is lower if compared to the other analyzed thresholds. The Oltrepò Pavese area is, then, more susceptible to shallow landsliding for long-duration [64] events. For short and medium events [64], the amount of rainfall able to trigger shallow landslides is higher, thus reducing the proneness of the territory in correspondence of such events.

According to the achieved results of this paper, the main relevance of this work and of the reconstructed thresholds are as follows: (i) empirically and physicallybased thresholds for a representative area of the Italian Apennines; (ii) different physicallybased thresholds according to different soil hydrological conditions and considering rainfall scenarios already measured in the study area; (iii) implementation of a physicallybased slope model allowing to couple the monitoring of soil hydrological responses toward atmosphere-soil interface, the modeling of slope hydrological responses, and the slope stability analysis; (iv) robust evaluation of the threshold's predictive capability through a different dataset with respect to that used in the reconstruction of the models; and (vi) determination of advantages and constraints in the use of empirically or physicallybased thresholds.

5. Conclusions

The reconstruction of reliable thresholds with a high predictive capability becomes very important, especially if their implementation in an earlywarning system is proposed. In regard to the Oltrepò Pavese hilly area, which can be assumed representative of the geological, geomorphological, and land-use settings prone to shallow landslides also of the whole Northern Apennines, empirically and physicallybased thresholds were estimated. These were evaluated by quantifying their predictive capability through the comparison between modeled and real triggering or non-triggering conditions, identified in a validation dataset covering a five-year time span.

The role played by the soil hydrological conditions at the beginning of a rainfall event is fundamental in making this rainfall able to trigger or not trigger shallow landslides. The lower the pore-water pressure is at the beginning of an event, the higher the amount of rainfall required to trigger shallow failures is. When shallow landslides occur as a consequence of rain fallen on previously saturated soil (nil pore-water pressure), as in the study area, physicallybased thresholds provide a better reliability in discriminating the event which could or could not trigger slope failures. Besides a good capability in identifying correctly the triggering conditions, empirical thresholds, based only on rainfall data and neglecting the antecedent soil hydrological conditions, provide a significant number of false positives. These are events similar to the ones that provoked observed shallow failures, but with initial soil conditions drier than those corresponding to the real triggering events.

Main conclusions of this work can be summarized as follows:

- The antecedent soil hydrological conditions have a primary role on predisposing or preventing shallow slope instability during a rainfall event. This should be taken into account, especially for those contexts characterized by seasonal hydrological behaviors and, in particular, during periods when initial pore-water pressure conditions are more favorable to lead the triggering of shallow landslides. In fact, in the Oltrepò Pavese area, cold and wet months between October and April are the most susceptible periods of the year, due to the permanence of saturated or close-to-saturation soil conditions;
- The most promising approach for developing an early warning system based on rainfall thresholds seems to be the reconstruction of physicallybased thresholds for the typical initial pore-water pressure conditions leading to slope instabilities. These tools can be supported further by the monitoring of the soil hydrological behaviors and slope stability analysis in correspondence of different rainfall scenarios. However, to confirm the better effectiveness of the physicallybased thresholds than the empirical ones, it is required a comparison of the threshold exceedance against existing early warning criteria and further landslide occurrence for future time span, as suggested in [12];
- Physicallybased models of a representative testsite and hydrological monitoring data could be not always available for a susceptible area. In such a context, empirical thresholds can represent a precautionary approach that allows us to identify the triggering conditions in a reliable way, in the awareness that they can give many false positives,especially for rainfall events similar to those provoking shallow landslides, but occurring in dry periods;
- Physicallybased thresholds are reconstructed based on physical simulation of slope stability, according to well-defined geotechnical, mechanical, and hydrological soil parameters. To take

into account the intrinsic variability of these parameters, also within small areas, probabilistic models will be applied to reconstruct this type of threshold in future developments.

Author Contributions: Conceptualization, M.B. (Massimiliano Bordoni) and C.M.; methodology, M.B. (Massimiliano Bordoni), B.C.,and C.M.; software, M.B. (Massimiliano Bordoni) and B.C.; validation, M.B. (Massimiliano Bordoni), and B.C.; formal analysis, M.B. (Massimiliano Bordoni), B.C., L.L., and V.V.; investigation, M.B. (Massimiliano Bordoni), B.C., L.L., R.V., M.B. (Marco Bittelli), V.V., and C.M.; resources, M.B. (Massimiliano Bordoni), R.V., M.B. (Marco Bittelli), and C.M.; data curation, M.B. (Massimiliano Bordoni), R.V., M.B. (Marco Bittelli), and C.M.; writing—original draft preparation, M.B. (Massimiliano Bordoni); writing—review and editing, M.B. (Massimiliano Bordoni), B.C., L.L., R.V., M.B. (Marco Bittelli), V.V., and C.M.; supervision, C.M.; project administration, C.M.

Funding: This work has been in the frame of the ANDROMEDA project, which has been supported by Fondazione Cariplo, grant n°2017-0677.

Acknowledgments: We thank Marco Tumiati for the assistance on the executions of the laboratory geotechnical tests of the soil horizons of the testsite.

Conflicts of Interest: The authors declare no conflicts of interest.

References

1. Zizioli, D.; Meisina, C.; Valentino, R.; Montrasio, L. Comparison between different approaches to modelling shallow landslide susceptibility: A case history in OltrepòPavese, Northern Italy. *Nat. Hazards Earth Syst. Sci.* **2013**, *13*, 559–573. [CrossRef]
2. Lacasse, S.; Nadim, F.; Kalsnes, B. Living with landslide risk. *Geotech. Eng J. Seags Agssea* **2010**, *41*.
3. Wieczorek, G.F.; Glade, T. Climatic factors influencing occurrence of debris flows. In *Debris Flow Hazards and Related Phenomena*; Jakob, M., Hungr, O., Eds.; Springer: Berlin, Germany, 2005; pp. 325–362.
4. Guzzetti, F.; Peruccacci, S.; Rossi, M.; Stark, C.P. The rainfall intensity-duration control of shallow landslides and debris flows: An update. *Landslides* **2008**, *5*, 3–17. [CrossRef]
5. Segoni, S.; Piciullo, L.; Gariano, S.L. A review of the recent literature on rainfall thresholds for landslides occurrence. *Landslides* **2018**, *15*, 1483–1501. [CrossRef]
6. Reichenbach, P.; Cardinali, M.; DeVita, P.; Guzzetti, F. Hydrological thresholds for landslides and floods in the Tiber River basin (central Italy). *Environ. Geol.* **1998**, *35*, 146–159. [CrossRef]
7. Crozier, M.J. The climate-landslide couple: A southern hemisphere perspective. In *Rapid Mass Movement as a Source of Climatic Evidence for the Holocene*; Matthews, J.A., Brunsden, D., Frenzel, B., Gläser, B., Weiß, M.M., Eds.; Gustav Fischer: Stuttgart, Germany, 1997; pp. 333–354.
8. Melillo, M.; Brunetti, M.T.; Perruccacci, S.; Gariano, S.L.; Roccati, A.; Guzzetti, F. A tool for the automatic calculation of rainfall thresholds for landslide occurrence. *Environ. Mod. Soft.* **2018**, *105*, 230–243. [CrossRef]
9. Caine, N. The rainfall intensity-duration control of shallow landslides and debris flows. *Geogr. Ann. Ser. A. Phys. Geogr.* **1980**, *62*, 23–27.
10. Aleotti, P. A warning system for rainfall-induced shallow failures. *Eng. Geol.* **2004**, *73*, 247–265. [CrossRef]
11. Gabet, E.J.; Burbank, D.W.; Putkonen, J.K.; Pratt-Sitaula, B.A.; Oiha, T.; Putkonen, J. Rainfall thresholds for landsliding in the Himalayas of Nepal. *Geomorphology* **2004**, *63*, 131–143. [CrossRef]
12. Godt, J.W.; Baum, R.L.; Chleborad, A.F. Rainfall characteristics for shallow landsliding in Seattle, Washington, USA. *Earth Surf. Process. Landf.* **2006**, *31*, 97–110. [CrossRef]
13. Brunetti, M.T.; Peruccacci, S.; Rossi, M.; Luciani, S.; Valigi, D.; Guzzetti, F. Rainfall thresholds for the possible occurrence of landslides in Italy. *Nat. Hazards Earth Syst. Sci.* **2010**, *10*, 447–458. [CrossRef]
14. Martelloni, G.; Segoni, S.; Fanti, R.; Catani, F. Rainfall thresholds for the forecasting of landslide occurrence at regional scale. *Landslides* **2012**, *9*, 485–495. [CrossRef]
15. Staley, D.M.; Kean, J.W.; Cannon, S.H.; Schmidt, K.M.; Laber, J.L. Objective definition of rainfall intensity duration thresholds for the initiation of post-fire debris flows in southern California. *Landslides* **2013**, *10*, 547–562. [CrossRef]
16. Segoni, S.; Rossi, G.; Rosi, A.; Catani, F. Landslides triggered by rainfall: A semiautomated procedure to define consistent intensity-duration thresholds. *Comput. Geosci.* **2014**, *63*, 123–131. [CrossRef]

17. Galanti, Y.; Barsanti, M.; Cevasco, A.; D'AmatoAvanzi, G.; Giannecchini, R. Comparison of statistical methods and multi-time validation for the determination of the shallow landslide rainfall thresholds. *Landslides* **2018**, *15*, 937–952. [CrossRef]
18. Gao, L.; Zhang, L.M.; Cheung, R.W.M. Relationships between natural terrain landslide magnitudes and triggering rainfall based on a large landslide inventory in Hong Kong. *Landslides* **2018**, *15*, 727–740. [CrossRef]
19. Dikshit, A.; Sarkar, R.; Pradhan, B.; Acharya, S.; Dorji, K. Estimating Rainfall Thresholds for Landslide Occurrence in the Bhutan Himalayas. *Water* **2019**, *11*, 1616. [CrossRef]
20. Gariano, S.L.; Sarkar, R.; Dikshit, A.; Dorji, K.; Brunetti, M.T.; Peruccacci, S.; Melillo, M. Automatic calculation of rainfall thresholds for landslide occurrence in Chukha Dzongkhag, Bhutan. *Bull. Eng. Geol. Environ.* **2019**, *78*, 4325–4332. [CrossRef]
21. Nikolopoulos, E.I.; Crema, S.; Marchi, L.; Marra, F.; Guzzetti, F.; Borga, M. Impact of uncertainty in rainfall estimation on the identification of rainfall thresholds for debris flow occurrence. *Geomorphology* **2014**, *221*, 286–297. [CrossRef]
22. Peres, D.J.; Cancelliere, A.; Greco, R.; Boogard, T.A. Influence of uncertain identification of triggering rainfall on the assessment of landslide early warning thresholds. *Nat. Hazards Earth Syst. Sci.* **2018**, *18*, 633–646. [CrossRef]
23. Lu, N.; Godt, J.W. *Hillslope Hydrology and Stability*; Cambridge University Press: NewYork, NY, USA, 2013.
24. Bordoni, M.; Meisina, C.; Valentino, R.; Lu, N.; Bittelli, M.; Chersich, S. Hydrological factors affecting rainfall-induced shallow landslides: From the field monitoring to a simplified slope stability analysis. *Eng. Geol.* **2015**, *193*, 19–37. [CrossRef]
25. Bogaard, T.; Greco, R. Invited perspectives: Hydrological perspectives on precipitation intensity-duration thresholds for landslide initiation: Proposing hydrometeorological thresholds. *Nat. Hazards Earth Syst. Sci.* **2018**, *18*, 31–39. [CrossRef]
26. Frattini, P.; Crosta, G.; Sosio, R. Approaches for defining thresholds and return periods for rainfall-triggered shallow landslides. *Hydrol. Process.* **2009**, *23*, 1444–1460. [CrossRef]
27. Alvioli, M.; Guzzetti, F.; Rossi, M. Scaling properties of rainfall induced landslides predicted by a physically based model. *Geomorphology* **2014**, *213*, 38–47. [CrossRef]
28. DeVita, P.; Napolitano, E.; Godt, J.W.; Baum, R. Deterministic estimation of hydrological thresholds for shallow landslide initiation and slope stability models: Case study from the Somma-Vesuvius area of southern Italy. *Landslides* **2013**, *10*, 713–728. [CrossRef]
29. DeVita, P.; Fusco, F.; Tufano, R.; Cusano, D. Seasonal and event-based hydrological and slope stability modeling of pyroclastic fall deposits covering slopes in Campania (Southern Italy). *Water* **2018**, *10*, 1140. [CrossRef]
30. Fusco, F.; DeVita, P.; Mirus, B.B.; Baum, R.L.; Allocca, V.; Tufano, R.; DiClemente, E.; Calcaterra, D. Physically based estimation of rainfall thresholds triggering shallow landslides in volcanic slopes of southern Italy. *Water* **2019**, *11*, 1915. [CrossRef]
31. Peres, D.J.; Cancelliere, A. Estimating return period of landslide triggering by Monte Carlo simulation. *J. Hydrol.* **2016**, *541*, 256–271. [CrossRef]
32. Mirus, B.B.; Becker, R.E.; Baum, R.L.; Smith, J.B. Integrating real-time subsurface hydrologic monitoring with empirical rainfall thresholds to improve landslide early warning. *Landslides* **2018**, *15*, 1909–1919. [CrossRef]
33. Yang, Z.; Cai, H.; Shao, W.; Huang, D.; Uchimura, T.; Lei, X.; Tian, H.; Qiao, J. Clarifying the hydrological mechanisms and thresholds for rainfall-induced landslide: In situ monitoring of big data to unsaturated slope stability analysis. *Bull. Eng. Geol. Environ.* **2019**, *78*, 2139–2150. [CrossRef]
34. Wei, X.; Fan, W.; Cao, Y.; Chai, X.; Bordoni, M.; Meisina, C.; Li, J. Integrated experiments on field monitoring and hydro-mechanical modeling for determination of a triggering threshold of rainfall-induced shallow landslides. A case study in Ren River catchment, China. *Bull. Eng. Geol. Environ.* **2019**. [CrossRef]
35. Corominas, J.; VanWesten, C.; Frattini, P.; Cascini, L.; Malet, J.P.; Fotopoulou, S.; Catani, F.; VanDenEeckhaut, M.; Mavrouli, O.; Agliardi, F.; et al. Recommendations for the quantitative analysis of landslide risk. *Bull. Eng. Geol. Environ.* **2014**, *73*, 209–263. [CrossRef]
36. Brunetti, M.; Buffoni, L.; Maugeri, M.; Nanni, T. Precipitation intensity trends in Northern Italy. *Int. J. Climatol.* **2000**, *20*, 1017–1031. [CrossRef]
37. Coppola, E.; Giorgi, F. An assessment of temperature and precipitation change projections over Italy from recent global and regional climate model simulations. *Int. J. Climatol.* **2010**, *30*, 11–32. [CrossRef]

38. Braga, G.; Braschi, G.; Calculli, S.; Caucia, F.; Cerro, A.; Colleselli, F.; Grisolia, M.; Piccio, A.; Rossetti, R.; Setti, M.; et al. I fenomeni franosi nell'Oltrepo Pavese: Tipologia e cause. *Geol. E Idrogeol.* **1985**, *20*, 621–666.
39. Meisina, C. Swelling-shrinking properties of weathered clayey soils associated with shallow landslides. *Quat. J. Eng. Geol. Hydrogeol.* **2004**, *37*, 77–94. [CrossRef]
40. Meisina, C. Characterisation of weathered clayey soils responsible for shallow landslides. *Nat. Hazards Earth Syst. Sci.* **2006**, *6*, 825–838. [CrossRef]
41. Cruden, D.M.; Varnes, D.J. Landslide types and processes. In *Landslides: Investigation and Mitigation Sp. Rep. 247*; Turner, A.K., Schuster, R.L., Eds.; Transportation Research Board, National Research Council, National Academy Press: Washington, DC, USA; pp. 36–75.
42. Thornthwaite, C.W.; Mather, J.R. Instructions and tables for computing potential evapotranspiration and the water balance. *Publ. Climatol. Lab. Climatol. Drexel Inst. Technol.* **1957**, *10*, 185–311.
43. Barrow, C.J. *World Atlas of Desertification*; United Nations Environment Programme: London, UK, 1992.
44. Melillo, M.; Brunetti, M.T.; Peruccacci, S.; Gariano, S.L.; Guzzetti, F. An algorithm for the objective reconstruction of rainfall events responsible for landslides. *Landslides* **2015**, *12*, 311–320. [CrossRef]
45. Peruccacci, S.; Brunetti, M.T.; Luciani, S.; Vennari, C.; Guzzetti, F. Lithological and seasonal control of rainfall thresholds for the possible initiation of landslides in central Italy. *Geomorphology* **2012**, *139–140*, 79–90. [CrossRef]
46. VanGenuchten, M.T. A closed-form equation for predicting the hydraulic conductivity of unsaturated soils. *Soil Sci. Soc. Am. J.* **1980**, *44*, 892–898. [CrossRef]
47. Bordoni, M.; Bittelli, M.; Valentino, R.; Chersich, S.; Meisina, C. Improving the estimation of complete field soil water characteristic curves through field monitoring data. *J. Hydrol.* **2017**, *552*, 283–305. [CrossRef]
48. Baum, R.L.; Savage, W.Z.; Godt, J.W. *TRIGRS—A Fortran Program for Transient Rainfall Infiltration and Grid-Based Regional Slope-Stability Analysis, Version 2.0, Open-File Report 2008-1159*; Savage, W.Z., Ed.; US Geological Survey: Denver, CO, USA, 2008.
49. Baum, R.L.; Godt, J.W.; Savage, W.Z. Estimating the timing and location of shallow rainfallinduced landslides using a model for transient, unsaturated infiltration. *J. Geophys. Res.* **2010**, *115*, F03013. [CrossRef]
50. Sorbino, G.; Sica, C.; Cascini, L. Susceptibility analysis of shallow landslides source areas using physically based models. *Nat. Hazards* **2010**, *53*, 313–332. [CrossRef]
51. Park, D.W.; Nikhil, N.V.; Lee, S.R. Landslide and debris flow susceptibility zonation using TRIGRS for the 2011 Seoul landslide event. *Nat. Hazards Earth Syst. Sci.* **2013**, *13*, 2833–2849. [CrossRef]
52. Schilirò, L.; Esposito, C.; ScarasciaMugnozza, G. Evaluation of shallow landslide-triggering scenarios through a physically based approach: An example of application in the southern Messina area (northeastern Sicily, Italy). *Nat. Hazards Earth Syst. Sci.* **2013**, *15*, 2091–2109. [CrossRef]
53. Alvioli, M.; Baum, R.L. Parallelization of the TRIGRS model for rainfall-induced landslides using the message passing interface. *Environ. Model. Soft.* **2016**, *81*, 122–135. [CrossRef]
54. Weidner, L.; Oommen, T.; Escobar-Wolf, R.; Sajinkumar, K.S.; Samuel, R.A. Regional-scale back-analysis using TRIGRS: An approach to advance landslide hazard modeling and prediction in sparse data regions. *Landslides* **2018**, *15*, 2343–2356. [CrossRef]
55. Srivastava, R.; Yeh, T.-C.J. Analytical solutions for one dimensional, transient infiltration toward the water table in homogeneous and layered soils. *Water Resour. Res.* **1991**, *27*, 753–762. [CrossRef]
56. Iverson, R.M. Landslide triggering by rain infiltration. *Water Resour. Res.* **2000**, *36*, 1897–1910. [CrossRef]
57. Gardner, W.R. Some steady-state solutions of the unsaturated moisture flow equation with application to evaporation from a water table. *Soil Sci.* **1958**, *85*, 228–232. [CrossRef]
58. Ghezzehei, T.A.; Kneafsey, T.J.; Su, G.W. Correspondence of the Gardner and van Genuchten/Mualem relative permeability function parameters. *Water Resour. Res.* **2007**, *43*, 10–W10417. [CrossRef]
59. Comegna, L. *Regional Analysis of Rainfall-Induced Landslides. The Case of Camaldoli Hill, Naples: Test Case nr. 1 October 2004*; Test Case Nr. 2—September 2005; Centro euro-Mediterraneo per I Cambiamenti Climatici CMCC: Lecce, Italy, 2008.
60. Simunek, J.; VanGenucthen, M.T.; Sejna, M. Development and applications of the HYDRUS and STANMOD software packages and related codes. *Vadose Zonej.* **2008**, *7*, 587–600. [CrossRef]
61. Hargreaves, G.L.; Hargreaves, G.H.; Riley, J.P. Irrigation water requirements for Senegal River basin. *J. Irrig. Drain. Eng. ASCE* **1985**, *111*, 265–275. [CrossRef]

62. Schilirò, L.; Cevasco, A.; Esposito, C.; ScarasciaMugnozza, G. Shallow landslide initiation on terraced slopes: Inferences from a physically based approach. *Geomat. Nat. Haz. Risks* **2018**, *9*, 295–324. [CrossRef]
63. Gariano, S.L.; Brunetti, M.T.; Iovine, G.; Melillo, M.; Peruccacci, S.; Terranova, O.; Vennari, C.; Guzzetti, F. Calibration and validation of rainfall thresholds for shallow landslide forecasting in Sicily, southern Italy. *Geomorphology* **2015**, *228*, 653–665. [CrossRef]
64. Alpert, P.; Ben-Gai, T.; Baharan, A.; Benjamini, Y.; Yekutieli, D.; Colacino, M.; Diodato, L.; Ramis, C.; Homar, V.; Romero, R.; et al. The paradoxical increase of Mediterranean extreme daily rainfall in spite of decrease in total values. *Geophys. Res. Lett.* **2002**, 29. [CrossRef]
65. Piciullo, L.; Mads-Peter, D.; Devoli, G.; Herve, C.; Calvello, M. Adapting the EDuMaP method to test the performance of the Norwegian early warning system for weather induced landslides. *Nat. Hazards Earth Syst. Sci.* **2017**, *17*, 817–831. [CrossRef]
66. Raggi, M.; Viaggi, D.; Bartolini, F.; Furlan, A. The role of policy priorities and target in the spatial location of participation in agri-environmental schemes in Emilia-Romagna (Italy). *Land Use Pol.* **2015**, *47*, 78–89. [CrossRef]
67. Tohari, A.; Nishigaki, M.; Komatsu, M. Laboratory rainfall-induced slope failure with moisture content measurement. *J. Geotech. Geoenviron. Eng.* **2007**, *133*, 575–587. [CrossRef]
68. Montrasio, L.; Valentino, R. A model for triggering mechanisms of shallow landslides. *Nat. Hazards Earth Syst. Sci.* **2008**, *8*, 1149–1159. [CrossRef]
69. Bittelli, M.; Valentino, R.; Salvatorelli, F.; RossiPisa, P. Monitoring soil–water and displacement conditions leading to landslide occurrence in partially saturated clays. *Geomorphology* **2012**, *173–174*, 161–173. [CrossRef]
70. Smethurst, J.A.; Clarke, D.; Powrie, D. Factors controlling the seasonal variation in soil water content and pore-water pressures within a lightly vegetated clay slope. *Geotechnique* **2012**, *62*, 429–446. [CrossRef]
71. Fressard, M.; Maquaire, O.; Thiery, Y.; Davidson, R.; Lissak, C. Multi-method characterisation of an active landslide: Case study in the Pays d'Auge plateau (Normandy, France). *Geomorphology* **2016**, *270*, 22–39. [CrossRef]
72. Cevasco, A.; Pepe, G.; Brandolini, P. The influences of geological and land use settings on shallow landslides triggered by an intense rainfall event in a coastal terraced environment. *Bull. Eng. Geol. Environ.* **2014**, *73*, 859–875. [CrossRef]
73. Cevasco, A.; Diodato, N.; Revellino, P.; Fiorillo, F.; Grelle, G.; Guadagno, F.M. Storminess and geo-hydrological events affecting small coastal basins in a terraced Mediterranean environment. *Sci. Total Environ.* **2015**, *532*, 208–219. [CrossRef]
74. Peruccacci, S.; Brunetti, M.T.; Gariano, S.L.; Melillo, M.; Rossi, M.; Guzzetti, F. Rainfall thresholds for possible landslide occurrence in Italy. *Geomorphology* **2017**, *290*, 39–57. [CrossRef]
75. Melillo, M.; Brunetti, M.T.; Peruccacci, S.; Gariano, S.L.; Guzzetti, F. Rainfall thresholds for the possible landslide occurrence in Sicily (Southern Italy) based on the automatic reconstruction of rainfall events. *Landslides* **2016**, *13*, 165–172. [CrossRef]
76. Palladino, M.R.; Viero, A.; Turconi, L.; Brunetti, M.T.; Peruccacci, S.; Melillo, M.; Luino, F.; Deganutti, A.M.; Guzzetti, F. Rainfall thresholds for the activation of shallow landslides in the Italian Alps: The role of environmental conditioning factors. *Geomorphology* **2018**, *303*, 53–67. [CrossRef]
77. Innes, J.L. Debris flows. *Prog. Phys.Geograp.* **1983**, *7*, 469–501. [CrossRef]

Article

Using a Tank Model to Determine Hydro-Meteorological Thresholds for Large-Scale Landslides in Taiwan

Guan-Wei Lin [1], Hsien-Li Kuo [1,*], Chi-Wen Chen [2], Lun-Wei Wei [3] and Jia-Ming Zhang [1]

1 Department of Earth Sciences, National Cheng Kung University, No. 1, University Road, Tainan City 701, Taiwan; gwlin@mail.ncku.edu.tw (G.-W.L.); star691487@gmail.com (J.-M.Z.)

2 Multi-hazard Risk Assessment Division, National Research Institute for Earth Science and Disaster Resilience, 3-1, Tennodai, Tsukuba, Ibaraki 305-0006, Japan; cwchen@bosai.go.jp

3 Department of Geosciences, National Taiwan University, No.1, Section 4, Roosevelt Road, Taipei 10617, Taiwan; wlw1105@gmail.com

* Correspondence: kuo1195@gmail.com; Tel.: +886-06-275-7575 (ext. 65424)

Received: 3 December 2019; Accepted: 15 January 2020; Published: 16 January 2020

Abstract: Rainfall thresholds for slope failures are essential information for establishing early-warning systems and for disaster risk reduction. Studies on the thresholds for rainfall-induced landslides of different scales have been undertaken in recent decades. This study attempts to establish a warning threshold for large-scale landslides (LSLs), which are defined as landslides with a disturbed area more massive than 0.1 km^2. The numerous landslides and extensive rainfall records make Taiwan an appropriate area to investigate the rainfall conditions that can result in LSLs. We used landslide information from multiple sources and rainfall data captured by 594 rain gauges to create a database of 83 rainfall events associated with LSLs in Taiwan between 2001 and 2016. The corresponding rainfall duration, cumulative event rainfall, and rainfall intensity for triggering LSLs were determined. This study adopted the tank model to estimate conceptual water depths (S_1, S_2, S_3) in three-layer tanks and calculated the soil water index (SWI) by summing up the water depths in the three tanks. The empirical SWI and duration (SWI–D) threshold for triggering LSLs occurring during 2001–2013 in Taiwan is determined as SWI = 155.20 − 1.56D and $D \geq 24$ h. The SWI–D threshold for LSLs is higher than that for small-scale landslides (SSLs), those with a disturbed area smaller than 0.1 km^2. The LSLs that occurred during 2015–2016 support this finding. It is notable that when the SWI and S_3 reached high values, the potential of LSLs increased significantly. The rainfall conditions for triggering LSLs gradually descend with increases in antecedent SWI. Unlike the rainfall conditions for triggering SSLs, those for triggering LSLs are related to the long duration–high intensity type of rainfall event.

Keywords: soil water index; large-scale landslide; SWI–D threshold; early warning system

---

1. Introduction

In the past two decades, the frequency of occurrence of extreme rainfall events and large-scale natural hazards has increased significantly worldwide [1–6], causing substantial economic losses and human casualties. In past studies [7–9], the characteristics of a large-scale landslide have been reported, including (1) extremely high movement velocity, (2) a large collapse volume, and (3) deep excavations into bedrock. Nevertheless, discriminating large-scale landslides (LSLs) from small-scale landslides (SSLs) requires many in situ observations, which remain difficult to accomplish extensively. In practical terms, the mass movement velocity and excavation depth are both difficult to observe, so the disturbed area or volume is mainly treated as a scale indicator of a landslide [10]. In the study, large-scale landslides (LSLs) are defined as landslides with a disturbed area more massive than 0.1 km^2. Although the occurrence frequency of LSLs is much lower than that of SSLs, LSLs induce rapid alterations of the topography, causing calamities on a far greater scale than do SSLs. Moreover, the Earth's surface processes in mountainous areas are significantly affected by LSLs.

For better evaluation of landslide hazards induced by rainfall-triggered LSLs, it is essential to comprehend the circumstances that induce failure and the mass movement following collapse [11,12]. Accurate landslide information on occurrence time, size, and location are beneficial for comprehending when, where, and how slopes may collapse following heavy rainfall [13]. Rainfall is well known as one of the significant factors in landslide occurrence, so in-depth knowledge of the effects of rainfall conditions is required. At present, Taiwan has an early-warning system for debris flows based on the relationship between rainfall intensity and effective rainfall [14]. The effective rainfall contributing to debris-flow occurrence includes the cumulative rainfall during the considered rainfall event and its 7-day antecedent rainfall before the rainfall event. However, there is no early-warning system for massive landslides in Taiwan. To control damage, the rainfall conditions that induce LSLs must be determined and used to define a rainfall threshold as a criterion of early-warning for the prevention and mitigation of disasters.

Rainfall parameters, including duration, intensity, cumulative rainfall, and antecedent rainfall, have been utilized in many previous studies to identify the essential rainfall conditions for shallow landslide occurrence [15–19]. Among the characteristics of rainfall, the cumulative rainfall represents the total height of precipitation on the ground surface, but it may not reflect the intratelluric water content, which involves the processes of infiltration, drainage, and even evapotranspiration. Kuo, et al. [20] adopted the traditional dual-factor analysis, i.e., rainfall intensity versus rainfall duration (*I-D*), cumulative rainfall versus rainfall duration (*R-D*), and rainfall intensity versus cumulative rainfall (*I-R*), to investigate in preliminary terms the rainfall thresholds for triggering LSLs. They reported that the cumulative rainfall might be the deterministic factor in triggering LSLs. However, the complicated relationship between the meteorological trigger and the hydrological cause was not considered in the study of Kuo, et al. [20]. Bogaard and Greco [21] have proposed analyzing the precipitation thresholds for landslides and debris flows from a hydro-meteorological point of view. In their study, the soil water index (SWI) is treated as a proxy for both meteorological trigger and hydrological cause.

The SWI proposed by Sugawara, et al. [22] is derived from a three-layer tank model. The value of the SWI is estimated to represent the depth of the remaining water in the three-layer tank. Similarly, Segoni, et al. [23] discovered that the performance of a regional scale landslide warning system could be improved by using soil moisture data instead of antecedent rainfall. The influences of infiltration and drainage on water content within slopes are considered when calculating the SWI. Currently, the Japan Meteorological Agency (JMA) adopts the SWI as the conceptual soil water content affected by antecedent rainfall as well as event rainfall [24]. Furthermore, the tank model has been successfully applied to discuss the influence of water infiltration on deep-seated landslides [25,26]. The estimation of groundwater supply caused by infiltration using the tank model can be considered as an indicator of pore water pressure changes in the deep layer of a slope. Chen, et al. [27] first proposed the SWI–*D* curve as an empirical rainfall threshold for shallow landslides in Taiwan. They noted that the SWI can be used as the indicator of the antecedent rainfall condition and recommended establishing a suitable warning system in Taiwan.

For landslide early warning systems, the thresholds for LSLs would be different from those for SSLs. Thus, different disaster alerts and evacuation strategies would be produced [21,28,29]. It would be worthwhile to create a regional warning threshold for landslides of different scales for Taiwan. However, the rainfall thresholds for landslides having a sizable disturbed area (i.e., exceeding 0.1 km^2) have rarely been determined for Taiwan in the past due to the limited number of cases. In this study, an LSL dataset containing information on landslide and rainfall parameters is created to carry out statistical analysis of multiple rainfall parameters. Then SWI and rainfall duration can be used to determine the critical threshold for triggering LSLs. Moreover, this study attempts to construct a multi-threshold model different from the single threshold for shallow landslides constructed in the past to provide a new landslide warning model, which can be used at different stages or for landslides of different scales. The threshold will provide invaluable information for helping disaster management authorities to alert the general public and prepare for prevention and disaster mitigation.

2. Study Area

Taiwan is located at the convergent plate boundary where the Philippine Sea plate subducts below the Eurasian plate at a velocity of 80 mm/year; hence, it experiences a high orogenic uplift rate (5–7 mm/year) and frequent earthquakes [30]. This high orogenic uplift rate is responsible for an active mountain belt with many summits higher than 3000 m above sea level (a.s.l.) [31]. Almost 48% of the mountain range in Taiwan is higher than 1000 m a.s.l., and the montane slopes are frequently steeper than 45° and have thin (<1 m) regolith cover [32,33]. The mountains in Taiwan have a steep slope and significant relief, and the rock formations are highly fractured and fragile. These geological settings are unfavorable to slope stability. Taiwan is also situated in the East Asian monsoon belt and in the region of subtropical climates, so Taiwan has a humid and warm climate. The annual rainfall is 2500 mm, and on average, 3–5 typhoon strikes occur yearly [34,35]. Torrential precipitation during the summer seasons often triggers geological hazards. In short, the geological and topographic conditions make Taiwan a high-risk region for slope failure (Figure 1).

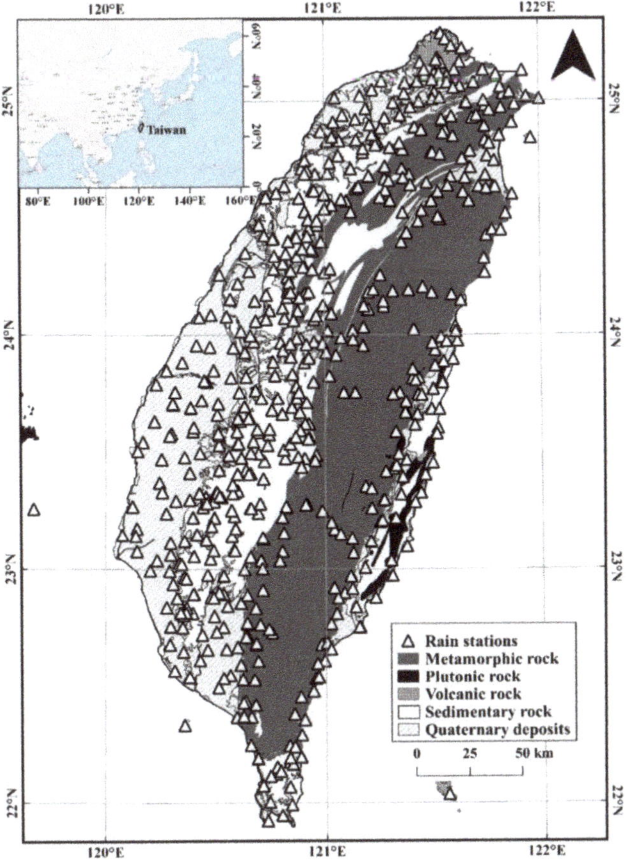

Figure 1. Distribution of rain gauges around Taiwan and geological map.

3. Data and Study Methods

3.1. Landslide Data

The main parameters of landslides include disturbed area, location, time of occurrence, failure mechanism, and type. In this study, we collected landslide data from the annual inventory of the Forestry Bureau of Taiwan. The information on the disturbed area and location of each landslide was integrated into the annual landslide inventory, but the failure mechanisms and types of landslides were not investigated. To delimit landslides of different scales, the criterion of a disturbed area of 0.1 km^2 was adopted in this study to separate LSLs from SSLs [20].

Since the mass movement of an LSL may generate ground motion, such ground motion can be recorded by nearby seismic stations [36,37]. In the frequency domain, the natural energy of landslide-induced ground-motion (called a landslide-quake) is mainly below 5 Hz, and the distribution pattern of energy in a spectrogram is triangular due to a gradual increase–decay process over time [36]. The triangular pattern in the spectrogram is the particular property that discriminates landslide-quakes from those of earthquakes and other background noise [20]. The Soil and Water Conservation Bureau (SWCB) of Taiwan extracts the occurrence times of LSLs triggered by heavy rainfall from seismic records through identifying landslide-quakes. The occurrence times of 83 LSLs triggered by rainfall over a period of 16 years (2001–2016) were observed from the landslide-induced seismic records of the SWBC and used to locate their sources using a locating approach proposed by Chen et al. [38]. Manconi, et al. [37] proposed a similar approach to detect, locate, and estimate the volumes of rockslides by analyzing waveforms acquired from broadband regional seismic networks in the eastern Italian Alps [37]. The types and failure mechanisms of the LSLs are not mentioned in the SWCB reports. However, according to some in situ investigations, the most recurrent types of LSLs in Taiwan are rock slides and debris avalanches. We collected the occurrence times and locations of LSLs from the reports of the SWCB. According to the SWCB reports, the identification error of landslide-quakes might occur due to the interference from local tremors or anthropogenic noise. A double-check conducted jointly with the analysis of remote-sensing imagery should be implemented to avoid misdetection. This study carefully compared the locations of 83 LSLs with the annual landslide inventory of the Forestry Bureau to create an LSL dataset containing the information on LSL location, disturbed area, and time (accuracy in minutes) (Figure 2 and Table 1). These 83 LSLs occurred in Taiwan during the typhoon season: 1 in June, 12 in July, 63 in August, and seven in September. This study used 75 LSLs during the period between 2001 and 2013 to analyze rainfall conditions and used 8 LSLs that occurred in 2015 and 2016 to verify the results.

The alignment of the LSLs with the geological map showed that of these 83 LSLs, 16 were located in the Western Foothills, where the lithology mainly consists of sedimentary rocks. Ten LSLs were situated in the Hsuehshan Range, where the rock formation mainly consists of alternating meta-sandstone and shale. Forty-seven LSLs occurred on the west flank of the Central Range, where the strata mainly consist of argillite and slate. Nine LSLs occurred in the eastern flank of the Central Range, where the lithology mainly comprises schist and marble. Only one LSL occurred in the Coastal Range, where the strata mainly comprise sedimentary rocks and igneous rock. The slope gradients of these LSLs were mainly distributed in the range between 20° and 40°. The LSLs primarily occurred on slopes with elevations ranging from 500 to 2000 m, but the distributions of the highest and lowest elevations of these LSLs showed that their average vertical displacement was greater than 500 m [20].

Most of the 83 LSLs occurred in metamorphic rock areas, indicating that metamorphic rock slopes were likely to be massively unstable, which could be attributed to the active tectonics in Taiwan's mountainous area inducing intense rock deformation and displacement. In contrast, the sedimentary rock areas in Taiwan have a relatively more moderate relief than the metamorphic rock area. Although massively unstable slopes still develop on sedimentary rocks, the number of LSLs was significantly lower than that on metamorphic rocks. Although the difference in LSL numbers between sedimentary rock slopes and metamorphic rock slopes seems to indicate that the geological and topographic features would influence the evolution of a massively unstable slope, the limited number of LSLs and SSLs considered in the study should be noted.

We also collected data on 174 SSLs occurring from 2006 to 2013 from the annual reports of the SWCB. The SSLs were investigated carefully with fieldwork, particularly in the cases of events that caused damage to public utilities or private property. The reports contained detailed information on the location, disturbed area, and approximate occurrence time of each SSL. These 174 SSLs occurred during the typhoon season: 2 in May, 32 in June, 36 in July, 39 in August, 24 in September, and 41 in October. The distribution of the slope gradients of the SSLs was similar to that of the LSLs. Unlike the LSLs, a large portion of the SSLs took place on slopes with elevations ranging from 750 to 1250 m. The occurrence times were estimated based on real-time videos and interviews with residents. Unfortunately, the accuracy of the time points was not mentioned in the reports. We have carefully double-checked the landslide data to exclude any SSLs that were not triggered by rainfall.

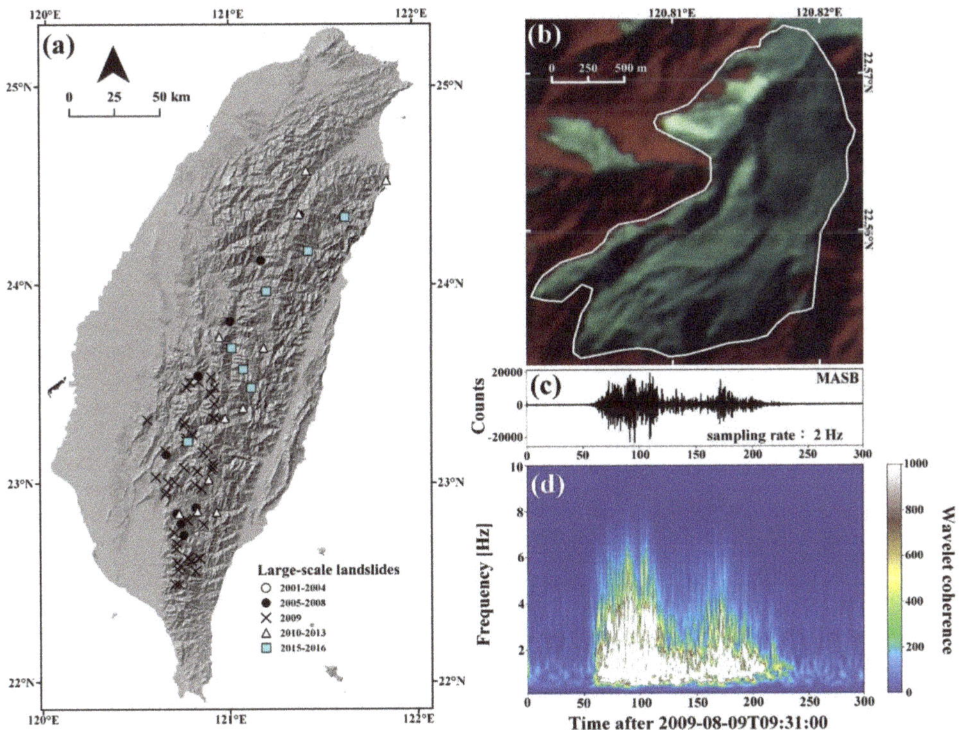

Figure 2. (**a**) Distribution of the 83 large-scale landslides (LSLs) that occurred between 2001 and 2016. (**b**) Example of satellite image of an LSL occurring in 2009 with a disturbed area of 2.3 km^2. (**c**) Original seismic waveform induced by the LSL. (**d**) Spectrogram of the vertical component of the seismic waveform induced by the LSL.

3.2. Rainfall Data

There are 594 rain gauge stations installed by the Central Weather Bureau (CWB) around Taiwan (Figure 1). Among these, 328 rainfall gauge stations were established in mountainous areas above 100 m a.s.l. The density of rain gauges is approximately one per 73 km^2. All rainfall gauges record hourly rainfall intensity. Due to the lack of rain gauge stations in the vicinity of landslide sites, we converted the records of the three nearest rain gauge stations into the representative rainfall data for each landslide site. This rainfall conversion involved conducting a deterministic interpolation using inverse distance weighting (IDW). IDW interpolation determines rainfall values on a landslide site using a weighted combination of a set of rainfall gauges. The weight is a function of the inverse distance from the landslide site to each rain gauge station. The interpolated rainfall should be a locational-dependent variable. The rainfall data of the nearest rainfall gauge stations will have the most significant influence in the interpolation. Chen and Liu [39] have proposed that a scan radius of 10–30 km would be the optimal parameter for IDW in interpolating rainfall data in Taiwan. In this study, we adopted a scan radius of 10 km from each landslide to collect rainfall data. If fewer than three rain gauge stations corresponded to this principle, we then adopted the record of the nearest rain gauge station. To determine the duration of a rainfall event, many previous studies have determined the length of a rainfall event using different non-rainfall intervals [40]. In this study, the starting-time of a rainfall event is defined as the time when the hourly rainfall exceeds 1 mm. The ending-time of the rainfall event is the time when the hourly rainfall becomes zero, but that level must be maintained for at least 24 h.

Table 1. Landslide data of the 83 LSLs.

ID	Date/Time (UTC)	Longitude (°E)	Latitude (°N)	Area (km^2)	ID	Date/Time (UTC)	Longitude (°E)	Latitude (°N)	Area (km^2)
1	31 Jul 2001/17:36	121.344	23.627	0.49	43	08 Aug 2009/13:56	120.661	22.960	0.11
2	18 Sep 2001/16:24	121.203	23.379	0.16	44	08 Aug2009/18:28	120.656	22.948	0.12
3	04 Aug 2003/11:57	120.738	22.391	0.18	45	10 Aug 2009/18:42	120.762	22.823	0.55
4	04 Aug 2003/04:18	120.966	23.429	0.13	46	09 Aug 2009/11;00	120.772	22.816	0.13
5	02 Jul 2004/19:03	121.491	24.241	0.1	47	08 Aug 2009/10:40	120.858	22.797	0.16
6	01 Jul 2004/19:36	121.248	23.596	0.15	48	08 Aug 2009/07:35	120.703	22.754	0.49
7	30 Jun 2004/23:51	121.333	23.174	0.11	49	08 Aug 2009/18:19	120.716	22.700	0.56
8	24 Aug 2004/14:54	120.763	23.566	0.22	50	08Aug 2009/19:19	120.712	22.673	0.64
9	20 Jul 2005/21:55	120.817	22.881	0.12	51	09 Aug 2009/03:55	120.719	22.603	0.63
10	21 Jul 2005/06:33	120.718	22.850	0.18	52	08 Aug 2009/20:15	120.733	22.586	0.73
11	18 Jul 2005/19:42	120.737	22.800	0.13	53	08 Aug 2009/00:04	120.724	22.565	0.39
12	20 Jul 2005/18:15	120.752	22.742	0.13	54	08 Aug 2009/17:05	120.708	22.494	0.94
13	09 Jun 2006/16:53	121.171	24.125	0.12	55	08 Aug 2009/00:35	120.727	22.494	0.12
14	15 Sep 2008/02:45	121.383	24.353	0.14	56	08 Aug 2009/21:42	120.909	23.100	0.25
15	18 Jul 2008/21:30	121.006	23.819	0.1	57	08 Aug 2009/17:53	120.911	23.079	0.19
16	18 Jul 2008/15:29	120.829	23.544	0.11	58	08 Aug 2009/17:21	120.902	23.072	0.28
17	18 Jul 2008/23:55	120.660	23.147	0.12	59	08 Aug 2009/02:20	120.847	22.975	0.11
18	08 Aug 2009/22:52	120.901	23.537	0.12	60	08 Aug 2009/23:15	120.772	22.627	0.15
19	08 Aug 2009/05:35	120.832	23.516	0.5	61	08 Aug 2009/18:16	120.831	22.626	0.72
20	08 Aug 2009/18:11	120.786	23.512	1.12	62	08 Aug 2009/23:41	120.837	22.625	0.12
21	08 Aug 2009/21:30	120.921	23.488	0.12	63	08 Aug 2009/09:00	120.793	22.611	0.62
22	08 Aug 2009/01:20	120.768	23.487	0.14	64	09 Aug 2009/09:31	120.813	22.560	2.31
23	09 Aug 2009/19:36	120.559	23.320	0.39	65	19 Sep 2010/23:24	120.728	22.850	0.15
24	08 Aug 2009/21:11	120.899	23.456	0.15	66	30 Aug 2011/09:13	121.183	23.685	0.11
25	08 Aug 2009/20:27	120.919	23.404	0.12	67	31 Aug 2011/09:37	120.976	23.331	0.11
26	08 Aug 2009/08:00	120.915	23.334	0.41	68	30 Aug 2011/07:10	120.929	22.859	0.12
27	08 Aug 2009/03:27	120.912	23.329	0.4	69	03 Aug 2012/01:00	121.377	24.359	0.19
28	08 Aug 2009/11:35	120.949	23.327	0.22	70	02 Aug 2012/19:00	120.946	23.740	0.25
29	10 Aug 2009/04:22	120.759	23.309	1.52	71	01 Aug 2012/18:39	121.417	24.576	0.12
30	08 Aug 2009/23:14	120.754	23.293	0.56	72	02 Aug 2012/10:00	121.853	24.525	0.12
31	10 Aug 2009/03:54	120.799	23.247	0.2	73	29 Aug 2013/19:48	120.825	22.862	0.21
32	09 Aug 2009/02:52	120.767	23.231	0.81	74	22 Aug 2013/19:05	121.073	23.383	0.18
33	09 Aug2009/00:34	120.767	23.215	2.24	75	13 Jul 2013/14:27	120.886	23.023	0.4
34	08 Aug 2009/16:15	120.881	23.180	0.14	76	09 Aug 2015/14:45	121.012	23.685	0.11
35	08 Aug 2009/22:16	120.656	23.166	2.5	77	09 Aug 2015/02:00	121.199	23.969	0.06
36	10 Aug 2009/11:06	120.857	23.157	0.34	78	08 Aug 2015/19:00	120.776	23.213	0.21
37	08 Aug 2009/03:55	120.754	23.082	0.33	79	16 Sep 2016/23:06	121.075	23.577	0.18
38	08 Aug 2009/06:25	120.825	23.062	0.39	80	08 Jul 2106/07:48	121.427	24.172	0.04
39	08 Aug 2009/23:02	120.604	23.034	0.13	81	16 Sep 2016/02:33	121.626	24.342	0.04
40	08 Aug 2009/07:15	120.704	23.012	0.23	82	28 Sep 2016/20:45	121.427	24.173	0.02
41	08 Aug 2009/06:28	120.671	23.008	0.15	83	29 Sep 2016/01:45	121.117	23.485	0.01
42	08 Aug 2009/08:10	120.813	22.997	0.19					

To confirm the rainfall threshold for triggering landslides, the rainfall conditions corresponding to the occurrence time of each landslide are necessary. Consequently, we counted the average rainfall intensity (I, mm/h), rainfall duration (D, h), and cumulative event rainfall (E, mm) from the starting-time of a rainfall event until the time point when the landslide occurred. If a landslide occurred after the peak hourly rainfall, the calculation of average rainfall intensity for the landslide would involve the value of the maximum hourly rainfall.

3.3. Soil Water Index

The soil water index (SWI) is a conceptual model which uses a three-layer tank model to estimate the depth of remaining water in three simulated soil layers during a rainfall event [41] (Figure 3). During a massive rainfall event, the water continues to infiltrate into the ground surface, and the moisture of the soil layers increases, which is strongly related to the potential for slope failure disasters. However, it is not an easy task to obtain the actual water contents in the soil layers if there are not enough hydrological instruments. Determining the physical or statistical relationship between rainfall, surface runoff, and groundwater is a compromise method for assessing underground water storage [42,43]. The tank model is a simple concept that uses three tanks, which represent reservoirs in a watershed. It considers rainfall as the input and generates output as the surface runoff, subsurface flow, intermediate flow, and sub-base flow. The tank model also explains the phenomena of infiltration, percolation, and water storage in the tanks. Thus, the SWI is established as a rainfall-runout model with some fixed parameters to estimate the permeation of water in the soil layers. This method was used to assess and predict potential landslides and to construct early warning systems in Japan [22].

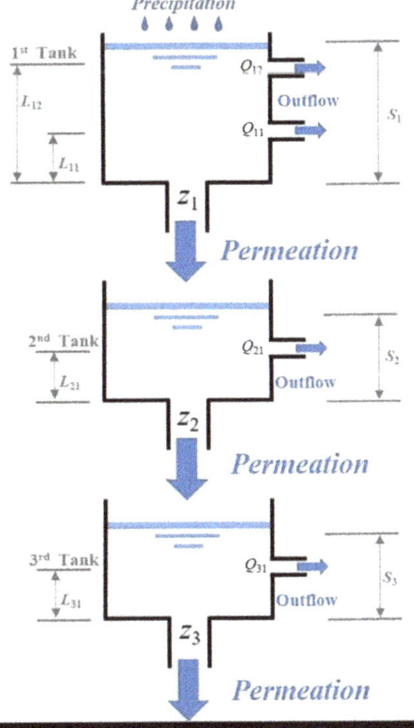

Figure 3. Schematic layout of the soil water index (SWI) tank model. The SWI represents the summation of water depth in the three tanks.

In this model, the SWI is defined as the total storage water height S_k, which is the sum of three tanks, and the formula can be written as [22]: s

$$\text{SWI}\ (t + \Delta t) = \sum_k S_k(t + \Delta t) \tag{1}$$

where t represents the time in hours; Δt expresses the passed time in hours; (k = 1,2,3) represents the tanks from top to bottom. Every S_k, the remaining water (mm) for each tank, is computed every hour (Δt = 1(h)) by:

$$S_k\,(t+\Delta t)=\begin{cases} S_k(t)-\left[\sum\limits_l Q_{kl}(t)+Z_k(t)\right]+R(\Delta t), & (k=1) \\ S_k(t)-\left[\sum\limits_l Q_{kl}(t)+Z_k(t)\right]+[Z_{k-1}(t+\Delta t)-Z_{k-1}(t)], & (k\geq 2) \end{cases} \tag{2}$$

where $R(\Delta t)$ represents the hourly rainfall amount in mm. Q_{kl} means the leakage height from the lth side tube of each tank (the top tank has two side tubes, and the others have one). Z_k is the water that permeates from the base tube of the kth tank. Z_{k-1} is the water that permeates from the base tube of the (k − 1) th tank. Both Q_{kl} and Z_k vary with t and can be calculated as follows:

$$Q_{kl}(t)=\begin{cases} a_{kl}\{S_k(t)-L_{kl}\} & (S_k(t)>L_{kl}) \\ 0 & (S_k(t)\leq L_{kl}) \end{cases} \tag{3}$$

where a_{kl} and b_k are the coefficients of seepage for the side holes and the base holes of the kth tank, respectively. L_{kl} represents the height (mm) of the leakage water flowing through the lth side hole of the kth tank.

In the three-layer tank model, the sum of Q_{11} and Q_{12} represents surface runoff, Q_{21} represents intermediate flow, and Q_{31} represents baseflow, respectively. In addition, Z_1 represents the infiltration amount from the first tank. Z_2 and Z_3 represent the percolation amounts from the second and third layers. S_1, S_2, and S_3 denote the depths of water storage in the first, second, and third tanks. Some previous studies performed statistical analysis of the relationship between river discharge and precipitation, and the constants a_{kl}, b_k, and L_{kl} (Table 2) were determined and used in the SWI model [22]. It may be true that the discharging rate and saturation capability would vary between distinct geological and topographic settings. However, the SWI is representative of conceptual water content. Furthermore, the variations of time series of the SWI using parameters adjusted with different areas have similar trends [41]. Thus, the Japanese government adopted the constant parameters developed by Okada, et al. [41] for the whole nation regardless of the various geological conditions [44]. Chen, et al. [27] applied the SWI model to calculate the rainfall characteristics for triggering shallow landslides in Taiwan and used the previous rainfall data over one month to inspect the effect of the antecedent rainfall. In this study, the SWI values of large-scale landslides and other rainfall factors were calculated and combined to find the hydrological conditions for triggering large-scale landslides. Figure 4 displays a paradigm for obtaining the time-varying SWI and its conditions for triggering an LSL. The hourly rainfall records of the three nearest rainfall gauge stations were used to estimate the representative hourly rainfall for the LSL (Figure 4a,b). Then the time-varying rainfall was interpolated to the landslide site using the IDW method and adopted to calculated S_1, S_2, and S_3 (Figure 4c). The time-varying SWI could be obtained by summing S_1, S_2, and S_3. The total remaining depths of the three tanks represent the water stored underground [45]. The concept of the tank model is easily understandable. In addition, the SWI can be used as a proxy for both meteorological trigger and hydrological cause. For the three-layer tank model, outputs through the outlets of the first tank, second tank, and third tank represent surface runoff, intermediate runoff, and baseflow [46,47]. Since most of the 83 LSLs were found to have depths of tens of meters by the SWCB, the remaining water depths in the deepest tank might be the critical hydrological causes for triggering LSLs. In this study, we calculated the time-varying values of the SWI for one month before the targeted rainfall event and in the period of the rainfall event.

Table 2. Parameters for calculating SWI.

Tank	First	Second	Third
Outflow height (mm)	$L_{11} = 15$ $L_{12} = 60$	$L_{21} = 15$	$L_{31} = 15$
Outflow coefficient (1/h)	$a_{11} = 0.1$ $a_{12} = 0.15$	$a_{21} = 0.05$	$a_{31} = 0.01$
Coefficient of permeability (1/h)	$b_1 = 0.12$	$b_1 = 0.05$	$b1 = 0.01$

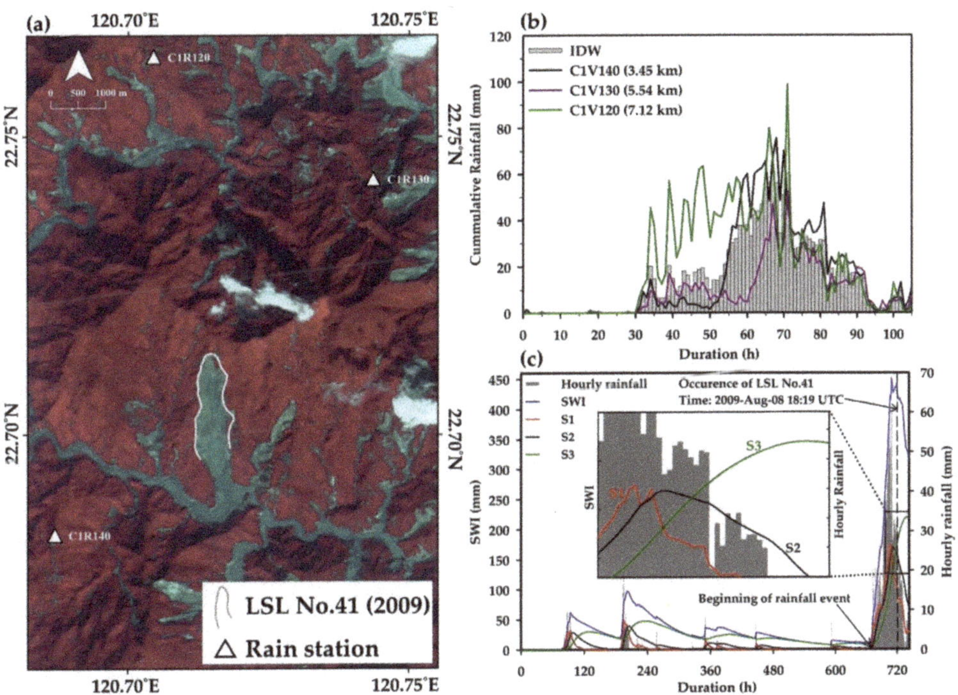

Figure 4. Example of three selected rain gauge stations and an LSL site. (**a**) Distribution of three rain gauges and the LSL No.41. (**b**) Rainfall records at the three stations and the estimated hourly rainfall using the inverse distance weighting (IDW) method. (**c**) Example of change in the SWI for the LSL.

4. Results and Discussion

4.1. Rainfall Conditions and SWI for Triggering Large-Scale Landslides

To define the rainfall threshold for landslide initiation, a detailed analysis of the rainfall conditions for the 83 landslides considered in the study was performed. Approximately half (41) of the LSLs collapsed when the cumulative event rainfall (E) exceeded 1000 mm (Figure 5). Moreover, 12 of these events happened after rainfall accumulations of more than 1500 mm in total. The amounts of hourly rainfall at the occurrence time points ranged from 0 to 91.8 mm, which included 31 events with values lower than 20 mm and 18 zero-value events (hourly rainfall was equal to zero at the occurrence time of the landslide). The duration (D) analysis showed that 63 of the LSLs occurred when the rainfall duration was longer than 48 h, and only one case had a duration time of less than 24 h. The results of the analysis of cumulative event rainfall, hourly rainfall, and duration indicated that hourly rainfall at the occurrence time was not a compelling factor in triggering large-scale landslides. Accordingly, average rainfall intensity (I) is usually adopted in the analysis of rainfall thresholds instead of hourly rainfall at the occurrence time of the landslide. On the other hand, the cumulative event rainfall and duration may have more remarkable effects on the conditions for triggering LSLs.

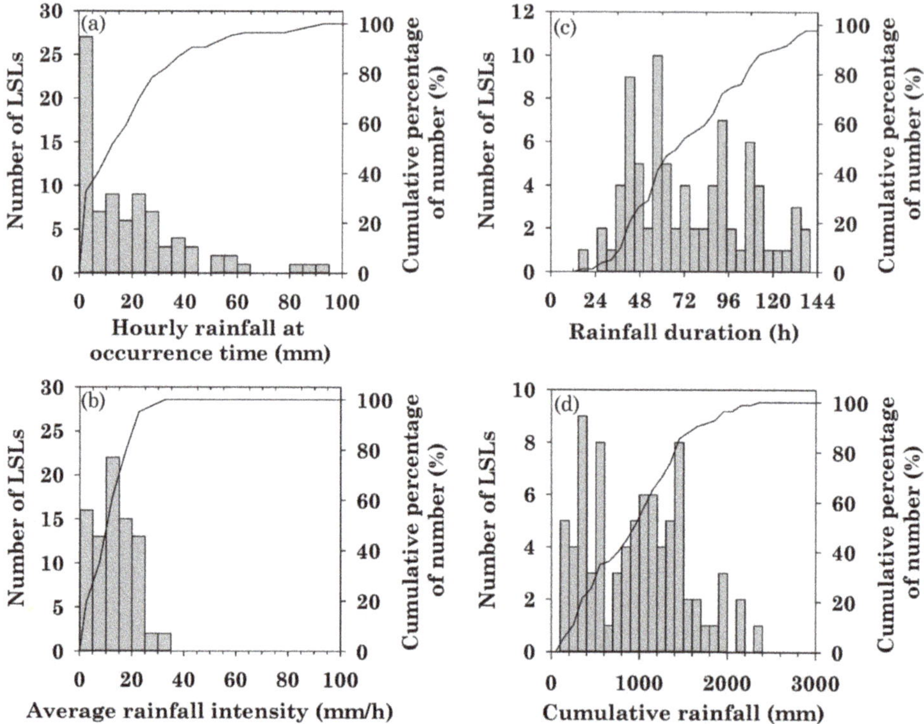

Figure 5. Statistics of triggered LSL number for each rainfall factor.

The SWI results contained the precipitation during the event as well as the antecedent precipitation of 30 days preceding the event. The average antecedent SWI was 16.3 mm, ranging from 1.2 mm to 56.2 mm. The SWI at the occurrence times of the LSLs was 311.9 mm on average, and the maximum and minimum values were 706.3 mm and 70.7 mm, respectively (Figure 6). The average S_3 value was 131.7 mm, which is higher than the average S_1 value of 78.1 mm. The values of S_1, S_2, and S_3 represent the depths of water content in the three simulated layers. SWI may not directly express the real water content in the deep rock formation. However, we need to note that the cumulative rainfall or rainfall intensity used in traditional rainfall analysis does not indicate the real water content in the rock formation, either. The statistics of the above two rainfall factors even neglect hydrological factors such as infiltration and drainage. In SWI calculations, infiltration and drainage are considered, and we can infer that the water entering the deep part of the soil is more closely related to the water content inside the rock formation than the cumulative rainfall is. Therefore, this study holds that replacing the cumulative rainfall factor with the SWI (including the values of S_1, S_2, and S_3) to construct the rainfall threshold for landslides is an enhanced approach [21,44].

The SWI analysis also revealed that 62 LSLs had S_3 values higher than the S_1 values. Under the condition that SWI is high, the situation $S_3 > S_1$ means more water remaining in the deeper layer but not in the shallow layer. Also, the average value of the ratios of S_3 to SWI at the occurrence times of LSLs was 0.46. In contrast, the average of the ratios of S_1 to SWI at the occurrence times of LSLs was only 0.22. Because S_3 represents the water content of the deeper soil layer, the results indicated that the deeper water might have a higher relationship with the occurrence of LSLs. The hydrographs of SWI demonstrated that most of the LSLs collapsed when the rising trend became smooth and even started to fall (Figure 7). According to the SWI model, the falling and smooth trend of the SWI hydrograph indicates the declination of rainfall events and the lack of recharging of the water in upper slope materials. This phenomenon responded well to the characteristic of cumulative rainfall and rainfall duration for triggering LSLs.

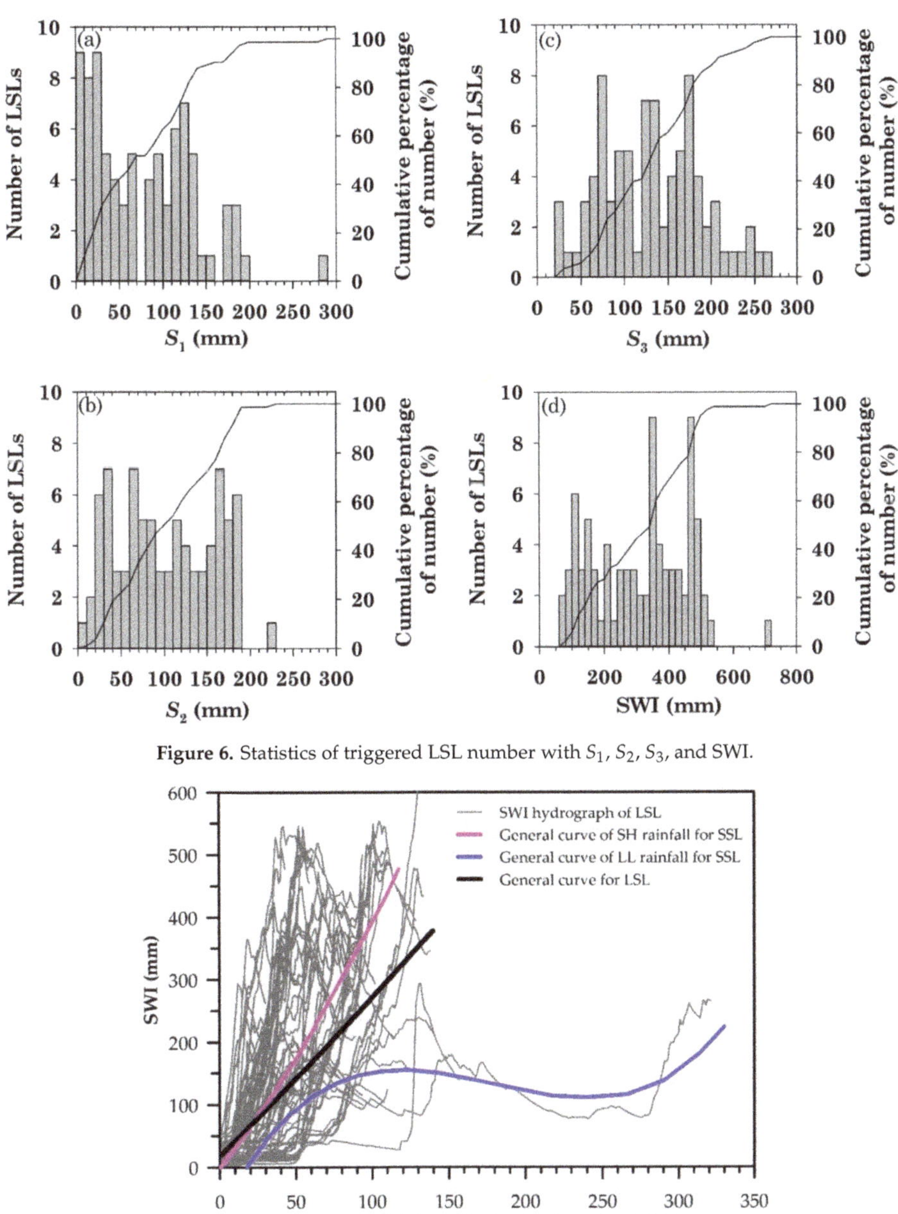

Figure 6. Statistics of triggered LSL number with S_1, S_2, S_3, and SWI.

Figure 7. Hourly changes in the SWI from the beginning of rainfall to landslide occurrence. The two general curves for small-scale landslides (SSL) were adopted from Chen, et al. [27].

4.2. Soil Water Index–Rainfall Duration (SWI–D) Threshold for Large-Scale Landslides (LSLs) and Verification

Unlike other consolidated approaches, we have defined soil water index–rainfall duration (SWI–D) thresholds, instead of the popular average rainfall intensity–rainfall duration thresholds. In this study, the SWI–D rainfall threshold curve at 5% exceedance probability was estimated by the method proposed by Brunetti et al. [48]. This threshold was expected to leave 5% of the data points below the threshold line. In general, a threshold requires verification with a certain number of cases. The verification cases are usually randomly selected from the total cases and excluded from the database used to build the threshold. Since the number of LSLs is inconsistent every year and the number of samples used to establish the threshold is limited, the random sampling method was not used to select the verification cases in this study; instead, the LSLs occurring in the last two years of the study period were selected as the threshold verification cases. Figure 8 depicts the SWI–D conditions associated with LSLs and SSLs in Taiwan and the threshold lines. The threshold for LSLs is determined as SWI = 155.20 − 1.56D and $D \geq 24$ h. Figure 9 presents the ratios of water depth in the third tank (S_3) to the SWIs for LSLs and SSLs. The 50th-percentile of S_3/SWI is 0.41, which indicates that when S_3 occupies more than 40% of the SWI, there is a higher potential of LSL initiation. This result implies the importance of the water content of the deeper layer.

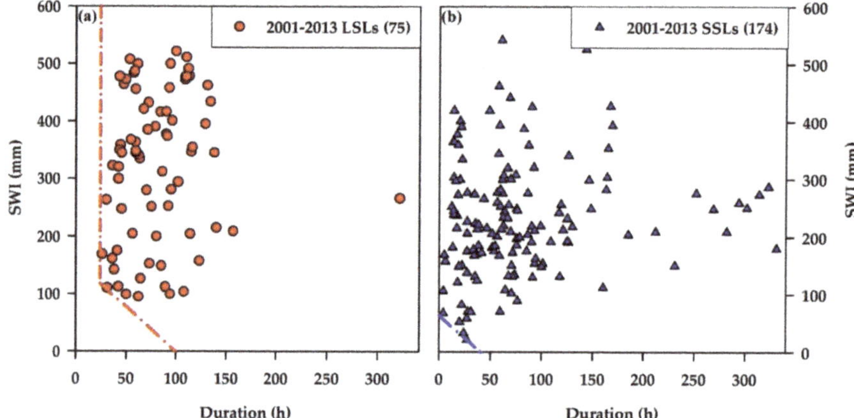

Figure 8. Soil water index–rainfall duration (SWI–D) thresholds for (**a**) LSLs and (**b**) SSLs.

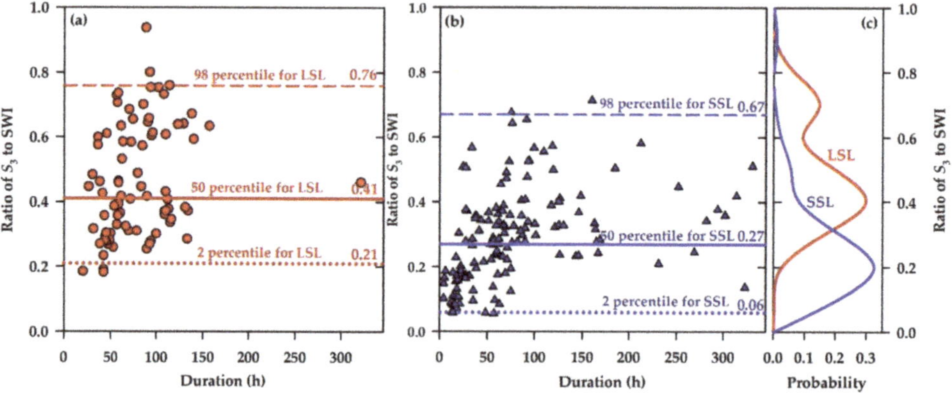

Figure 9. (**a**) Data distribution of the ratios of S_3 to SWI for (**a**) LSLs and (**b**) SSLs. (**c**) Probability distribution of the ratios of S_3 to SWI for LSLs and SSLs. Duration was calculated from the beginning of rainfall to landslide occurrence.

The SWI–*D* threshold built with LSL data for 2001–2013 was validated with eight LSLs triggered by heavy rainfall in 2015 and 2016 (Figure 10). Six of them accorded with the SWI–*D* conditions for 2001–2013, verifying that the SWI can be treated as an indicator for triggering LSLs. However, two cases were lower than the SWI–*D* threshold. Even so, the two lower values were relatively close to the threshold line. The verification confirmed the advantage of using the SWI. For instance, the SWI often increases rapidly before landslide occurrence, and this phenomenon can be used for a warning system.

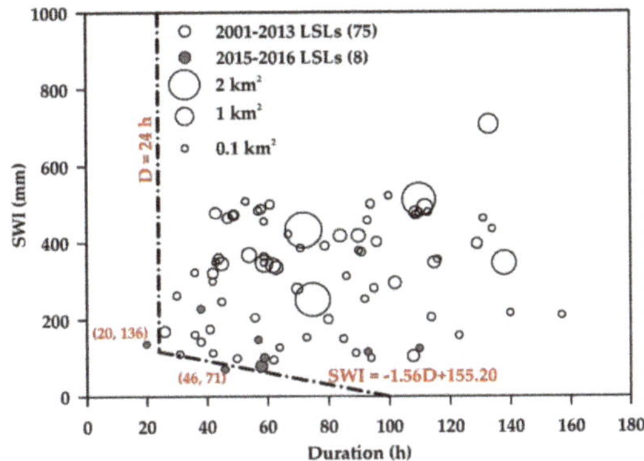

Figure 10. Validation of the SWI–*D* threshold using the data for six LSLs occurring during 2015–2016. Dashed line represents the SWI–*D* threshold.

4.3. Comparison with Small-Scale Landslides

For comparison, we used an inventory dataset including 174 small-scale landslides (<0.1 km^2) provided by the Soil and Water Conservation Bureau, Taiwan. The inventory reports occurrence times and rainfall records. One evident contrast between the rainfall conditions for SSLs and those for LSLs in Taiwan is that the LSLs primarily occur with higher cumulative rainfall. However, SSLs are more likely with shorter but intense rainfall events.

For further comparison, Figure 8 shows the SWI–*D* conditions for triggering SSLs and LSLs. Although the variations of the SWI–*D* conditions for the two types of landslide are challenging to distinguish, we note that the SWI–*D* threshold for LSLs is much higher than that for SSLs. The slope materials require a more substantial amount of water content to evolve into a massive landslide. Figure 9 displays the variation of the ratio of S_3 to SWI for SSLs and LSLs. We also noted that each percentile of S_3/SWI for LSLs was significantly higher than that for SSLs. For the 50th percentile, SSLs can potentially occur when S_3 occupies 27% of the SWI; however, LSLs can potentially occur when S_3 occupies 41% of the SWI. A rainfall event that raises the SWI and S_3 to high values is critical for triggering LSLs. Therefore, identifying changes in the SWI is conducive to determining the lowest rainfall thresholds for landslides of different scales. Based on the data for 2001–2016, Figure 7 shows the hourly changes of the SWI from the beginning of the rainfall events to LSL occurrences. We used non-parametric median regressions to determine the general trend of the SWI for triggering these LSLs. In addition to an LSL with a duration greater than 200 h, the general SWI curve for the remaining 82 LSLs was obtained by calculating the 50th-percentile of SWIs per each 1-hour increasing interval of duration. Figure 7 also displays the general trends of SSLs in Taiwan, reported by Chen, et al. [27]. Saito, et al. [3] classified rainfall conditions for triggering SSLs into two types, shorter duration–high intensity (SH) and long duration–low intensity (LL). The general trend of the time-varying changes in the SWI for the LSLs in Taiwan is located between the SH type and the LL type. Among the 83 LSLs, only one LSL, occurring in 2006, was triggered by an LL type rainfall event. Comparing SSLs with LL types, the rainfall conditions for triggering LSLs are associated with high average intensity. Furthermore, comparing SSLs with SH types, rainfall conditions for triggering LSLs are associated with a slightly longer duration.

4.4. *Effect of Antecedent SWI*

The antecedent SWI values for one month until the beginnings of rainfall events were calculated for comparison with the rainfall conditions for triggering LSLs. Based on the approaches proposed by Chen, et al. [27] for separating landslides into three categories by two values of antecedent SWI, 14.7 mm (the average value) and 29.4 mm (twofold the average value), Table 3 displays the values of the 98th percentiles of average intensity, cumulative event rainfall, and duration for these three categories. Figure 11 shows that the rainfall conditions for triggering LSLs gradually descend with increases in antecedent SWI. With increases in antecedent SWI, the decline in rainfall intensity and cumulative rainfall for triggering LSLs is more evident than that for SSLs. This finding reveals the significant effect of water content before a rainfall event on the occurrence of an LSL. Figure 11 also shows that when antecedent SWI is less than 29.4 mm, the 98th percentiles of rainfall intensity and cumulative rainfall for triggering LSLs are higher than those for SSLs. The 98th percentile of duration for LSLs exceeds 100 h, even though it is shorter than that for SSLs. When antecedent SWI is higher than 29.4 mm, the decrease in the duration for triggering LSLs is smaller than that for SSLs. This finding indicates that long-duration rainfall can be considered as one of the main conditions for triggering LSLs.

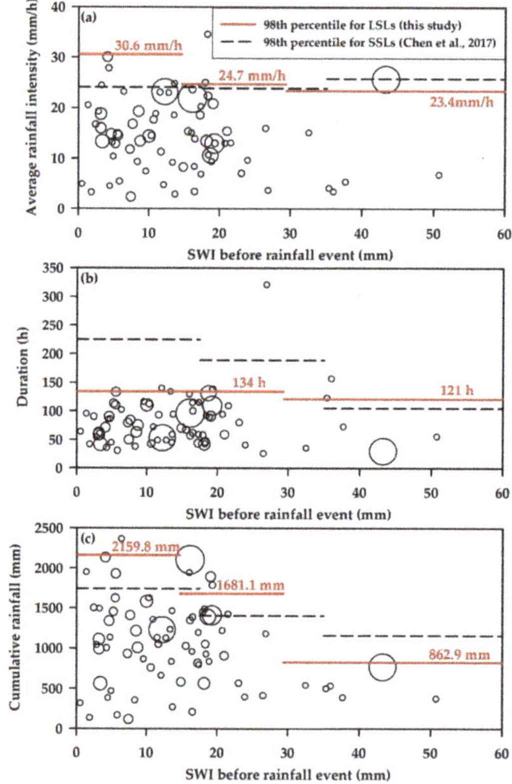

Figure 11. Variations of (**a**) average rainfall intensity, (**b**) duration, and (**c**) cumulative rainfall for triggering LSLs during 2001–2016 with the antecedent SWI. Red lines represent the 98th percentile of rainfall parameters for triggering LSLs in different categories of antecedent SWI. Dashed lines represent the 98th percentile of rainfall parameters for triggering SSLs in different categories of antecedent SWI reported by Chen, et al. [27].

Table 3. The 98th percentile of duration, cumulative amount, average intensity of rainfall triggering LSLs in different categories of antecedent SWI.

Rainfall Conditions	SWI ≤ 14.7	14.7 < SWI < 29.4	SWI ≥ 29.4
Duration (h)	134	134	121
Cumulative rainfall (mm)	2159.8	1681.1	826.9
Average rainfall intensity (mm/h)	30.6	24.7	23.4

5. Conclusions

Due to the limited number of large-scale landslides (LSLs) in a specific region and the difficulty of obtaining their occurrence times, it is still not easy to establish effective rainfall thresholds for LSLs. Exact time and location information on the LSLs in Taiwan provides the opportunity to develop regional rainfall thresholds for triggering LSLs. In addition to the rainfall thresholds determined by general rainfall factors (intensity, duration, and cumulative rainfall), the conceptual water depth estimated by the three-layer tank model can be used as one of the hydro-meteorological conditions that cause landslides. This study analyzed rainfall factors and evaluated conceptual water depths, including S_1, S_2, S_3, and the Soil Water Index (SWI), in three-layer tanks to assess hydro-meteorological thresholds for LSLs from 2001 to 2016 in Taiwan. The analysis of rainfall duration (D), cumulative event rainfall (E), and hourly rainfall (I) indicated that the hourly rainfall at the occurrence time is not a crucial factor for triggering an LSL. In fact, the cumulative event rainfall and duration may be the determining factors for triggering an LSL. The average antecedent SWI is 16.3 mm, and the average SWI when LSLs occurred is 311.9 mm. Sixty-Two events occurred when S_3 was higher than S_1, which indicated that the deeper water might have a higher relationship with the initiation of LSLs. The triggering rainfall for 75 LSLs from 2001 to 2013 was used to make an SWI–D threshold (SWI = 155.20 − 1.56D and $D \geq 24$ h) and was tested with 8 LSLs from 2015 to 2016. A substantial amount of water content within deeper materials is required for initiation of an LSL. The result verified that the SWI can be treated as an indicator of possible LSL initiation. This study also compared the rainfall conditions for LSLs with those for small-scale landslides (SSLs) and revealed that the antecedent rainfall for triggering LSLs is higher than that for triggering SSLs. Unlike the rainfall conditions for triggering SSLs, which are associated with the shorter duration–high intensity type, the rainfall conditions for triggering LSLs are related to long duration–high intensity rainfall events. Through understanding the different hydro-meteorological conditions for LSLs and SSLs using the tank model, the results of this study provide the potential to develop an enhanced landslide early warning model considering disasters of various scales.

Author Contributions: G.-W.L. devised the research, the main conceptual ideas, and the proof outline. H.-L.K. and G.-W.L. performed the analytic calculations. G.-W.L., H.-L.K., C.-W.C., L.-W.W., and J.-M.Z. contributed to the interpretation of the results. G.-W.L. took the lead in writing the manuscript. All authors provided critical feedback and helped shape the research, analysis, and manuscript. All authors have read and agreed to the published version of the manuscript.

Funding: The authors gratefully acknowledged the financial support of the Ministry of Science and Technology of Taiwan and the Soil and Water Conservation Bureau, Council of Agriculture, Executive Yuan of Taiwan.

Acknowledgments: The source of all seismic and rainfall information included in this paper was the Institute of Earth Sciences, Academia Sinica of Taiwan, and the Seismology Center, Central Weather Bureau (CWB), Taiwan.

Conflicts of Interest: The authors declare no conflict of interest

References

1. Tu, J.Y.; Chou, C. Changes in precipitation frequency and intensity in the vicinity of Taiwan: Typhoon versus non-typhoon events. *Environ. Res. Lett.* **2013**, *8*, 014023. [CrossRef]
2. Westra, S.; Alexander, L.V.; Zwiers, F.W. Global Increasing Trends in Annual Maximum Daily Precipitation. *J. Clim.* **2013**, *26*, 3904–3918. [CrossRef]

3. Saito, H.; Korup, O.; Uchida, T.; Hayashi, S.; Oguchi, T. Rainfall conditions, typhoon frequency, and contemporary landslide erosion in Japan. *Geology* **2014**, *42*, 999–1002. [CrossRef]
4. Gariano, S.L.; Guzzetti, F. Landslides in a changing climate. *Earth Sci. Rev.* **2016**, *162*, 227–252. [CrossRef]
5. Kelly, M.J. Trends in Extreme Weather Events since 1900—An Enduring Conundrum for Wise Policy Advice. *J. Geogr. Nat. Disasters* **2016**, *6*, 1000155.
6. Froude, M.J.; Petley, D.N. Global fatal landslide occurrence from 2004 to 2016. *Nat. Hazards Earth Syst. Sci.* **2018**, *18*, 2161–2181. [CrossRef]
7. Chigira, M.; Kiho, K. Deep-seated rockslide-avalanches preceded by mass rock creep of sedimentary rocks in the Akaishi Mountains, central Japan. *Eng. Geol.* **1994**, *38*, 221–230. [CrossRef]
8. Lin, C.W.; Liu, S.H.; Lee, S.Y.; Liu, C.C. Impacts of the Chi-Chi earthquake on subsequent rainfall-induced landslides in central Taiwan. *Eng. Geol.* **2006**, *86*, 87–101. [CrossRef]
9. Kirschbaum, D.; Stanley, T.; Zhou, Y. Spatial and temporal analysis of a global landslide catalog. *Geomorphology* **2015**, *249*, 4–15. [CrossRef]
10. Brunetti, M.T.; Guzzetti, F.; Rossi, M. Probability distributions of landslide volumes. *Nonlinear Process. Geophys.* **2009**, *16*, 179–188. [CrossRef]
11. Cruden, D.M.; Varnes, D.J. Chapter 3-Landslide types and process. In *Landslides: Investigation and Mitigation*; Turner, A.K., Schuster, R.L., Eds.; National Academy Press: Washington, DC, USA, 1996; pp. 36–75.
12. Hung, C.; Liu, C.H.; Lin, G.W.; Leshchinsky, B. The Aso-Bridge coseismic landslide: A numerical investigation of failure and runout behavior using finite and discrete element methods. *Bull. Eng. Geol. Environ.* **2018**, *78*, 2459–2472. [CrossRef]
13. Guzzetti, F.; Gariano, S.L.; Peruccacci, S.; Brunetti, M.T.; Marchesini, I.; Rossi, M.; Melillo, M. Geographical landslide early warning systems. *Earth Sci. Rev.* **2019**, *200*, 102973. [CrossRef]
14. Jan, C.D.; Lee, M.H. A debris-flow rainfall-based warning model. *J. Chin. Soil Water Conserv.* **2004**, *35*, 273–283.
15. Caine, N. The rainfall intensity-duration control of shallow landslides and debris flows. *Geogr. Ann. Phys. Geogr.* **1980**, *62*, 23–27.
16. Guzzetti, F.; Peruccacci, S.; Rossi, M.; Stark, C.P. Rainfall thresholds for the initiation of landslides in central and southern Europe. *Meteorol. Atmos. Phys.* **2007**, *98*, 239–267. [CrossRef]
17. Chen, C.W.; Saito, H.; Oguchi, T. Rainfall intensity–duration conditions for mass movements in Taiwan. *Prog. Earth Planet. Sci.* **2015**, *2*, 14. [CrossRef]
18. Segoni, S.; Piciullo, L.; Gariano, S.L. A review of the recent literature on rainfall thresholds for landslide occurrence. *Landslides* **2018**, *15*, 1483–1501. [CrossRef]
19. Gariano, S.L.; Sarkar, R.; Dikshit, A.; Dorji, K.; Brunetti, M.T.; Peruccacci, S.; Melillo, M. Automatic calculation of rainfall thresholds for landslide occurrence in Chukha Dzongkhag, Bhutan. *Bull. Eng. Geol. Environ.* **2019**, *78*, 4325–4332. [CrossRef]
20. Kuo, H.L.; Lin, G.W.; Chen, C.W.; Saito, H.; Lin, C.W.; Chen, H.; Chao, W.A. Evaluating critical rainfall conditions for large-scale landslides by detecting event times from seismic records. *Nat. Hazards Earth Syst. Sci.* **2018**, *18*, 2877–2891. [CrossRef]
21. Bogaard, T.; Greco, R. Invited perspectives: Hydrological perspectives on precipitation intensity-duration thresholds for landslide initiation: Proposing hydro-meteorological thresholds. *Nat. Hazards Earth Syst. Sci.* **2018**, *18*, 31–39. [CrossRef]
22. Sugawara, M.E.; Ozaki, I.; Watanabe, I.; Katsuyama, Y. Tank model and its application to Bird Creek, Wollombi Brook, Bikin River, Kitsu River, Sanaga River and Nam Mune. In *Research Notes of the National Research Center for Disaster Prevention*; Science and Technology Agency: Tokyo, Japan, 1974; pp. 1–64.
23. Segoni, S.; Rosi, A.; Lagomarsino, D.; Fanti, R.; Casagli, N. Brief communication: Using averaged soil moisture estimates to improve the performances of a regional-scale landslide early warning system. *Nat. Hazards Earth Syst. Sci.* **2018**, *18*, 807–812. [CrossRef]
24. Oku, Y.; Yoshino, J.; Takemi, T.; Ishikawa, H. Assessment of heavy rainfall-induced disaster potential based on an ensemble simulation of Typhoon Talas (2011) with controlled track and intensity. *Nat. Hazards Earth Syst. Sci.* **2014**, *14*, 2699–2709. [CrossRef]
25. Lin, S.E.; Chan, Y.H.; Kuo, C.Y.; Chen, R.F.; Hsu, Y.J.; Chang, K.J.; Lee, S.P.; Wu, R.Y.; Lin, C.W. The Use of a Hydrological Catchment Model to Determine the Occurrence of Temporal Creeping in Deep-seated Landslides. *J. Chin. Soil Water Conserv.* **2017**, *48*, 153–162. (In Chinese)

26. Nie, W.; Krautblatter, M.; Leith, K.; Thuro, K.; Festl, J. A modified tank model including snowmelt and infiltration time lags for deep-seated landslides in alpine environments (Aggenalm, Germany). *Nat. Hazards Earth Syst. Sci.* **2017**, *17*, 1595–1610. [CrossRef]
27. Chen, C.W.; Saito, H.; Oguchi, T. Analyzing rainfall-induced mass movements in Taiwan using the soil water index. *Landslides* **2017**, *14*, 1031–1041. [CrossRef]
28. Mercogliano, P.; Segoni, S.; Rossi, G.; Sikorsky, B.; Tofani, V.; Schiano, P.; Catani, F.; Casagli, N. Brief communication "A prototype forecasting chain for rainfall induced shallow landslides". *Nat. Hazards Earth Syst. Sci.* **2013**, *13*, 771–777. [CrossRef]
29. Dikshit, A.; Sarkar, R.; Pradhan, B.; Acharya, S.; Dorji, K. Estimating rainfall thresholds for landslide occurrence in the Bhutan Himalayas. *Water* **2019**, *11*, 1616. [CrossRef]
30. Willett, S.D.; Fisher, D.; Fuller, C.; Yeh, E.C.; Lu, C.Y. Erosion rates and orogenic-wedge kinematics in Taiwan inferred from fission-track thermochronometry. *Geology* **2003**, *31*, 945–948. [CrossRef]
31. Ho, C.S. An Introduction to the geology of Taiwan: Explanatory TEXT of the geological map of Taiwan. In *Central Geological Survey*, 2nd ed.; Ministry of Economic Affairs: Taipei, Taiwan, 1988; p. 192.
32. Li, Y.H. Denudation of Taiwan island since the Pliocene epoch. *Geology* **1976**, *4*, 105–107. [CrossRef]
33. Hovius, N.; Stark, C.P.; Chu, H.T.; Lin, J.C. Supply and Removal of Sediment in a Landslide-Dominated Mountain Belt: Central Range, Taiwan. *J. Geol.* **2000**, *108*, 73–89. [CrossRef]
34. Shieh, S.L.; Wang, S.T.; Cheng, M.D.; Yeh, T.C.; Chiou, T.K. *User's Guide for Typhoon Forecasting in the Taiwan Area (VII)*; Central Weather Bureau: Taipei, Taiwan, 1998; p. 171.
35. Wu, C.C.; Kuo, Y.H. Typhoons affecting Taiwan: Current understanding and future challenges. *Bull. Am. Meteorol. Soc.* **1999**, *80*, 67–80. [CrossRef]
36. Dammeier, F.; Moore, J.R.; Haslinger, F.; Loew, S. Characterization of alpine rockslides using statistical analysis of seismic signals. *J. Geophys. Res. Earth Surf.* **2011**, *116*, F04024. [CrossRef]
37. Manconi, A.; Picozzi, M.; Coviello, C.; De Santis, F.; Elia, L. Real-time detection, location, and characterization of rockslides using broadband regional seismic networks. *Geophys. Res. Lett.* **2016**, *43*, 6960–6967. [CrossRef]
38. Chen, C.H.; Chao, W.A.; Wu, Y.M.; Zhao, L.; Chen, Y.G.; Ho, W.Y.; Lin, T.L.; Kuo, K.H.; Chang, J.M. A seismological study of landquakes using a real-time broad-band seismic network. *Geophys. J. Int.* **2013**, *194*, 885–898. [CrossRef]
39. Chen, F.W.; Liu, C.W. Estimation of the spatial rainfall distribution using inverse distance weighting (IDW) in the middle of Taiwan. *Paddy Water Environ.* **2012**, *10*, 209–222. [CrossRef]
40. Melillo, M.; Brunetti, M.T.; Peruccacci, S.; Gariano, S.L.; Guzzetti, F. An algorithm for the objective reconstruction of rainfall events responsible for landslides. *Landslides* **2015**, *12*, 311–320. [CrossRef]
41. Okada, K.; Makihara, Y.; Shimpo, A.; Nagata, K.; Kunitsugu, M.; Saito, K. Soil water index. *Tenki* **2001**, *47*, 36–41.
42. Xie, Z.; Su, F.; Liang, X.; Zeng, Q.; Hao, Z.; Guo, Y. Applications of a surface runoff model with Horton and Dunne runoff for VIC. *Adv. Atmos. Sci.* **2003**, *20*, 165–172.
43. Kirchner, J.W. Catchments as simple dynamical systems: Catchment characterization, rainfall-runoff modeling, and doing hydrology backward. *Water Resour. Res.* **2009**, *45*, W02429. [CrossRef]
44. Osanai, N.; Shimizu, T.; Kuramoto, K.; Kojima, S.; Noro, T. Japanese early-warning for debris flows and slope failures using rainfall indices with Radial Basis Function Network. *Landslides* **2010**, *7*, 325–338. [CrossRef]
45. Song, J.H.; Her, Y.; Park, J.; Lee, K.D.; Kang, M.S. Simulink Implementation of a Hydrologic Model: A Tank Model Case Study. *Water* **2017**, *9*, 639. [CrossRef]
46. Paik, K.; Kim, J.H.; Kim, H.S.; Lee, D.R. A conceptual rainfall-runoff model considering seasonal variation. *Hydrol. Process.* **2005**, *19*, 3837–3850. [CrossRef]
47. Jang, T.; Kim, H.; Kim, S.; Seong, C.; Park, S. Assessing irrigation water capacity of land use change in a data-scarce watershed of Korea. *J. Irrig. Drain. Eng.* **2011**, *138*, 445–454. [CrossRef]
48. Brunetti, M.T.; Peruccacci, S.; Rossi, M.; Luciani, S.; Valigi, D.; Guzzetti, F. Rainfall thresholds for the possible occurrence of landslides in Italy. *Nat. Hazards Earth Syst. Sci.* **2010**, *10*, 447–458. [CrossRef]

Article

Temporal Probability Assessment and Its Use in Landslide Susceptibility Mapping for Eastern Bhutan

Abhirup Dikshit [1], Raju Sarkar [2,3], Biswajeet Pradhan [1,4,*], Ratiranjan Jena [1], Dowchu Drukpa [5] and Abdullah M. Alamri [6]

1 Centre for Advanced Modelling and Geospatial Information Systems (CAMGIS), Faculty of Engineering & IT, University of Technology Sydney, Sydney, NSW 2007, Australia; abhirupdikshit@gmail.com (A.D.); ratiranjan.jena@student.uts.edu.au (R.J.)
2 Department of Civil Engineering, Delhi Technological University, Bawana Road, Delhi 110042, India; rajusarkar@dce.ac.in
3 Department of Civil Engineering and Architecture, College of Science and Technology, Royal University of Bhutan, Rinchending 21101, Bhutan
4 Department of Energy and Mineral Resources Engineering, Sejong University, Choongmu-gwan, 209, Neungdongro Gwangjin-gu, Seoul 05006, Korea
5 Department of Geology and Mines, Ministry of Economic Affairs, Royal Govt. of Bhutan, Thimphu 11001, Bhutan; ddukpa@moea.gov.bt
6 Department of Geology & Geophysics, College of Science, King Saud University, P.O. Box 2455, Riyadh 11451, Saudi Arabia; alamri.geo@gmail.com
* Correspondence: biswajeet.pradhan@uts.edu.au

Received: 25 November 2019; Accepted: 15 January 2020; Published: 17 January 2020

Abstract: Landslides are one of the major natural disasters that Bhutan faces every year. The monsoon season in Bhutan is usually marked by heavy rainfall, which leads to multiple landslides, especially across the highways, and affects the entire transportation network of the nation. The determinations of rainfall thresholds are often used to predict the possible occurrence of landslides. A rainfall threshold was defined along Samdrup Jongkhar–Trashigang highway in eastern Bhutan using cumulated event rainfall and antecedent rainfall conditions. Threshold values were determined using the available daily rainfall and landslide data from 2014 to 2017, and validated using the 2018 dataset. The threshold determined was used to estimate temporal probability using a Poisson probability model. Finally, a landslide susceptibility map using the analytic hierarchy process was developed for the highway to identify the sections of the highway that are more susceptible to landslides. The accuracy of the model was validated using the area under the receiver operating characteristic curves. The results presented here may be regarded as a first step towards understanding of landslide hazards and development of an early warning system for a region where such studies have not previously been conducted.

Keywords: shallow landslide; landslide susceptibility; temporal probability; Bhutan

1. Introduction

Rainfall triggered landslides are one of the most devastating naturally occurring disasters across the world [1]. The global dataset of landslide hazards in the 2004–2016 period extracted from Reference [2] showed that almost 75% of the world's fatal landslides occurred in the Himalayan region. Bhutan is no exception to this, and is a part of one of the world's highly landslide-prone regions in the world [3]. The damage caused by landslides in this country has led to casualties and loss of land, affecting people's livelihoods and disrupting the transportation network, which is key to the country's economy. Most of the landslides in the Bhutan Himalayas are triggered by rainfall, especially during the monsoon period [4,5]. Therefore, it is imperative to identify the areas that could be affected by landslides, in order to reduce the probability of damage in the future. The key to achieving this is through a detailed landslide hazard assessment that will help civic authorities to curtail landslide damage through effective land use management.

Landslide hazard may be defined as the probability of a damaging landslide in a spatial ("where") and temporal ("when") context, along with the magnitude ("how large") of the event [6,7]. Landslide susceptibility is defined as the likelihood of landslide occurrence ("where") in an area depending on local terrain conditions [8]. It may be regarded as the first step towards analyzing hazard and risk. Various spatial assessment models for landslide susceptibility have been developed [9–11]. However, compared to spatial assessment, there have been fewer attempts to carry out temporal probability assessment (studies have been conducted in Nilgiris, India [12], Hoa Binh, Vietnam [13], and Cameroon [14]). The two main techniques used to assess temporal probability for future landslide occurrences are (i) analysis of potential slope failure and (ii) statistical analysis of past landslide events [13,15]. The first technique involves evaluation of the current slope conditions and determination of the probability for future slope instability, which may be difficult to apply in large study areas [16]. Statistical analysis of past landslide events may be done directly using records of the landslides identified in the study area or, alternatively, it may be performed indirectly by using information related to recurrence of the landslide-triggering events [17]. Direct analysis requires a long time span of historical landslide data which is extremely difficult to obtain, especially in underdeveloped countries. Therefore, an indirect approach analyzing the frequency of occurrence of rainfall was used in this study to determine temporal probability. Even though this approach did not require complete multi-temporal landslide inventory data, it required determination of the relationship between rainfall and landslide incidences. After the calculation of rainfall thresholds, the landslide temporal probability was computed based on the number of times precipitation exceeded the threshold value [16]. As the frequency of rainfall-induced landslides only evaluates how often landslides might occur, it therefore needs to be integrated with spatial probability (susceptibility) and temporal probability to develop a landslide hazard map [17,18].

The prediction of landslide incidences using rainfall thresholds has been successfully carried out for various regions, including Italy [19–21], New Zealand [22], Malaysia [23], and the Himalayan arc [24–26]. The calculation of rainfall thresholds for landslide triggering can be determined using three main approaches: (i) physically based models [27], (ii) empirical rainfall threshold models [28], and (iii) statistically based models [29]. The physically based models are linked to the physical attributes of the study region and can be difficult to apply in cases of unavailability of an extensive dataset. Empirical models calculate rainfall thresholds based on past rainfall events which led to landslide incidences. The threshold is usually obtained by drawing lower-bound lines to the rainfall conditions that resulted in landslides plotted in Cartesian, semi-logarithmic, or logarithmic coordinates [30]. Statistical models use statistical tools like Bayesian inference or logistic regression to calculate thresholds [31].

In the case of Bhutan, studies to date have focused on the southwest region covering the Phuentsholing–Thimphu Highway (known as the Asian Highway), which connects the national capital Thimphu with neighboring countries. These studies have primarily focused on rainfall estimation and spatial assessment, using various techniques such as a probabilistic approach [5,32], a semi-automatic algorithm approach [26], an empirical approach [4], and machine learning models [33]. The other major highway, Samdrup Jongkhar–Trashigang (S-T), situated in the eastern part of the country, has been neglected, and a landslide study in this region is yet to be conducted. The main aim of this study was to assess landslide susceptibility utilizing temporal rainfall for the S-T highway. The two objectives in this study were (i) to estimate the temporal probability, and (ii) to estimate the landslide susceptibility using a multi-criterion decision-making approach. We addressed three major research themes in the current study: (i) determination of rainfall threshold, probability estimation of the threshold being exceeded, and landslide probability after the threshold has exceeded; (ii) susceptibility of the region with respect to landslides; and (iii) validation of the thresholds and susceptibility map. For this, the rainfall thresholds were determined based on the relation between daily rainfall and past landslide events that occurred between 2014 and 2017. The thresholds were validated using the rainfall records of 2018. Thereafter, the exceedance probability of the threshold was calculated and the temporal probability of landslides was determined using a Poisson model. Finally, a landslide susceptibility map was developed using the analytic hierarchy process (AHP), utilizing the determined threshold values. This study was the first attempt to conduct such an elaborate study for the eastern region of Bhutan. The results from the present work can be understood as a preliminary step towards setting up an operational landslide early warning system so that damage to the transportation corridor can be reduced and human lives can be saved.

2. Study Area

The study area was the Samdrup Jongkar–Trashigang (S-T) highway, which is a 180 km stretch of road located in the eastern part of Bhutan, which covers 1880 km^2 (Figure 1). The region was selected as it connects eight districts (known as "dzongkhags" in Bhutanese) and is a major route for the people residing around the highway. The highway is critical as it is the only transportation network connecting eight dzongkhags in East Bhutan, and it is a lifeline for the people residing in these areas. The transportation corridor has a history of severe slope failures in the form of frequent landslides, rockfalls, and mudflow. The region falls in a sub-tropical zone, where heavy rainfall is frequent. The region receives its maximum rainfall during the monsoon seasons (June–September). The rainfall pattern in this region can be described as low-intensity and long-duration with occasional intermittent outbursts.

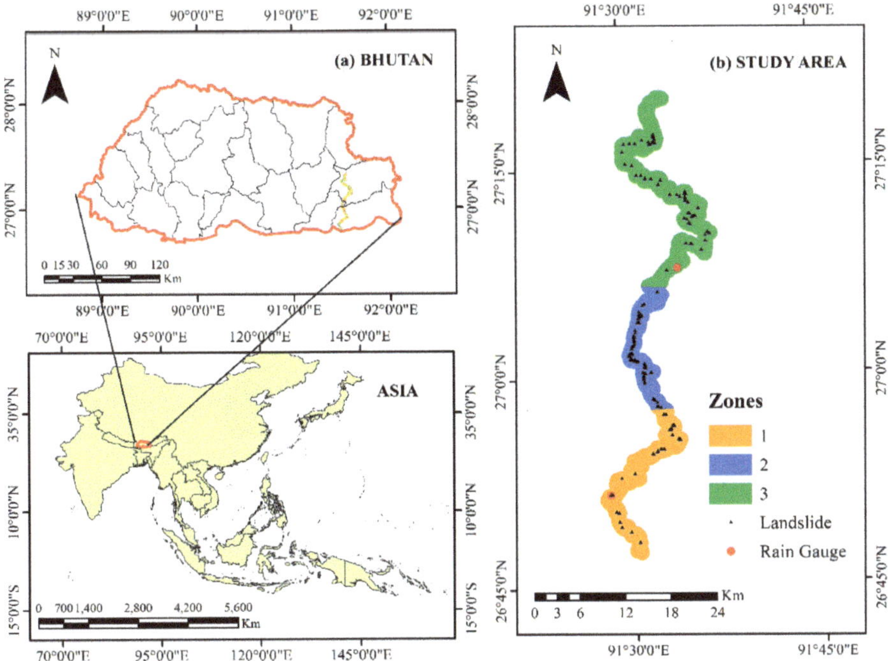

Figure 1. Location of (**a**) Bhutan; (**b**) spatial distribution of landslides in the study area considered in this analysis. The zones were categorized based on spatial coverage of rain gauges, rain gauge location, and elevation difference.

The area includes five different geological groups, which are Baxa Group, Daling Shumar Group, and the Greater, Lesser, and Sub-Himalayan zones. The geology varies from the Baxa Group in the south to the Greater Himalayan Zone in the north. The rocks found in the region are predominantly dark grey to green fine-grained phyllites, slate varying from dark brown to black, and fine-grained, medium- to thick-bedded quartzite with thin to very thin grey-black fine-grained phyllite interbeds [34,35].

The average thickness of the quartzite is about 100 m, but the individual bands of quartzite range from 10 cm to 2 m. The orientation of the latter is 48° NW, with an average dip 40° towards the slope direction of the slide. The quartzite in the landslide area undergoes brittle deformation with many irregular joints [36]. Figure 2 depicts examples of the damage caused by landslides along the highway.

Figure 2. Landslide damage: (**a**) landslide at 68.1 km along the Samdrup Jongkar–Trashigang (S-T) road (N 26.903°, E 91.505°) (5 July 2016); (**b**) landslide at 93.8 km along the S-T road (N 27.112°, E 91.544°) (17 July 2016) [36].

3. Data

Data from a total of 347 landslides which occurred from 2014 to 2018 were collected from Border Road Organization, Government of India under Project DANTAK (Figure 1b). The data included the dates and geographic coordinates of landslide location based on field observations, interactions with locals, and media reports. The types of movement in the region based on Cruden and Varnes' [37] classification are: debris slides, rock falls, earth flows, and rotational landslides. The field visit conducted in November 2017 revealed the landslides to be shallow, with depths ranging up to few meters, and able to be mapped as single points. The yearly distribution (Figure 3a) of landslides shows that the majority of the landslides occurred in 2017 and 2018, whereas the monthly distribution (Figure 3b) shows that 89% of landslides occurred between the months of June and September. It is often difficult to determine the rainfall conditions responsible for failures, due to lack of rain gauge density and high distances between rain gauges and landslide points [26,38]. As multiple landslides can occur during a rainfall event, subsequent landslides for a single rainfall event after the initial failure were not considered for the threshold analysis. Thus, in this study, we defined a landslide event as "single landslide-triggering event", in which the landslide events after the initial failure were not considered for threshold estimation [16,26]. A rainfall event was defined as a period of continuous rainfall separated by dry (without rainfall) period. Using all these criteria, the total numbers of rainfall and landslide events during the threshold determination time frame (2014–2017) were 477 and 104, respectively. The zones were divided based on spatial coverage of rain gauges [4,26]. A buffer radius of 15 km around each rain gauge was selected to divide the region into zones. In terms of landslide events for respective zones, half of the landslide events occurred in Zone 2 (49) followed by Zone 3 (40) and Zone 1 (15).

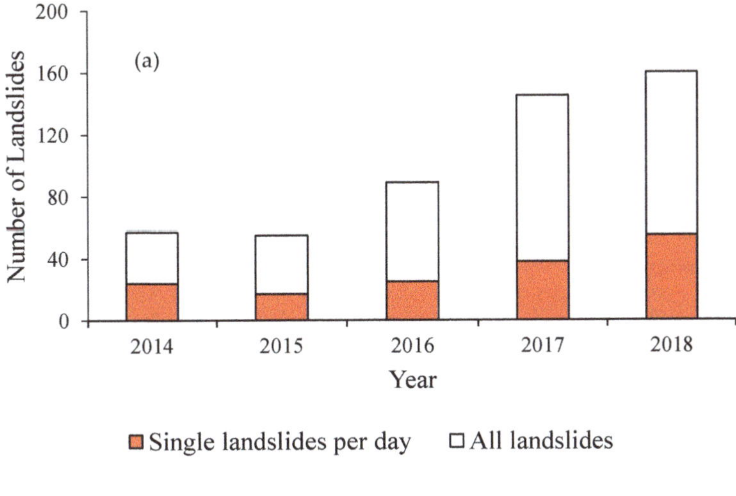

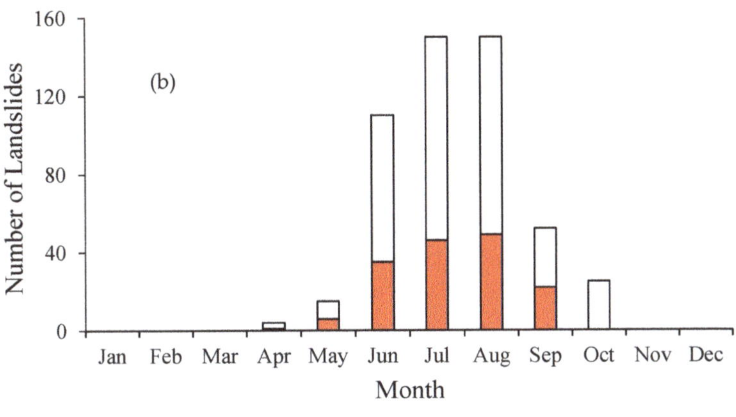

Figure 3. Number of landslides (**a**) per year and (**b**) per month between 2014 and 2018.

The daily rainfall data used for this study (Figure 4) was collected from two rain gauges at Deothang and Kanglung, managed by the National Centre for Hydrology and Meteorology, Bhutan (http://www.hydromet.gov.bt). The average cumulative yearly rainfalls for Deothang and Kanglung for 2014–2018 were 3495 mm and 1020 mm, respectively, of which 89% occurred during the monsoon season (June–September). The higher precipitation observed in Deothang is due to its location on the windward side of the mountain.

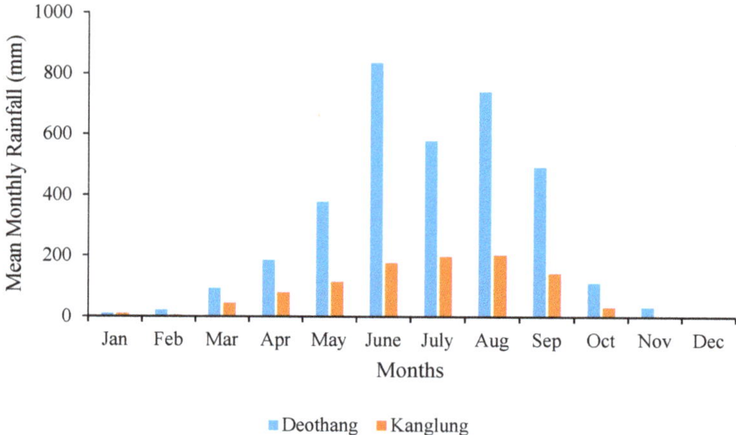

Figure 4. Average monthly rainfall in the study region for both rain gauges (2014–2018).

4. Landslide Temporal Probability Assessment

4.1. Rainfall Threshold Estimation

A rainfall threshold determines the minimum rainfall conditions necessary for landslide initiation in a specific region [39], and various researchers have attempted to quantify thresholds using several approaches [39–41]. A recent review article [40] on rainfall threshold estimation explained in detail the various approaches currently in use, along with their merits and demerits. Of the various techniques, empirically based approaches are widely used because of the simplicity and ease with which they can provide an accurate approximation of minimum precipitation conditions. Various rainfall thresholds using different parameters have been developed, such as ID (intensity–duration) [21], ED (cumulative event rainfall duration) [4], and AD (antecedent rainfall duration) [16,42]. The most commonly used rainfall variables for threshold estimation are daily rainfall, antecedent rainfall, and cumulative rainfall [40]. However, the choice of rainfall variable with which to determine thresholds is primarily dependent on the type of landslides in the region [20].

For the S-T highway, monsoonal rainfall occurs with interruptions and can be characterized mostly low-intensity and long-duration events along with occasional extreme events, making the choice of antecedent rainfall appropriate [30]. Antecedent rainfall is a significant factor for landslide triggering, especially in less impermeable soils, as it lowers soil suction and increases pore water pressure [13]. The use of antecedent rainfall was based on analysis of historical landslide pattern, the field visit, and previous studies conducted in other regions of Bhutan. Although estimation of the number of days to be considered to analyze the effect of antecedent rainfall was a challenge, it has been widely accepted that antecedent rainfall over 15 days to 30 days plays a crucial role for landslide initiation in the Himalayas [43]. The calculation of the antecedent period prior to landslide incidence is usually based on a trial-and-error approach, ranging from 3 days to 120 days [30,44,45].

For this study, the correlation between daily and antecedent rainfall conditions was analyzed for six different time periods (3, 5, 7, 10, 15, and 30 days) (Figure 5a–f). The analysis of the various antecedent rainfall time periods was conducted using the method proposed by Zezere et al. [42]. Blue dots represent the daily rainfall, whereas the orange points depict the antecedent rainfall values for respective time periods. The best discrimination between daily and antecedent conditions, according to the method suggested by Reference [42], was observed for 30 day antecedent rainfall and was accepted as the metric for threshold calculation.

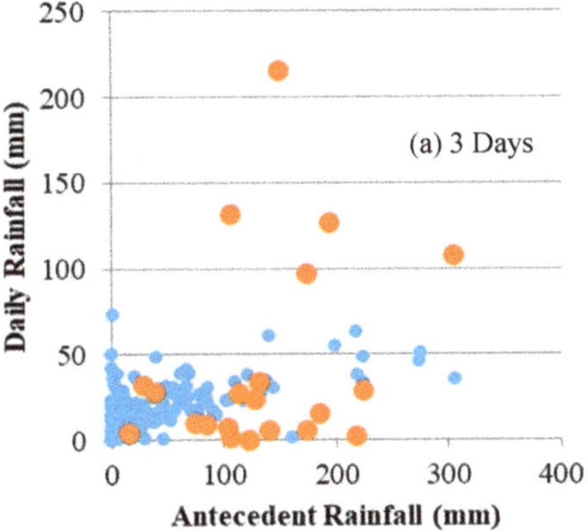

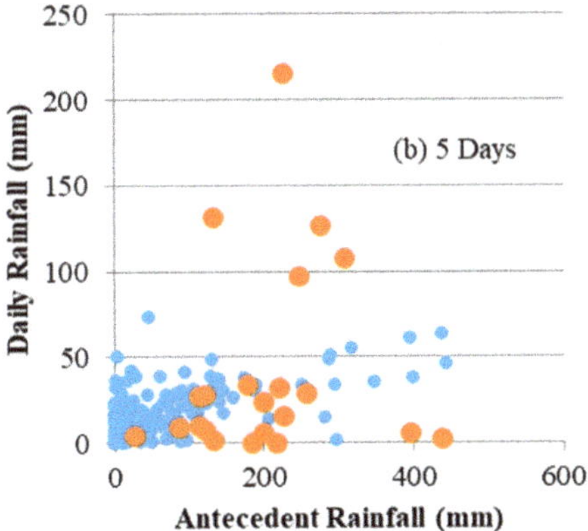

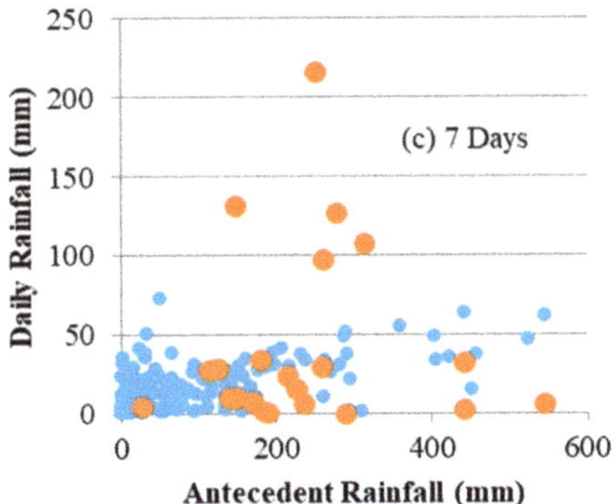

Figure 5. *Cont.*

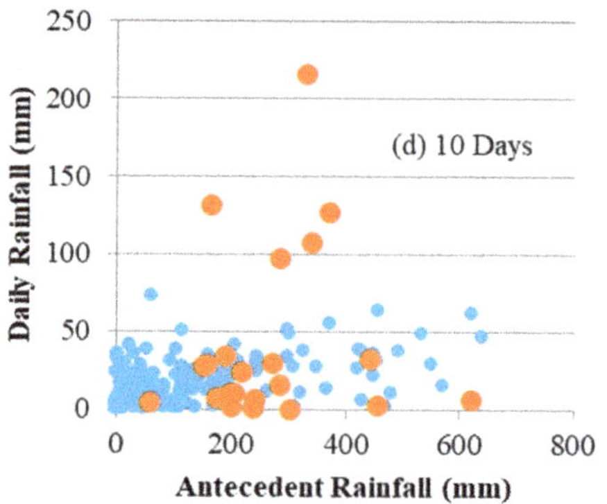

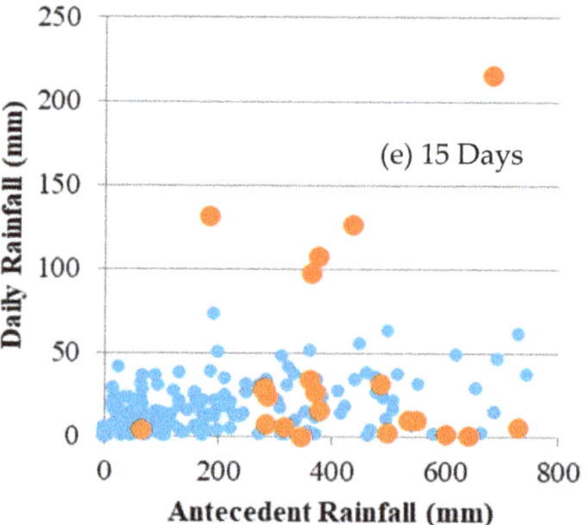

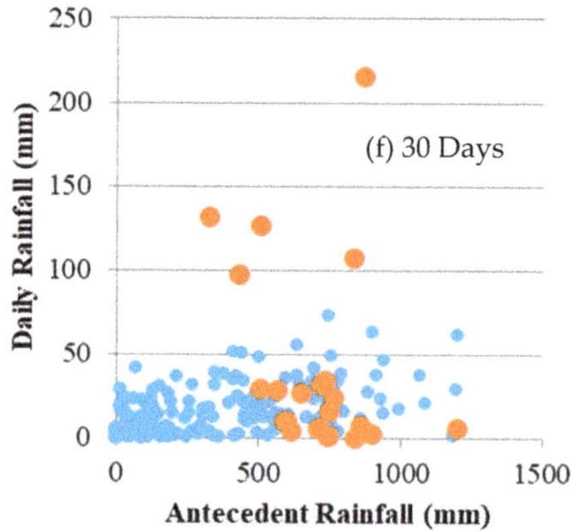

Figure 5. Relationship between daily (blue) and antecedent (orange) rainfall in 2014–2017.

The threshold determination was performed using a scatter chart for daily (R_{TH}) and 30 day antecedent rainfall (R_{30ad}), and was calculated for all three zones. The graph was generated using the rainfall and landslide data from 2014–2017. The threshold equation was obtained by using the lower end points in the scattered graph [13,16]. The threshold equations of various zones are depicted in Figure 6.

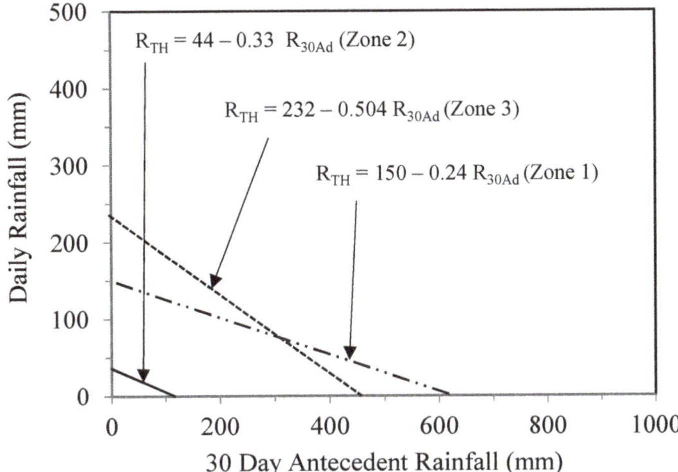

Figure 6. Threshold equation between daily rainfall and 30 day antecedent rainfall for all the three zones.

4.2. Validation of Rainfall Threshold

The significance of any landslide study is determined by the validation of the results obtained. One review of rainfall threshold studies [40] emphasized the importance of validation of rainfall thresholds for conducting landslide studies. The rainfall threshold validation was performed using rainfall and landslide data from 1 January to 31 December 2018. During this period, a total of 52 landslide events occurred, out of which 80% of landslides (41) happened during the monsoon season. The threshold equation for Zone 1 was $R_{TH} = 150 - 0.24R_{30ad}$, and its validation for the monsoon of 2018 is depicted in Figure 7. The threshold exceedance axis depicts the value of R_{TH} with respect to R_{30AD} value for each day, wherein the positive values indicates landslide occurrence. This figure shows that a heavy measure of rainfall occurred, exceeding the threshold. The abrupt increase in the magnitude of daily rainfall or constant rise in 30 day antecedent rainfall is shown by the rise in the threshold curve.

During the validation period, the threshold was exceeded nine times, out of which seven times landslides occurred. No landslides were reported on 2 June and 7 June, even though the threshold was exceeded. This observation can be attributed to the fact that a landslide event does not generally happen with the increase in threshold curve, and sometimes happens a couple of days later as the result of a difference in pore pressure because of changes in the measure of antecedent rainfall. From 1 October to 30 December, there was no threshold exceedance, and no landslides occurred during that period. Similar validations were carried out for Zone 2 and Zone 3 threshold equations using 2018 rainfall and landslide data. These results indicate that the threshold model performed well for landslide forecasting in 2018.

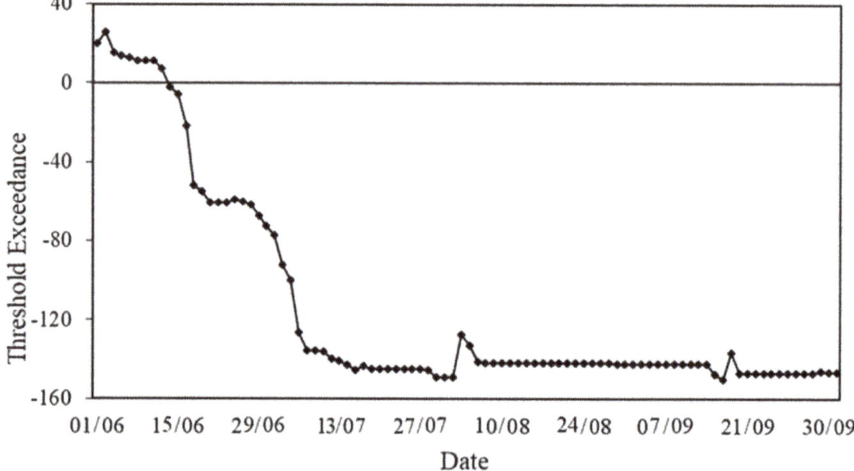

Figure 7. Validation of the threshold equation for Zone 1. The positive values on the Y axis represent exceedance and negative values denote non-exceedance of the threshold.

4.3. Temporal Probability of Landslide Initiation

The temporal probability of landslide incident is determined as the product of annual exceedance probability (AEP) and probability of landslide occurrence [16]. AEP is defined as the probability of the threshold being exceeded in a given year [46], and is determined using the Poisson probability model defined as [16]

$$P(N(t) = n) = e^{-\lambda t}\frac{(-\lambda t)^n}{n!} \quad (1)$$

where N(t) represents the number of landslide incidences during time t and λ is landslide occurrence rate. The exceedance probability for time t is calculated as [13]

$$P[N(t) \geq 1] = 1 - \exp\,(-t/\mu) \quad (2)$$

where μ is the mean recurrence interval between subsequent landslides determined from landslide inventory.

The determination of temporal probability was based on the following assumptions: (i) probability of landslide incidence is correlated to the probability of rainfall threshold being reached or exceeded, and (ii) landslides will not or will seldom occur when precipitation value is less than the threshold value [13,16,47].

The annual temporal probabilities for different zones of the study region are depicted in Figure 8, and their distribution along with the threshold equations is presented in Table 1. For Zone 1, the threshold value was exceeded 55 times during the simulation period, and out of these 55 cases, landslides were triggered in 16. The estimated probability P[L|(R > RT)] for Zone 1 was 0.29. Similarly, for Zone 2 and 3, the threshold value was exceeded 76 and 64 times in the 4 year period, leading to 31 and 21 landslides being triggered and contributing temporal probability values of 0.41 and 0.33, respectively. The probability of having one or more rainfall events in any given year varied from 0.29 to 0.41. The highest probability values were obtained for Zone 2, followed by Zone 3 and Zone 1. This variation was also observed in the number of landslide incidences for each zone. These precipitation events were also capable of triggering multiple landslides during the monsoon period.

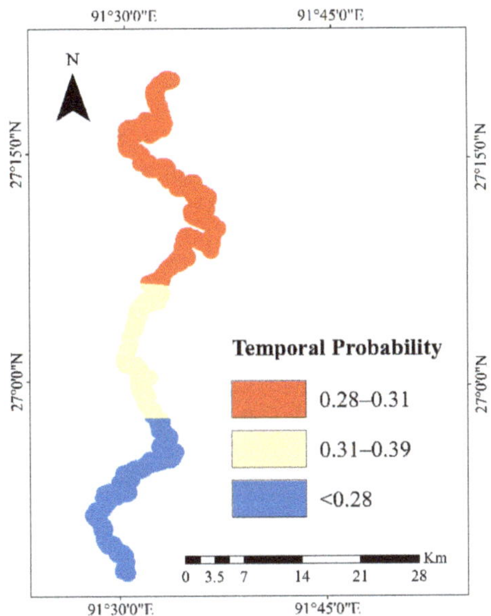

Figure 8. Annual exceedance probability (AEP) of landslide occurrences along the Samdrup Jongkhar–Trashigang Highway.

Table 1. Temporal Probability of Landslide Occurrence.

Area	Threshold Equation (R_T)	Number of Threshold Exceedances	$P(R > R_T)$	Landslide Frequency	$P[L \mid (R > R_T)]$	Temporal Probability
Zone 1	$150 - 0.24R_{30ad}$	55	0.99	16	0.29	0.28
Zone 2	$232 - 0.50R_{30ad}$	76	0.99	31	0.40	0.39
Zone 3	$44 - 0.30R_{30ad}$	64	0.99	21	0.32	0.31

5. Landslide Susceptibility Mapping

Landslide susceptibility can be defined as the probability of spatial occurrences of slope failures for a given set of geo-environmental conditions [48], and its determination is one of the crucial steps needed to understand identify potentially landslide-prone sections for any study region. Several studies around the world have been conducted towards the development of landslide susceptibility maps (LSM) using various methods [8]; however, there seems to be no consensus as to the best method for analysis [49]. Aleotti and Chowdhury [50] categorized LSM methods as either quantitative or qualitative. Qualitative models are mostly based on expert opinion, whereas quantitative models are data-driven, which makes them more reliable. The quantitative approaches include several kinds of techniques such, including statistical, deterministic, and other approaches [51–53]. In the case of statistical approaches, it is assumed that the parameters affecting landslide events in the past will be the same in future [54], and these analyses can be categorized into bivariate and multivariate [49]. In bivariate analysis, the factors affecting landslides are compared with landslide inventory data by providing weights based on landslide causative factors. The most frequently used methods in bivariate models are overlay, index-based, and weight-of-evidence analyses [8,51,55]. Bui et al. [6] performed a comparison between a bivariate approach (statistical index) and a multivariate approach (logistic regression) for Vietnam, and found equal forecasting capability. However, one of the main issues with the use of a quantitative approach is the assignment of weights to the landslide-affecting factors [56–58]. The use of GIS has been proven to be a powerful tool with which to validate the significance of factors, and it has been used for multi-criterion decision analysis [49,59]. The decision analysis technique combines primary- and secondary-level weights for every causative factor, where primary weights are similar to the bivariate approach and secondary weights are expert-opinion-based. For secondary weights, the analytic hierarchy process (AHP) has become popular and has been successfully applied for decision-making systems [60,61]. AHP uses a pairwise relative comparison between every landslide-causative factor. Generally, AHP consists of five key steps: (a) simplify the decision process into its component factors, (b) distribute the factors in a hierarchy process, (c) allocate numerical values to analyze the relative significance of each factor, (d) compose a comparison matrix, and (e) provide weights to every factor by calculating normalized principal eigenvectors [62].

To determine susceptibility, a variety of factors responsible for landslides in the study region were considered. Parameter selection depends on various factors, such as landslide type, data availability and reliability, and adopted methods [63]. For the present study, we used eight landslide-conditioning factors based on the characteristics of the area and prepared from various data sources (Table 2). Figure 9a–f represents all the maps used for the analysis, derived from the Shuttle Radar Topography Mission (SRTM) digital elevation model (DEM) with 30 × 30 m resolution, which was the only terrain data source available for this region. The factors with continuous values were reclassified into categories based on Jenks' natural breaks optimization method [64] and developed using ArcGIS 10.4.1.

Table 2. Parameters Used for Landslide Susceptibility Mapping.

Parameters	Data Source	Explanation	Scale
Slope (°)	SRTM	Derived from raster DEM	1:30,000
Average daily rainfall (mm)	Project DANTAK, Border Road Organization, Govt. of India		1:30,000
Proximity to road (m)	Topographical map	Shape file	1:30,000
Proximity to stream (m)	SRTM	Derived from Raster DEM using the [65] order greater than 5 in vector format	1:30,000
Aspect (N/E/S/W)	SRTM	Derived from raster DEM	1:30,000
Elevation (m)	SRTM	Derived from raster DEM	1:30,000
Landcover	Ministry of Agriculture and Forests, Royal Govt. of Bhutan	Vector data	1:30,000
Geology	Department of Geology and Mines, Royal Govt. of Bhutan	Geological map	1:30,000

The above-mentioned thematic layers were combined by using a weight-of-factors approach determined by AHP to develop the landslide susceptibility map. The use of AHP to develop landslide susceptibility maps has been successfully applied in various regions [61,66,67]. The weights required to carry out AHP were calculated by performing pairwise comparisons for each landslide factor and assigning values from 1 to 9 [63,68–70]. Table 3 shows the pairwise comparison and priority calculation, along with rankings of all indicators. These values were based on an expert's opinion and were placed in n × n matrix, where n is the number of factors.

Table 3. Parameter Wise Weights, Matrix, and Consistency Ratio as Determined Using AHP.

Parameters	Slope	Average Daily Rainfall	Proximity to Road	Proximity to Stream	Aspect	Elevation	Landcover	Geology	Weights (%)
Slope	1	0.33	4	2	2	2	3	3	19.2
Average daily rainfall	3	1	4	2	2	3	3	3	26.7
Proximity to road	0.25	0.25	1	0.33	0.33	0.5	0.5	0.33	4.1
Proximity to stream	0.5	0.5	3	1	0.5	1	2	2	11.2
Aspect	0.5	0.5	3	2	1	3	2	2	15.2
Elevation	0.5	0.33	2	1	0.33	1	0.5	0.5	7
Landcover	0.33	0.33	2	0.5	0.5	2	1	1	8
Geology	0.33	0.33	3	0.5	0.5	2	1	1	8.5

Consistency Ratio = 0.039.

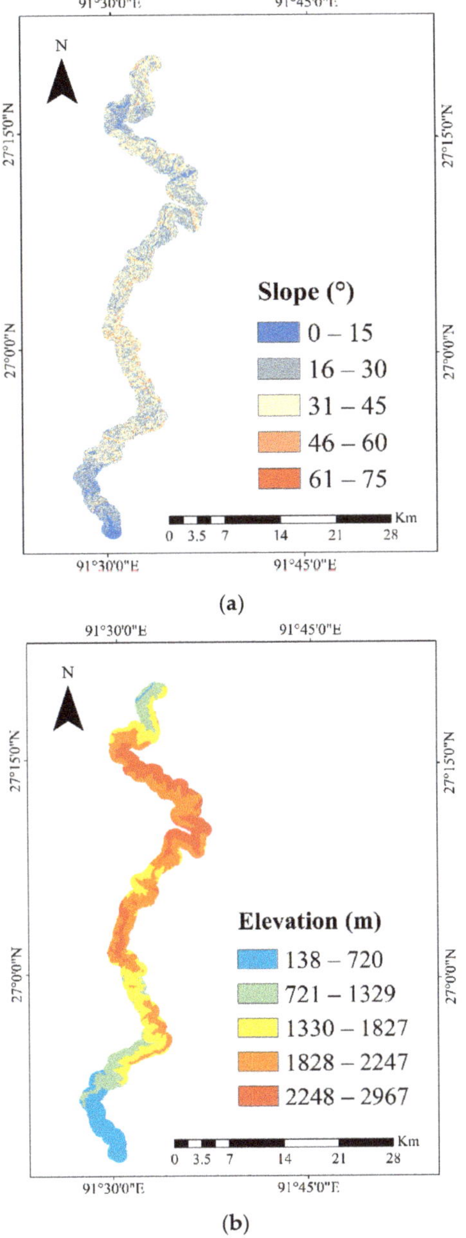

Figure 9. *Cont.*

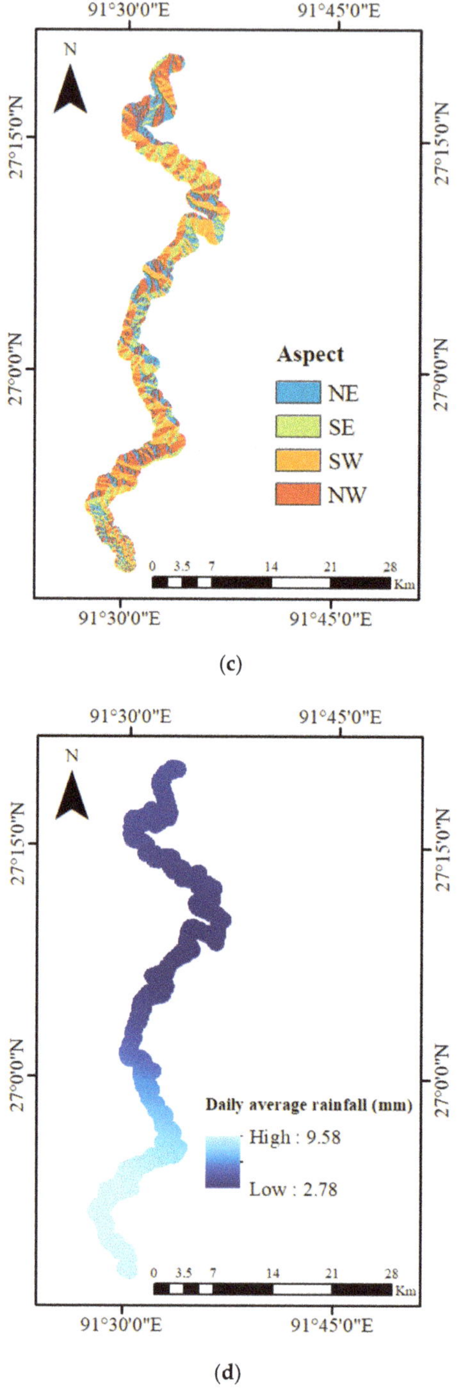

(c)

(d)

Figure 9. *Cont.*

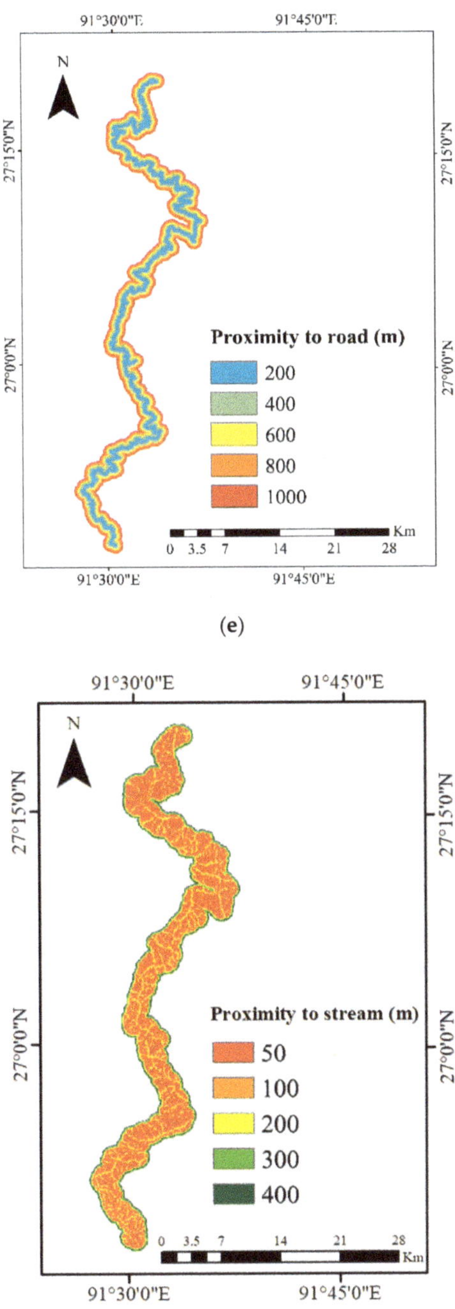

Figure 9. *Cont.*

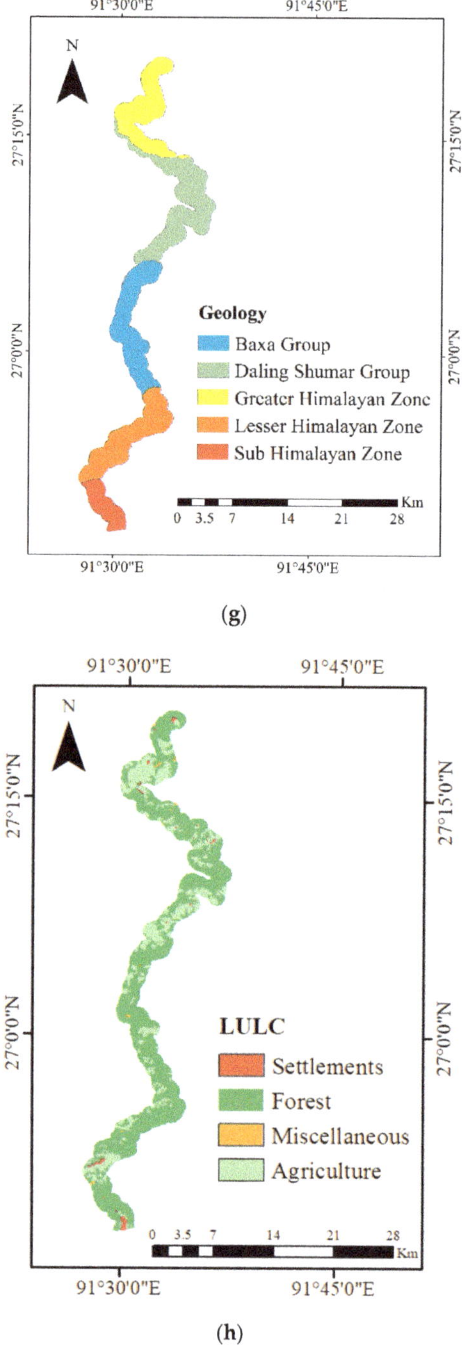

Figure 9. Landslide-conditioning factor maps: (**a**) slope, (**b**) elevation, (**c**) aspect, (**d**) mean daily rainfall, (**e**) proximity to road, (**f**) proximity to stream, (**g**) geology, and (**h**) land use and land cover (LULC).

The AHP reduced the inconsistencies formed due to the subjectivity of different experts' opinions by computing a consistency index (CI) and consistency ratio (CR), which were determined by

$$CI = (\lambda_{max} - n)/(n - 1) \tag{3}$$

$$CR = CI/RI \tag{4}$$

where λ_{max} represents the largest Eigenvector of the matrix and n represents the total causative factors (order of the matrix) used in the generation of the LSM. RI (random index) is the average value of CI for a randomly generated pairwise matrix and can be accepted only when CR values are less than 10% [71]. For the present study, the average consistency index was estimated for a sample size of 500 and its value was 0.039 (3.9%), which was considered acceptable. Several authors have calculated and estimated different RIs based on various simulation methods and the total number of matrices involved in the process (Table 4). However, we have used Satty's [71] RI values of n = 11 and up to 500 matrices, where the values are 0, 0, 0.58, 0.9, 1.12, 1.24, 1.32, 1.41, 1.45, 1.49, and 1.51.

Table 4. RI Values Obtained by Various Authors (Adopted from Reference [72]).

	[71]	[71]	[73]	[74]	[75]	[76]	[77]	[78]	[79]
	100	500	1000	2500		500		100,000	100,000
3	0.382	0.58	0.5799	0.52	0.5233	0.49	0.500	0.525	0.5245
4	0.946	0.90	0.8921	0.87	0.8860	0.82	0.834	0.882	0.8815
5	1.220	1.12	1.1159	1.10	1.1098	1.03	1.046	1.115	1.1086
6	1.032	1.24	1.2358	1.25	1.2539	1.16	1.178	1.252	1.2479
7	1.468	1.32	1.3322	1.34	1.3451	1.25	1.267	1.341	1.3417
8	1.402	1.41	1.3952	1.40		1.31	1.326	1.404	1.4056
9	1.350	1.45	1.4537	1.45		1.36	1.369	1.452	1.4499
10	1.464	1.49	1.4882	1.49		1.39	1.406	1.484	1.4854
11	1.576	1.51	1.5117			1.42	1.433	1.513	1.5141
12	1.476		1.5356	1.54		1.44	1.456	1.535	1.5365
13	1.564		1.5571			1.46	1.474	1.555	1.5551
14	1.568		1.5714	1.57		1.48	1.491	1.570	1.5713
15	1.586		1.5831			1.49	1.501	1.583	1.5838

The values of CR cannot be negative and can attain a maximum value of 0.3. Values of CR less than 0.1 are considered acceptable; if this is not achieved, new attempts are made until the value is acceptable [80]. However, the values of CR are dependent on the analysis type and the number of criteria being considered. In some cases, CR > 0.1 may not be considered critical, and values of CR ranging from 0.15 to 0.3 can also be considered acceptable. A matrix will be considered consistent, according to Saaty [71], if

$$\lambda max < n + 0.1((\lambda max) - n) \tag{5}$$

Finally, all the landslide-causative factors and classes were integrated by a method of weighted overlay in ArcGIS to generate the landslide susceptibility index (LSI).

$$LSI = \sum_{k=1}^{n} W_k W_{jk} \tag{6}$$

where W_k is the weight of the causative factor, W_{jk} is the rank value for factor class j of causative factor k, and n represents the total causative factors selected. Ranks of criteria were calculated based on priority or weight values. The highest priority considered was Rank 1, while the lowest priority considered was the last rank in the AHP.

Model Development and Validation

The landslide inventory data was randomly categorized into two datasets, i.e., training (70%) and testing (30%). The landslide susceptibility map based on 8 causative factors using AHP is depicted in

Figure 10. The map was divided into 4 classes: very low (0–0.24), low (0.25–0.49), moderate (0.5–0.74) and high (0.75–1) according to natural breaks to define the class intervals in the susceptibility map. From total of 242 landslides, more than 78.1% falls under the high zone. Whereas 11.98% falls under the moderate zone and combined 9.92% falls under the low and very low zones (Table 5).

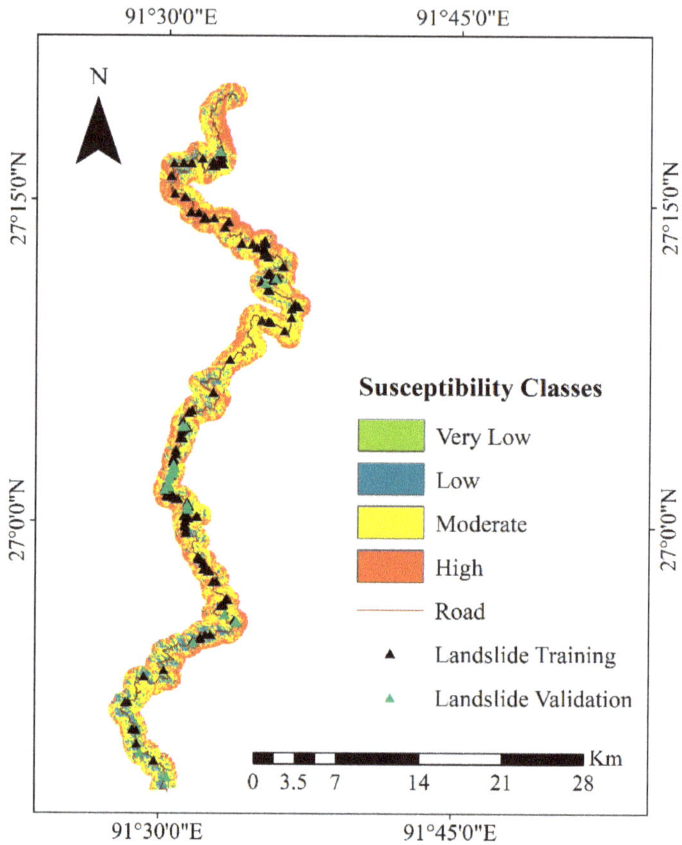

Figure 10. Landslide Susceptibility Map of the Study Region.

Table 5. Landslide Susceptibility Results for the Samdrup Jongkhar–Trashigang Highway Region.

Inventory		**Susceptibility**							
Zone	**Landslides**	**High**		**Moderate**		**Low**		**Very Low**	
		(%)	**(km²)**	**(%)**	**(km²)**	**(%)**	**(km²)**	**(%)**	**(km²)**
Zone 1	39	21.23	13.84	59.90	39.05	18.71	12.19	0.15	0.10
Zone 2	119	36.6	38.82	55.7	59.08	6.8	7.25	0.9	0.94
Zone 3	84	28.31	13.69	63.45	30.67	8.06	3.89	0.16	0.08

Validation of the susceptibility maps was performed for randomly selected data from the inventory data using receiver operating characteristics (ROC). This is an effective way to analyze the quality of predictive techniques [81]. The ROC curve is plotted between true positive rate (sensitivity) on the Y axis against false positive rate (specificity) on the X axis. The terms "sensitivity" and "specificity", which are used to plot ROC curves, are defined as follows.

$$\text{Sensitivity} = \frac{\text{TP}}{\text{TP} + \text{FN}} \tag{7}$$

$$\text{Specificity} = \frac{\text{TN}}{\text{FP} + \text{TN}} \tag{8}$$

where true positive (TP) is the number of actual landslides predicted correctly, and true negative (TN) is the total number of non-occurring landslides predicted correctly. False positive (FP) is the number of actual landslides inaccurately predicted as non-occurring landslides, and false negative (FN) is the number of non-occurring landslides inaccurately predicted as actual landslides. [82]. The area under the ROC curve (AUC) was also used to determine the quality of the prediction by analyzing the model's ability to forecast the occurrence or nonoccurrence of predefined events [83]. The results of the success rate curve of the AHP model had an AUC of 0.798, corresponding to a prediction accuracy of 79.8% (Figure 11).

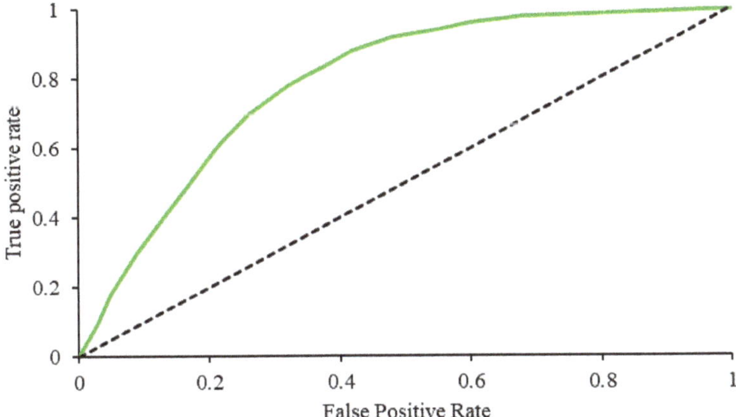

Figure 11. Receiver operating characteristic (ROC) curve of the susceptibility map by AHP.

The results of pairwise comparison, priority estimation, and ranking of all the criteria could be applied to other study areas for susceptibility assessment. Several high-resolution satellite image datasets are required to better understand the locations and perform these assessments. In cases of unavailability of high-resolution satellite images, Google Earth images could be useful for preparation of some indicators. The use of Google Earth images could also be helpful for accurate identification of landslide locations and conduction of extensive field studies. All the requirements for more accurate analysis depend on study location tectonics, data availability, and proper methods. Therefore, based on local geology and tectonic conditions, these results could be transferable and applicable in other locations in both small- and large-scale areas.

6. Conclusions

Landslides are the most frequently occurring natural hazards, especially in the Himalayan regions, which suffer from heavy monsoonal rainfall and subsequent landslides. In this study, the temporal probability of landslide events was determined using rainfall and landslide data from 2014–2017 along Samdrup Jongkhar–Trashigang highway in East Bhutan. The highway was divided into three zones based on land use, topography, and rain gauge coverage for the determination of temporal probability. Thereafter, a landslide susceptibility map was developed using AHP. The results of temporal probability were validated with landslide event dataset of 2018 to understand the applicability of the model. The conclusions from the study can be summarized as follows.

(1) Threshold determination was performed considering the antecedent rainfall duration approach. The selection of antecedent rainfall was based on previous studies which have highlighted its significance, especially in the Himalayan region and in Bhutan. A trial-and-error approach was adopted to determine the number of antecedent days required for landslide initiation, and a 30 day antecedent rainfall was adopted for threshold determination.
(2) The temporal probability of landslide occurrence was determined based on the Poisson model, and the validation results based on 2018 data revealed that the model performed well.
(3) The landslide susceptibility map of the region was developed using AHP classified into four categories (Very Low, Low, Moderate, and High), and the results showed that 78% of the region falls under into high-susceptibility zone. The performance of the model was assessed using the area under ROC and an accuracy of 79.8% was achieved. However, due to the dynamic nature of land use and rainfall patterns, the susceptibility map will need to be updated from time to time.

The present study on rainfall threshold estimation and the development of a susceptibility map for eastern Bhutan along Samdrup Jongkhar–Trashigang highway is an important study in the context of the Bhutan Himalayas, for which study on both these aspects is lacking. The current study can be regarded as a preliminary step towards risk management, which could be supported by conducting a hazard and vulnerability assessment of the region. The temporal probabilities determined can be integrated with the susceptibility map to obtain a landslide hazard map. However, the results might be improved by increasing the number of landslide events and using precipitation data with a higher temporal resolution. The results from the present study also could prove helpful to civic authorities in identifying key sections of the road which are most vulnerable to landslides, and undertaking strict measures to prevent slope failures, strengthening the transportation network and saving human lives.

Author Contributions: Conceptualization, A.D. and R.S.; methodology, A.D. and R.J.; data curation, D.D.; writing—original draft preparation, A.D. and R.S.; writing—review and editing, B.P. and A.M.A.; supervision, B.P. and R.S. All authors have read and agreed to the published version of the manuscript.

Funding: This research was supported by the Centre for Advanced Modelling and Geospatial Information Systems (CAMGIS) in the University of Technology Sydney (UTS) under Grants 321740.2232335 and 321740.2232357; Grant 321740.2232424, and Grant 321740.2232452. This research is also supported by Researchers Supporting Project number RSP-2019/14, King Saud University, Riyadh, Saudi Arabia. Also, the research was funded from BRACE project (NERC/GCRF NE/P016219/1).

Acknowledgments: The authors are thankful to National Center for Hydrology and Meteorology, Royal Government of Bhutan for providing rainfall data and the Border Roads Organization (Project DANTAK), Government of India for providing landslide data. Authors are also thankful to staff of College of Science and Technology, Royal University of Bhutan, who helped directly or indirectly while carrying out the present study. We also thank the editor and the reviewers for reviewing and suggesting valuable modifications, which has helped in improving the quality of this paper.

Conflicts of Interest: The authors declare no conflict of interest. The funders had no role in the design of the study; in the collection, analyses, or interpretation of data; in the writing of the manuscript, or in the decision to publish the results.

References

1. Petley, D. Global patterns of loss of life from landslides. *Geology* **2012**, *40*, 927–930. [CrossRef]
2. Froude, M.J.; Petley, D.N. Global fatal landslide occurrence from 2004 to 2016. *Nat. Hazards Earth Syst. Sci.* **2018**, *18*, 2161–2181. [CrossRef]
3. Stanley, T.; Kirschbaum, D.B. A heuristic approach to global landslide susceptibility mapping. *Nat. Hazards* **2017**, *87*, 145–164. [CrossRef]
4. Dikshit, A.; Sarkar, R.; Pradhan, B.; Acharya, S.; Dorji, K. Estimating Rainfall Thresholds for Landslide Occurrence in the Bhutan Himalayas. *Water* **2019**, *11*, 1616. [CrossRef]

5. Sarkar, R.; Dorji, K. Determination of the Probabilities of Landslide Events—A Case Study of Bhutan. *Hydrology* **2019**, *6*, 52. [CrossRef]
6. Bui, D.T.; Löfman, O.; Revhaug, I.; Dick, O. Landslide susceptibility analysis in the Hoa Binh province of Vietnam using statistical index and logistic regression. *Nat. Hazards* **2011**, *59*, 1413–1444. [CrossRef]
7. Qiu, H.; Cui, Y.; Hu, S.; Yang, D.; Pei, Y.; Yang, W. Temporal and spatial distributions of landslides in the Qinba Mountains, Shaanxi Province, China. *Geomat. Nat. Hazards Risk* **2019**, *10*, 599–621. [CrossRef]
8. Reichenbach, P.; Rossi, M.; Malamud, B.D.; Mihir, M.; Guzzetti, F. A review of statistically-based landslide susceptibility models. *Earth-Sci. Rev.* **2018**, *180*, 60–91. [CrossRef]
9. Althuwaynee, O.F.; Pradhan, B.; Lee, S. Application of an evidential belief function model in landslide susceptibility mapping. *Comput. Geosci.* **2012**, *44*, 120–135. [CrossRef]
10. Pradhan, B.; Lee, S. Landslide susceptibility assessment and factor effect analysis: Backpropagation artificial neural networks and their comparison with frequency ratio and bivariate logistic regression modelling. *Environ. Model. Softw.* **2010**, *25*, 747–759. [CrossRef]
11. Pradhan, B.; Lee, S.; Buchroithner, M.F. A GIS-based back-propagation neural network model and its cross-application and validation for landslide susceptibility analyses. *Comput. Environ. Urban Syst.* **2010**, *34*, 216–235. [CrossRef]
12. Jaiswal, P.; Van Westen, C.J.; Jetten, V. Quantitative landslide hazard assessment along a transportation corridor in southern India. *Eng. Geol.* **2010**, *116*, 236–250. [CrossRef]
13. Bui, D.T.; Pradhan, B.; Lofman, O.; Revhaug, I.; Dick, Ø.B. Regional prediction of landslide hazard using probability analysis of intense rainfall in the Hoa Binh province, Vietnam. *Nat. Hazards* **2013**, *66*, 707–730.
14. Afungang, R.N.; Bateira, C.V. Temporal probability analysis of landslides triggered by intense rainfall in the Bamenda Mountain Region, Cameroon. *Environ. Earth Sci.* **2016**, *75*, 1032. [CrossRef]
15. Saez, J.L.; Corona, C.; Stoffel, M.; Schoeneich, P.; Berger, F. Probability maps of landslide reactivation derived from tree-ring records: Pra Bellon landslide, southern French Alps. *Geomorphology* **2012**, *138*, 189–202. [CrossRef]
16. Jaiswal, P.; Van Westen, C.J. Estimating temporal probability for landslide initiation along transportation routes based on rainfall thresholds. *Geomorphology* **2009**, *112*, 96–105. [CrossRef]
17. Corominas, J.; Moya, J. A review of assessing landslide frequency for hazard zoning purposes. *Eng. Geol.* **2008**, *102*, 193–213. [CrossRef]
18. Das, I.; Stein, A.; Kerle, N.; Dadhwal, V.K. Probabilistic landslide hazard assessment using homogeneous susceptible units (HSU) along a national highway corridor in the northern Himalayas, India. *Landslides* **2011**, *8*, 293–308. [CrossRef]
19. Segoni, S.; Lagomarsino, D.; Fanti, R.; Moretti, S.; Casagli, N. Integration of rainfall thresholds and susceptibility maps in the Emilia Romagna (Italy) regional-scale landslide warning system. *Landslides* **2015**, *12*, 773–785. [CrossRef]
20. Martelloni, G.; Segoni, S.; Fanti, R.; Catani, F. Rainfall thresholds for the forecasting of landslide occurrence at regional scale. *Landslides* **2012**, *9*, 485–495. [CrossRef]
21. Giannecchini, R.; Galanti, Y.; Avanzi, G.D. Critical rainfall thresholds for triggering shallow landslides in the Serchio River Valley (Tuscany, Italy). *Nat. Hazards Earth Syst. Sci.* **2012**, *12*, 829–842. [CrossRef]
22. Schmidt, M.; Glade, T. Linking global circulation model outputs to regional geomorphic models: A case study of landslide activity in New Zealand. *Clim. Res.* **2003**, *25*, 135–150. [CrossRef]
23. Althuwaynee, O.F.; Pradhan, B.; Ahmad, N. Estimation of rainfall threshold and its use in landslide hazard mapping of Kuala Lumpur metropolitan and surrounding areas. *Landslides* **2015**, *12*, 861–875. [CrossRef]
24. Dikshit, A.; Satyam, D.N. Estimation of rainfall thresholds for landslide occurrences in Kalimpong, India. *Innov. Infrastruct. Solut.* **2018**, *3*, 24. [CrossRef]
25. Gabet, E.J.; Burbank, D.W.; Putkonen, J.K.; Pratt-Sitaula, B.A.; Ojha, T. Rainfall thresholds for landsliding in the Himalayas of Nepal. *Geomorphology* **2004**, *63*, 131–143. [CrossRef]
26. Gariano, S.L.; Sarkar, R.; Dikshit, A.; Dorji, K.; Brunetti, M.T.; Peruccacci, S.; Melillo, M. Automatic calculation of rainfall thresholds for landslide occurrence in Chukha Dzongkhag, Bhutan. *Bull. Eng. Geol. Environ.* **2019**, *78*, 4325–4332. [CrossRef]
27. Crosta, G.B.; Frattini, P. Distributed modelling of shallow landslides triggered by intense rainfall. *Nat. Hazards Earth Syst. Sci.* **2003**, *3*, 81–93. [CrossRef]

28. Jemec, M.; Komac, M. Rainfall patterns for shallow landsliding in perialpine Slovenia. *Nat. Hazards* **2013**, *67*, 1011–1023. [CrossRef]
29. Bui, D.T.; Pradhan, B.; Löfman, O.; Revhaug, I.; Dick, O.B. Spatial prediction of landslide hazards in Hoa Binh province (Vietnam): A comparative assessment of the efficacy of evidential belief functions and fuzzy logic models. *Catena* **2012**, *96*, 28–40.
30. Kanungo, D.; Sharma, S. Rainfall thresholds for prediction of shallow landslides around Chamoli-Joshimath region, Garhwal Himalayas, India. *Landslides* **2014**, *11*, 629–638. [CrossRef]
31. Frattini, P.; Crosta, G.; Sosio, R. Approaches for defining thresholds and return periods for rainfall-triggered shallow landslides. *Hydrol. Process.* **2009**, *23*, 1444–1460. [CrossRef]
32. Sarkar, R.; Dikshit, A.; Hazarika, H.; Yamada, K.; Subba, K. Probabilistic Rainfall Thresholds for Landslide Occurrences in Bhutan. *Int. J. Recent Technol. Eng.* **2019**, *8*. [CrossRef]
33. Sameen, M.I.; Sarkar, R.; Pradhan, B.; Drukpa, D.; Alamri, A.M.; Park, H.-J. Landslide spatial modelling using unsupervised factor optimisation and regularised greedy forests. *Comput. Geosci.* **2019**. [CrossRef]
34. Long, S.; McQuarrie, N.; Tobgay, T.; Grujic, D.; Hollister, L. Geologic Map of Bhutan. *J. Maps* **2011**, *7*, 184–192. [CrossRef]
35. McQuarrie, N.; Long, S.; Tobgay, T.; Nesbit, J.; Gehrels, G.; Ducea, M. Documenting basin scale, geometry and provenance through detrital geochemical data: Lessons from the Neoproterozoic to Ordovician Lesser, Greater, and Tethyan Himalayan strata of Bhutan. *Gondwana Res.* **2013**, *23*, 1491–1510. [CrossRef]
36. Phuntso, Y.; Wangda, U.; Tenzin, T. *Monitoring of Landslide at Arong, Moshi and Phongmey Using GPS*; Department of Geology and Mines, Ministry of Economic Affairs, Royal Government of Bhutan: Thimphu, Bhutan, 2018.
37. Cruden, D.M.; Varnes, D.J. Chapter 3—Landslide types and processes. In *Landslides: Investigation and Mitigation*; Transportation Research Board: Washington, DC, USA, 1996.
38. Gariano, S.L.; Brunetti, M.; Iovine, G.; Melillo, M.; Peruccacci, S.; Terranova, O.; Vennari, C.; Guzzetti, F. Calibration and validation of rainfall thresholds for shallow landslide forecasting in Sicily, southern Italy. *Geomorphology* **2015**, *228*, 653–665. [CrossRef]
39. Guzzetti, F.; Peruccacci, S.; Rossi, M.; Stark, C.P. Rainfall thresholds for the initiation of landslides in central and southern Europe. *Theor. Appl. Clim.* **2007**, *98*, 239–267. [CrossRef]
40. Segoni, S.; Piciullo, L.; Gariano, S.L. A review of the recent literature on rainfall thresholds for landslide occurrence. *Landslides* **2018**, *15*, 1483–1501. [CrossRef]
41. Segoni, S.; Rosi, A.; Lagomarsino, D.; Fanti, R.; Casagli, N. Brief communication: Using averaged soil moisture estimates to improve the performances of a regional-scale landslide early warning system. *Nat. Hazards Earth Syst. Sci.* **2018**, *18*, 807–812. [CrossRef]
42. Zêzere, J.L.; Trigo, R.M.; Trigo, I.F. Shallow and deep landslides induced by rainfall in the Lisbon region (Portugal): Assessment of relationships with the North Atlantic Oscillation. *Nat. Hazards Earth Syst. Sci.* **2005**, *5*, 331–344. [CrossRef]
43. Mathew, J.; Babu, D.G.; Kundu, S.; Kumar, K.V.; Pant, C.C. Integrating intensity–duration-based rainfall threshold and antecedent rainfall-based probability estimate towards generating early warning for rainfall-induced landslides in parts of the Garhwal Himalaya, India. *Landslides* **2014**, *11*, 575–588. [CrossRef]
44. Glade, T.; Crozier, M.; Smith, P. Applying Probability Determination to Refine Landslide-triggering Rainfall Thresholds Using an Empirical "Antecedent Daily Rainfall Model". *Pure Appl. Geophys.* **2000**, *157*, 1059–1079. [CrossRef]
45. Dahal, R.K.; Hasegawa, S. Representative rainfall thresholds for landslides in the Nepal Himalaya. *Geomorphology* **2008**, *100*, 429–443. [CrossRef]
46. Fell, R.; Ho, K.K.; Lacasse, S.; Leroi, E. A framework for landslide risk assessment and management. In *Landslide Risk Management*; CRC Press: Boca Raton, FL, USA, 2005; pp. 13–36.
47. Chleborad, A.F.; Baum, R.L.; Godt, J.W. Rainfall thresholds for forecasting landslides in the Seattle, Washington, area: Exceedance and probability. In *US Geological Survey Open-File Report*; US Geological Survey: Reston, VA, USA, 2006; Volume 1064, p. 31.
48. Guzzetti, F.; Reichenbach, P.; Cardinali, M.; Galli, M.; Ardizzone, F. Probabilistic landslide hazard assessment at the basin scale. *Geomorphology* **2005**, *72*, 272–299. [CrossRef]
49. Goetz, J.; Brenning, A.; Petschko, H.; Leopold, P. Evaluating machine learning and statistical prediction techniques for landslide susceptibility modeling. *Comput. Geosci.* **2015**, *81*, 1–11. [CrossRef]

50. Aleotti, P.; Chowdhury, R. Landslide hazard assessment: Summary review and new perspectives. *Bull. Int. Assoc. Eng. Geol.* **1999**, *58*, 21–44. [CrossRef]
51. Pawluszek, K.; Borkowski, A. Impact of DEM-derived factors and analytical hierarchy process on landslide susceptibility mapping in the region of Rożnów Lake, Poland. *Nat. Hazards* **2017**, *86*, 919–952. [CrossRef]
52. Pourghasemi, H.R.; Rahmati, O. Prediction of the landslide susceptibility: Which algorithm, which precision? *Catena* **2018**, *162*, 177–192. [CrossRef]
53. Bui, D.T.; Shahabi, H.; Omidvar, E.; Shirzadi, A.; Geertsema, M.; Clague, J.J.; Khosravi, K.; Pradhan, B.; Pham, B.T.; Chapi, K.; et al. Shallow Landslide Prediction Using a Novel Hybrid Functional Machine Learning Algorithm. *Remote Sens.* **2019**, *11*, 931. [CrossRef]
54. Guzzetti, F.; Carrara, A.; Cardinali, M.; Reichenbach, P. Landslide hazard evaluation: A review of current techniques and their application in a multi-scale study, Central Italy. *Geomorphology* **1999**, *31*, 181–216. [CrossRef]
55. Kavzoglu, T.; Sahin, E.K.; Colkesen, I. Selecting optimal conditioning factors in shallow translational landslide susceptibility mapping using genetic algorithm. *Eng. Geol.* **2015**, *192*, 101–112. [CrossRef]
56. Tien Bui, D.; Tuan, T.A.; Hoang, N.-D.; Thanh, N.Q.; Nguyen, D.B.; Van Liem, N.; Pradhan, B. Spatial prediction of rainfall-induced landslides for the Lao Cai area (Vietnam) using a hybrid intelligent approach of least squares support vector machines inference model and artificial bee colony optimization. *Landslides* **2017**, *14*, 447–458. [CrossRef]
57. Lagomarsino, D.; Tofani, V.; Segoni, S.; Catani, F.; Casagli, N. A Tool for Classification and Regression Using Random Forest Methodology: Applications to Landslide Susceptibility Mapping and Soil Thickness Modeling. *Environ. Model. Assess.* **2017**, *22*, 201–214. [CrossRef]
58. Segoni, S.; Tofani, V.; Rosi, A.; Catani, F.; Casagli, N. Combination of Rainfall Thresholds and Susceptibility Maps for Dynamic Landslide Hazard Assessment at Regional Scale. *Front. Earth Sci.* **2018**, *6*, 85. [CrossRef]
59. Ahmed, B. Landslide susceptibility mapping using multi-criteria evaluation techniques in Chittagong Metropolitan Area, Bangladesh. *Landslides* **2015**, *12*, 1077–1095. [CrossRef]
60. El Jazouli, A.; Barakat, A.; Khellouk, R. GIS-multicriteria evaluation using AHP for landslide susceptibility mapping in Oum Er Rbia high basin (Morocco). *Geoenviron. Disasters* **2019**, *6*, 3. [CrossRef]
61. Shahabi, H.; Hashim, M. Landslide susceptibility mapping using GIS-based statistical models and Remote sensing data in tropical environment. *Sci. Rep.* **2015**, *5*, 9899. [CrossRef]
62. Kayastha, P.; Dhital, M.; De Smedt, F. Application of the analytical hierarchy process (AHP) for landslide susceptibility mapping: A case study from the Tinau watershed, west Nepal. *Comput. Geosci.* **2013**, *52*, 398–408. [CrossRef]
63. Hasekioğulları, G.D.; Ercanoglu, M. A new approach to use AHP in landslide susceptibility mapping: A case study at Yenice (Karabuk, NW Turkey). *Nat. Hazards* **2012**, *63*, 1157–1179. [CrossRef]
64. Hong, H.; Naghibi, S.A.; Pourghasemi, H.R.; Pradhan, B. GIS-based landslide spatial modeling in Ganzhou City, China. *Arab. J. Geosci.* **2016**, *9*, 112. [CrossRef]
65. Crovelli, R.A. *Probability Models for Estimation of Number and Costs of Landslides*; US Geological Survey: Reston, VA, USA, 2000.
66. Palchaudhuri, M.; Biswas, S. Application of AHP with GIS in drought risk assessment for Puruliya district, India. *Nat. Hazards* **2016**, *84*, 1905–1920. [CrossRef]
67. Pourghasemi, H.R.; Pradhan, B.; Gokceoglu, C. Application of fuzzy logic and analytical hierarchy process (AHP) to landslide susceptibility mapping at Haraz watershed, Iran. *Nat. Hazards* **2012**, *63*, 965–996. [CrossRef]
68. Mokarram, M.; Zarei, A.R. Landslide Susceptibility Mapping Using Fuzzy-AHP. *Geotech. Geol. Eng.* **2018**, *36*, 3931–3943. [CrossRef]
69. Saaty, T.L. A scaling method for priorities in hierarchical structures. *J. Math. Psychol.* **1977**, *15*, 234–281. [CrossRef]
70. Saaty, T.L.; Vargas, L.G. Diagnosis with Dependent Symptoms: Bayes Theorem and the Analytic Hierarchy Process. *Oper. Res.* **1998**, *46*, 491–502. [CrossRef]
71. Saaty, T.L. *Fundamentals of Decision Making and Priority Theory with the Analytic Hierarchy Process*; RWS Publications: Pittsburgh, PA, USA, 2000; Volume 6.
72. Alonso, J.A.; Lamata, M.T. Consistency in the analytic hierarchy process: A new approach. *Int. J. Uncertain. Fuzziness Knowl.-Based Syst.* **2006**, *14*, 445–459. [CrossRef]

73. Golden, B.L.; Wang, Q. An Alternate Measure of Consistency. In *The Analytic Hierarchy Process*; Springer Science and Business Media: Berlin/Heidelberg, Germany, 1989; pp. 68–81.
74. Lane, E.F.; Verdini, W.A. A Consistency Test for AHP Decision Makers. *Decis. Sci.* **1989**, *20*, 575–590. [CrossRef]
75. Forman, E.H. Random indices for incomplete pairwise comparison matrices. *Eur. J. Oper. Res.* **1990**, *48*, 153–155. [CrossRef]
76. Noble, E.E.; Sanchez, P.P. A note on the information content of a consistent pairwise comparison judgment matrix of an AHP decision maker. *Theory Decis.* **1993**, *34*, 99–108. [CrossRef]
77. Tummala, V.R.; Wan, Y.-W. On the mean random inconsistency index of analytic hierarchy process (AHP). *Comput. Ind. Eng.* **1994**, *27*, 401–404. [CrossRef]
78. Aguarón, J.; Moreno-Jiménez, J.M. The geometric consistency index: Approximated thresholds. *Eur. J. Oper. Res.* **2003**, *147*, 137–145. [CrossRef]
79. Alonso, J.; Lamata, M. Estimation of the random index in the analytic hierarchy process. *Proc. Proc. Inf. Process. Manag. Uncertain. Knowl.-Based Syst.* **2014**, *1*, 317–322.
80. Jena, R.; Pradhan, B.; Beydoun, G.; Nizamuddin; Ardiansyah; Sofyan, H.; Affan, M. Integrated model for earthquake risk assessment using neural network and analytic hierarchy process: Aceh province, Indonesia. *Geosci. Front.* **2019**. [CrossRef]
81. Fan, W.; Wei, X.-S.; Cao, Y.-B.; Zheng, B. Landslide susceptibility assessment using the certainty factor and analytic hierarchy process. *J. Mt. Sci.* **2017**, *14*, 906–925. [CrossRef]
82. Sangchini, E.K.; Emami, S.N.; Tahmasebipour, N.; Pourghasemi, H.R.; Naghibi, S.A.; Arami, S.A.; Pradhan, B. Assessment and comparison of combined bivariate and AHP models with logistic regression for landslide susceptibility mapping in the Chaharmahal-e-Bakhtiari Province, Iran. *Arab. J. Geosci.* **2016**, *9*. [CrossRef]
83. Devkota, K.C.; Regmi, A.D.; Pourghasemi, H.R.; Yoshida, K.; Pradhan, B.; Ryu, I.C.; Dhital, M.R.; Althuwaynee, O.F. Landslide susceptibility mapping using certainty factor, index of entropy and logistic regression models in GIS and their comparison at Mugling–Narayanghat road section in Nepal Himalaya. *Nat. Hazards* **2013**, *65*, 135–165. [CrossRef]

Article

Rainfall Event–Duration Thresholds for Landslide Occurrences in China

Shuangshuang He [1], Jun Wang [1,2,*] and Songnan Liu [2]

[1] Nansen-Zhu International Research Centre, Institute of Atmospheric Physics, Chinese Academy of Sciences, Beijing 100029, China; heshuangshuang16@mails.ucas.ac.cn

[2] Collaborative Innovation Center on Forecast and Evaluation of Meteorological Disasters/Key Laboratory of Meteorological Disaster, Ministry of Education, Nanjing University of Information Science and Technology, Nanjing 210044, China; liusn113@126.com

* Correspondence: wangjun@mail.iap.ac.cn

Received: 13 November 2019; Accepted: 8 February 2020; Published: 12 February 2020

Abstract: A rainfall threshold for landslide occurrence at a national scale in China has rarely been developed in the early warning system for landslides. Based on 771 landslide events that occurred in China during 1998–2017, four groups of rainfall thresholds at different quantile levels of the quantile regression for landslide occurrences in China are defined, which include the original rainfall event–duration (E–D) thresholds and normalized (the accumulated rainfall is normalized by mean annual precipitation) (EMAP–D) rainfall thresholds based on the merged rainfall and the Climate Prediction Center Morphing technique (CMORPH) rainfall products, respectively. Each group consists of four sub-thresholds in rainy season and non-rainy season, and both are divided into short duration (<48 h) and long duration (≥48 h). The results show that the slope of the regression line for the thresholds in the events with long durations is larger than that with short durations. In addition, the rainfall thresholds in the non-rainy season are generally lower than those in the rainy season. The E–D thresholds defined in this paper are generally lower than other thresholds in previous studies on a global scale, and a regional or national scale in China. This might be due to there being more landslide events used in this paper, as well as the combined effects of special geological environment, climate condition and human activities in China. Compared with the previous landslide model, the positive rates of the rainfall thresholds for landslides have increased by 16%–20%, 10%–17% and 20%–38% in the whole year, rainy season and non-rainy season, respectively.

Keywords: landslide and debris flow; rainfall thresholds; China; quantile regression

1. Introduction

Landslides are one of the most devastating geo-hazards that cause heavy casualties and great economic losses around the world every year. China is one of the countries that are most prone to landslides due to its geological, geomorphological and climate conditions [1,2]. According to the National Geological Disaster Bulletin issued by the Ministry of Natural Resources of the People's Republic of China from 2007 to 2016, an average of 762 people are reported dead or missing per year due to landslides, debris flows and other geological disasters, resulting in direct economic losses of 4.2 billion Chinese yuan. Since precipitation is the main trigger factor for landslides, the frequency and intensity changes of precipitation, which are caused by the global warming [3], will probably lead to an increase in the potential occurrence of landslides in the future [4–6]. In addition, as China is currently experiencing economic booming and developing rapidly, the early warning of landslides is crucial in China.

Rainfall thresholds can be used to predict the possible occurrence of slope failures in an area [7]. Currently, many countries and regions have developed the early warning systems (EWSs) based on rainfall thresholds to provide geological hazard information for the public [8–10].

By using the global rainfall threshold, combined with the landslide and debris flow susceptibility in China, a real-time EWS has been established for landslides and debris flows in China [11]. The EWS was verified by using 106 events in China during 2016–2017, and the results showed that the early or delayed warning is effective for 69% of the total events, while 72% of landslide events in the rainy season (May to September) can be warned, but only 35% of landslide events in the non-rainy season can be warned in advance. Some landslide events which did not reach the critical value of rainfall triggering the landslide and debris flow cannot be predicted, which may be due to the fact that these rainfall thresholds are not applied for China [12]. Thus, it is necessary to establish rainfall thresholds suitable for China to improve the positive prediction rate for landslides.

In fact, some rainfall thresholds have been developed in China, but most of them were established with few events or just suitable for specific areas. For example, Huang et al. (2015) have calculated the rainfall thresholds with 50 landslide events in Huangshan region of Anhui Province [13]. Zhou and Tang (2013) have defined a rainfall threshold with 11 rainfall events inducing debris flows in the Wenchuan earthquake-stricken area in Sichuan Province [14]. Li et al. (2017) have established a rainfall intensity–duration threshold in China, but only with 60 landslide events happening during 2005–2011 [15]. By far, few studies have used a large amount of historical data of landslides to establish rainfall thresholds in China.

Rainfall thresholds can be established by using both physical and empirical methods. Rainfall infiltrates into the slope and changes the pore water pressure, reducing the shear stress and thus may result in the slope failure [16]. Based on the physical methods, these physical processes can be analyzed and the critical value of rainfall can be calculated out [17,18]. However, such methods require an extensive data collection of geological information that requires high spatial resolution, so it's not feasible for threshold calculation at large regional or global scales [19]. Based on the rainfall information of historical landslide events, the empirical methods can determine the low boundary with statistical methods for the rainfall conditions that result in slope failures. Several statistical methods have been applied to obtain the rainfall thresholds, such as Bayesian statistics [20], logistical regression [21], confidence level [22] and quantile regression [23,24]. Thus, the empirical methods are more suitable for larger areas.

Guzzetti et al. (2008) have classified the rainfall thresholds into four categories: (i) rainfall intensity–duration (I–D) thresholds, (ii) thresholds based on the total rainfall of the events, (iii) rainfall event–duration (E–D) thresholds and (iv) rainfall event–intensity (E–I) thresholds [25]. The I–D and E–D thresholds are the most commonly, world-widely used among them. The E–D thresholds are equivalent to the I–D thresholds, while the E–D thresholds can avoid unnecessary conversions. Therefore, the E–D thresholds are selected in this paper.

This paper aims to calculate the E–D and normalized (the accumulated rainfall is normalized by mean annual precipitation, MAP) (EMAP–D) thresholds in the whole of China, and to validate the thresholds for early warning of landslides. A landslide inventory with more landslide events covering the whole region of China will be used. The rest of this paper is organized as follows. Section 2 describes the landslides inventory, rainfall dataset and quantile regression method to calculate the rainfall thresholds, and the thresholds established in this study are shown in Section 3. In Section 4, the thresholds defined in this paper are compared with the existed thresholds and validated with those from the previous analysis [11]. Finally, the main conclusions are summarized in Section 5.

2. Data and Methods

2.1. Landslide Inventory

The landslide information is obtained from four sources: (i) China Geological Environment Information site (http://www.cigem.cgs.gov.cn/), which provide the reports of geological disasters occurred in China; (ii) a global landslide catalog developed by Kirschbaum et al. (2015 [1]), available at https://catalog.data.gov/dataset/global-landslide-catalog-export; (iii) literatures including annual reports published by related institutions and/or government departments and local news report related to landslides, and (iv) other reports online. 771 landslide events were collected in a period covering 20 years from 1998 to 2017.

For each landslide, the corresponding information includes location, time, type of failure and triggering reason. Not all of this information is complete for every single event, i.e., some information is unknown. There are four kinds of spatial resolutions with the collected landslide events, namely, S1—province, S2—prefecture-level region, S3—county or district, and S4—town, village, country or site. As is shown in Table 1, 24 landslides (accounting for 3.1% of the total landslide events) are collected at the resolution of S1, 87 (11.3%) at S2, 286 (37.1%) at S3 and 374 (48.5%) at S4. In addition, there are also two kinds of temporal resolutions with the collected data, which are T1—the occurrence dates for 623 (80.8%) landslide events are known, and T2—the occurrence time (hour) for 148 (19.2%) landslide events are known.

Table 1. Information of landslide inventory in China during 1998–2017. NE: number of landslide events; S1: province; S2: prefecture-level region; S3: county/district; S4: village/town/country/site; T1: day; T2: hour; L: landslides; DF: debris flow; RF: rock fall.

Year	NE	Spatial Accuracy				Temporal Accuracy		Landslide Types		
		S1	S2	S3	S4	T1	T2	L	DF	RF
1998	24	0	1	1	22	24	0	24	0	0
1999	19	0	0	0	19	19	0	19	0	0
2000	17	0	1	2	14	17	0	17	0	0
2001	7	0	2	0	5	7	0	7	0	0
2002	4	0	0	1	3	1	3	3	1	0
2003	35	5	13	12	5	35	0	35	0	0
2004	10	0	2	2	6	10	0	10	0	0
2005	11	0	1	2	8	8	3	9	2	0
2006	15	0	0	0	15	7	8	7	6	2
2007	65	10	13	10	32	57	8	45	17	3
2008	73	7	19	16	31	73	0	61	11	1
2009	80	1	5	26	48	59	21	60	14	6
2010	73	0	11	38	24	57	16	63	6	4
2011	39	0	0	25	14	34	5	28	6	5
2012	49	1	5	21	22	49	0	23	13	13
2013	58	0	6	38	14	58	0	36	5	17
2014	40	0	3	34	3	40	0	20	3	17
2015	29	0	1	22	6	29	0	14	5	10
2016	75	0	4	34	37	29	46	62	13	0
2017	48	0	0	2	46	10	38	38	10	0
Total	771	24	87	286	374	623	148	581	112	78
Percent	100	3.1%	11.3%	37.0%	48.6%	80.8%	19.2%	75.4%	14.5%	10.1%

In addition, information about the landslide types is also collected, which can be classified into debris-flow, rock fall and generic shallow landslide. Usually the mechanisms among them are considered to be similar, but in fact, it's not easy to distinguish them clearly, because sometimes the report uses imprecise language to describe the landslides. Thus, most of them are classified as generic shallow landslides (581, 75.36%). In fact, whether the landslide occurs for the first time or reactivates is very important in the landslide prediction, because the corresponding geological conditions would definitely change after the slope failure occurred. However, the collected landslide information does not contain this specific information, and the landslide model used here cannot simulate those processes. So, this paper mainly concentrates on the rainfall thresholds, and only the shallow landslides which occurred at the first time are considered. The above information of landslide inventory is listed in Table 1, including the number of landslide events with different spatio-temporal resolutions and landslide types in each year.

Landslide events used to reconstruct the rainfall thresholds are selected based on criteria as follows: (i) the location of the landslide has a spatial resolution of S3 or higher; (ii) the occurrence time of the landslide is known at least with a daily resolution (T1). The landslides which are not triggered by rainfall are excluded. Based on these requirements, 660 landslide events are selected to construct the rainfall thresholds.

The latitude and longitude of the landslide events can be acquired by map tool. Figure 1 shows the locations of landslide events and elevation distribution in China. The elevation data are derived from HydroSHEDS—Hydrological data and maps based on SHuttle Elevation Derivatives at multiple Scales, with a spatial resolution of 3 arc-seconds, which can be downloaded at http://hydrosheds.cr.usgs.gov [26]. Most of the landslides are distributed in transitional regions from high altitude to low altitude and hilly areas in the southern China, where heavy precipitation occurs frequently due to monsoon and typhoon activities. Landslides occurred in 1998–2015 (black triangle in Figure 1) are used to construct the rainfall thresholds, and those occurred in 2016–2017 (red triangle in Figure 1) are used for validation.

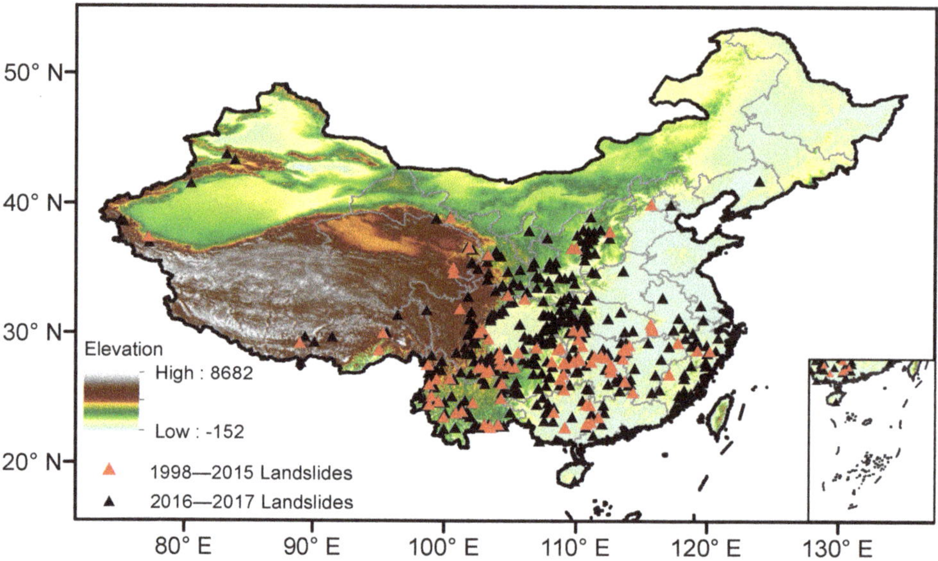

Figure 1. Location of landslides collected during 1998–2017 in China and the elevation (unit: m) distribution.

2.2. Rainfall Data

Three kinds of precipitation products are used to calculate the rainfall thresholds: (i) the merged precipitation product in China [27], and it is the main precipitation data used in this paper, (ii) satellite precipitation product produced by the NOAA Climate Prediction Center Morphing technique (CMORPH) [28], and (iii) CN05.1 [29] are also used in this paper. The hourly CMORPH and merged precipitation products are used to acquire the rainfall information when landslides occur, and daily CN05.1 is used to obtain the MAP for the normalization of the rainfall thresholds.

It has been pointed out that the development of satellite and ground-based radar technology has provided great support for the studies of rainfall thresholds [30]. The estimated precipitation from satellite-based data has a high spatio-temporal resolution, and the comprehensive quality evaluation in China is also good. In particular, the CMORPH performs better among the satellite precipitation products [31–33].

The CMORPH is developed by the National Oceanic and Atmospheric Administration (NOAA) Climate Prediction Center (CPC) in the United States. By using the geostationary satellite IR data to detect cloud systems and the associated motion characteristics, the cloud system advection vector is calculated, which is then used to deduct the instantaneous precipitation estimated by low orbiter satellite microwave observations. Finally, the continuous precipitation distribution is obtained. There are two versions of CMORPH: the original version (CMOPRH V0.x) and the new version (CMOPRH V1.0, used here). CMORPH V1.0 has provided near-real-time and bias-corrected products named CMORPH-RAW and CMORPH-CRT, respectively. CMORPH-RAW is satellite only precipitation and is provided in near-real-time so is suitable for real-time application. CMORPH-CRT is adjusted through matching the PDF of daily CMORPH-RAW against that for the CPC unified daily gauge analysis at each month over land, thus the CMORPH-CRT data is provided months delay. In this study, CMORPH-RAW with a temporal resolution of 30 min and a spatial resolution of −8 km, covering 60° S–60° N is used.

Based on the hourly precipitation observed by more than 30,000 automatic weather stations (AWS) in China and the satellite precipitation data retrieved from CMORPH, a merged rainfall product is developed through a two step merging algorithm of PDF (Probability Density Function) and OI (Optimal Interpolation). The merged rainfall combines the advantages of both the AWS and satellite products, so the spatio-temporal distributions of precipitation is more accurate and reasonable. The gauge stations used in this data are far more than which used in CMORPH-CRT, and it updates more rapidly than CMORPH-CRT, so that we use merged data in China rather than CMORPH-CRT. The spatial resolution of the merged rainfall is 0.1° × 0.1°, with a temporal resolution of 1 h. The dataset is available from 1 January 2008 to date, which can be downloaded online (available at http://data.cma.cn/data/detail/dataCode/SEVP_CLI_CHN_MERGE_CMP_PRE_HOUR_GRID_0.10/).

The CN05.1 dataset has been developed with observations at more than 2400 stations in China [29]. It is constructed by the "anomaly approach" [34], that is, first calculating a gridded climatology and then adding a gridded daily anomaly to the climatology to obtain the final dataset. The CN05.1 includes daily and monthly precipitation data and the period is from 1961 to 2017, with a spatial resolution of 0.25° × 0.25°. Here, the monthly data in 1981–2010 is adopted to calculate the MAP for 30 years.

As is shown in Figure 2, the MAP in China ranges from 24.62 to 2315.27 mm, which is high in the southeast of the Yangtze River and gradually decreases from the southeast to northwest, as it is known that heavy rainfall occurs frequently in the southern China during the summer monsoon season. Furthermore, typhoons from the Northwest Pacific Ocean land in these areas frequently from June to October and bring abundant rainfall. The high MAP in the southeastern China could partly explain why this region with low topographic relief is prone to landslides. As the MAP varies greatly in China, and the climate affects the meteorological conditions that can result in landslides, there is a need to reduce the effects of climate diversity. Commonly, due to the differences of rainfall thresholds in different climatic regions, the MAP as a climatic index is selected to normalize the rainfall thresholds [20,24,35,36]. Therefore, the accumulated rainfall is divided by MAP in this paper to acquire the EMAP–D thresholds.

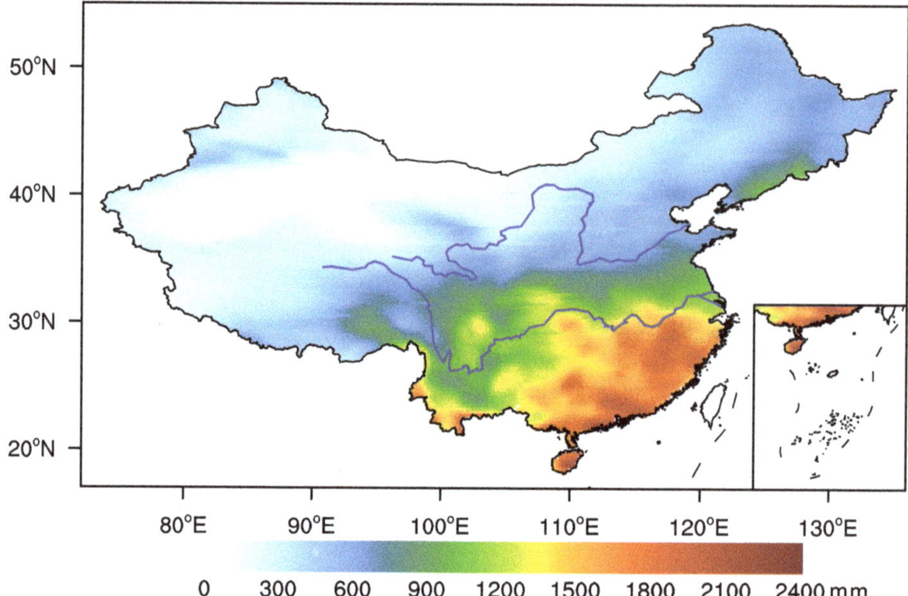

Figure 2. Mean annual precipitation (MAP) distribution in China during the period of 1981–2010.

2.3. Inventory Methodology

The formula of the E–D rainfall threshold is expressed as follow:

$$E = \alpha D^{\gamma} \tag{1}$$

where D is the duration (unit: h) from the beginning of the rainfall (Ts) to the occurrence time of the landslide (Te), E is the accumulated rainfall (unit: mm) during this duration, α and γ are constant parameters calculated by the regression method. Te means the time when the landslide event occurred; but for the landslide event with a temporal resolution of the day at S3, Te is the time when the rainfall ended on that day, and Ts is the time when the rainfall event began. A rainfall event is defined by a no-rainfall period. Saito et al. (2010) used 24 h to define the rainfall event in Japan [24]; Brunetti et al. (2010) used 48 h in May–September, and 96 h in October–April in Italy [37]; Segoni et al. (2014) did several runs using different no rain periods to get the lowest number of false alarms; the no rain period varies 18–36 h for different alert zones [22]. In this study, a rainfall event is defined as continuous rainfall separated from the preceding and the following events by a no-rainfall period of no less than 24 h.

Firstly, the rainfall information (E and D) relevant to landslides needs to be acquired. As CMORPH and merged rainfall data are both gridded data, the grid whose center is closest to the location of the landslide event is selected to reconstruct the rainfall information for landslides.

Then, Formula (1) is log-transformed to a linear equation:

$$\log10\,(E) = \log10\,(\alpha) + \gamma \log10\,(D) \tag{2}$$

For each landslide event, E–D values are plotted on the log10-log10 graph. In this process, it has been found that the rainfall durations of some landslide events are 0 h, which may be due to the fact that part of the landslides are not triggered by short-term rainfall, but by earthquake, snow melt or antecedent rainfall. Moreover, the AWSs coverage is not wide enough in China, especially in the mountain areas, therefore, such landslides are excluded when calculating rainfall thresholds.

Finally, based on the CMORPH products, 367 landslides are selected to calculate the rainfall thresholds; based on the merged rainfall products (available from 2008), 276 landslides are used to construct the rainfall thresholds.

To calculate the rainfall thresholds for landslides, the quantile regression [38] is adopted to determine the rainfall E–D threshold in this study. R software has a package called "quantreg", which implements the quantile regression [39,40]. It has been proposed that different levels of early warning information can be issued by the EWS with probability levels of 5%, 20% and 50%, categorized as "Null" (below the 5th percentile), "Alert" (5–20th percentiles), "Warning" (20–50th percentiles) and "Evacuation" (above the 50th percentiles), respectively. In this paper, the 10th, 20th, 30th, 40th, 50th, 60th, 70th, 80th and 90th percentile values are calculated for quantile regression lines at different levels, and the values of intercept α and slope β are returned by the linear regression method.

The main procedures for calculating the thresholds are summarized as follows.

(1) Collect information of landslides, including time, latitude and longitude of events.
(2) Screen the event that satisfies the criteria of spatio-temporal resolution.
(3) Define the rainfall event that is separated by a no-rainfall period of no less than 24 h, which means the period between two rainfall events is more than 24 h.
(4) Find the grid of the rainfall data whose center is closest to the location of the landslide event.
(5) Get the E and D of the landslide events and plot them as dots in the log10-log10 graph.
(6) Calculate the thresholds (including E–D and EMAP–D thresholds) by using the quantile regression.
(7) Compare with thresholds in other published literatures and validate the rainfall thresholds.

In addition, China is affected by the monsoon system. With the onset of the summer monsoon in May, the mainland of China becomes wet and rainy from south to north. When it comes to September, the summer monsoon retreats rapidly, followed by the establishment of the winter monsoon, and then the mainland of China is controlled by the Siberia high in winter. During this period, the soil is dry due to less rain. Considering the special climate conditions in China, the rainfall characteristics and soil moisture are different between the two periods in each year. Thus, the periods of May–September and October–April are selected as the rainy season and non-rainy season, respectively [41]. Moreover, as mentioned in Section 1, the existing landslide warning system shows different warning effects in rainy and non-rainy seasons [11], so the rainfall thresholds for the rainy and non-rainy seasons are calculated respectively in this paper.

3. Results

By using the landslide events occurred during 1998–2015 in China which satisfy the requirements mentioned above, that is, the spatial resolution at the levels of county, district or higher, and the temporal resolution at the level of day or finer, the D and E for each landslide event are inferred from the merged and CMOPRH rainfall data.

Figure 3 shows the distribution of the rainfall conditions for each landslide event in the log10-log10 graph. Roughly taking 48 h as a split point, the trend seems different in the two parts, which can be seen in both the CMORPH and merged rainfall data during the rainy season and non-rainy season. Thus, the duration can be divided into two periods: short duration ($1 \leq D < 48$ h) and long duration ($D > 48$). This may be because the mechanisms of landslide and debris flow triggered by long-term rainfall and short-term rainfall are different. In the long-term rainfall, the threshold is affected by the antecedent precipitation and soil moisture, and the evaporation is also an important factor [25]. In Figure 3a, for the merged data, in the rainy season, the range of the accumulated rainfall is 1.02–388.68 mm and the duration is 1–412 h (around 17 days); in the non-rainy season (Figure 3b), the range of the accumulated rainfall is 1.04–271.39 mm and the duration is 2–291 h (around 12 days). For the CMORPH data, in the rainy season (Figure 3c), the range of the accumulated rainfall is 1–394.8 mm and the duration is 1–341 h (around 14 days); in the non-rainy season (Figure 3d), the range of the accumulated rainfall is 1.2–179.4 mm and the duration is 1–393 h (around 16 days).

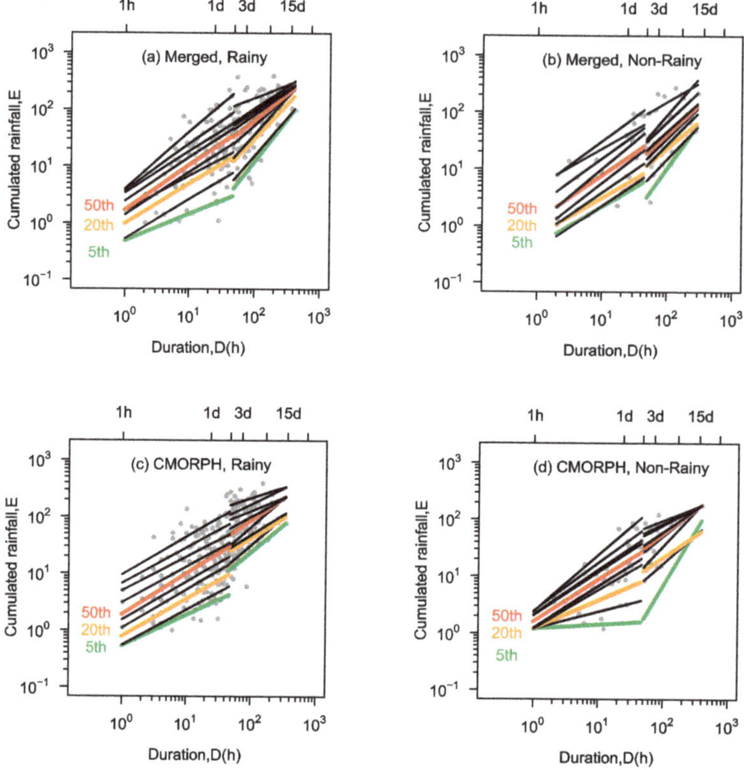

Figure 3. Event–duration (E–D) conditions of landslides calculated by the merged rainfall in the (**a**) rainy season and (**b**) non-rainy season, and by the Climate Prediction Center Morphing (CMORPH) rainfall in the (**c**) rainy season and (**d**) non-rainy season with different quantile regression lines (5%, 10%, 20%, 30%, 40%, 50%, 60%, 70%, 80% and 90% from bottom to top).

The coefficients of intercept α and slope γ at different levels of quantile regression lines obtained by combining precipitation and CMORPH precipitation are listed in Table 2. By using thresholds at different levels of quantile regression, different levels of warnings can be established in the EWS [42]. Generally, the threshold at the 5% quantile level is used as a lower safety threshold. If the threshold exceeds the value at the 5% quantile level, relevant tracking, monitoring and prediction should be carried out; if the threshold exceeds the value at the 20% quantile level, attention should be paid; and if the threshold exceeds the value at the 50% quantile level, people should be evacuated.

It can be found that for the thresholds at the quantile level of 50% or below, the slope of the regression line for the thresholds in the events with long durations is bigger than those with short durations, which means for long duration events much more rainfall is required to trigger landslide than. The possible reason might be that the evaporation plays an important role in the events with long durations, while for those with short durations, the evaporation can be ignored. Furthermore, the antecedent precipitation, soil moisture and climate condition also influence the rainfall thresholds in the initiation of landslides, and we will discuss this later in this article.

Table 2. Coefficients of intercept α and slope γ at different levels of quantile regression lines obtained by combining precipitation and CMORPH precipitation.

Season	Rainy Season				Non-Rainy Season			
Duration	D < 48 h		D ≥ 48 h		D < 48 h		D ≥ 48 h	
Coefficient	α	γ	α	γ	α	γ	α	γ
Merged-5%	0.49	0.47	0.01	1.48	0.45	0.68	0.01	1.61
10%	0.53	0.69	0.03	1.33	0.38	0.78	0.06	1.20
20%	0.98	0.70	0.11	1.22	0.65	0.67	0.28	0.96
30%	1.63	0.62	0.09	1.30	0.62	0.80	0.12	1.17
40%	1.41	0.75	0.46	1.03	0.72	0.90	0.19	1.13
50%	1.69	0.79	0.88	0.93	1.24	0.79	0.29	1.07
60%	2.50	0.76	1.81	0.83	1.10	0.96	0.36	1.05
70%	3.42	0.73	2.71	0.77	2.26	0.83	0.42	1.10
80%	3.77	0.82	2.83	0.78	4.98	0.66	0.19	1.33
90%	4.07	1.00	18.94	0.46	4.36	0.85	6.81	0.67
CMORPH-5%	0.53	0.53	0.28	0.96	1.20	0.07	0.00	1.94
10%	0.55	0.61	0.32	1.00	1.20	0.30	0.16	1.01
20%	0.78	0.65	1.55	0.71	1.20	0.49	0.58	0.78
30%	1.10	0.67	1.71	0.72	1.20	0.68	0.12	1.22
40%	1.54	0.66	0.59	1.02	1.26	0.73	0.74	0.92
50%	1.88	0.72	2.22	0.79	1.61	0.73	1.13	0.85
60%	3.10	0.62	5.02	0.65	2.06	0.76	1.61	0.79
70%	5.00	0.59	13.93	0.48	2.14	0.78	4.92	0.60
80%	6.73	0.62	12.76	0.56	2.47	0.84	6.54	0.55
90%	9.52	0.67	37.00	0.38	2.41	0.99	11.73	0.46

Considering that the MAP varies greatly in China (Figure 2), the accumulated rainfall is normalized by MAP. Figure 4 shows the scatter diagram of the relationship between EMAP and D on the logarithmic coordinates and the quantile regression lines at different levels. In addition, the corresponding coefficients of slope γ and intercept α for quantile regression lines at different levels are shown in Table 3. Similar to the E–D formula, split by 48 h, the formulas for both short-term and long-term rainfall thresholds are established, respectively. It can be seen that for the EMAP–D thresholds at the 50% quantile level or below established by the merged and CMORPH rainfall products, the slope of the regression line on rainfall threshold for the landslide and debris flow triggered by short-term rainfall is generally smaller than or close to that triggered by long-term rainfall, while the intercept is higher than that triggered by the long-term rainfall. The difference in the mechanisms for landslide and debris flow triggered by long-term rainfall and short-term rainfall is reflected in the rainfall thresholds.

Table 3. Same as Table 2, but for the coefficients of EMAP–D thresholds. (EMAP–D: E–D thresholds normalized by mean annual precipitation).

Season	Rainy Season				Non-Rainy Season			
Duration	D < 48 h		D ≥ 48 h		Duration		D < 48 h	
Coefficient	α	γ	α	γ	α	γ	α	γ
Merged-5%	0.00026	0.68	0.00002	1.38	0.00060	0.45	0.00001	1.51
10%	0.00054	0.76	0.00003	1.37	0.00075	0.38	0.00005	1.20
20%	0.00051	0.87	0.00010	1.20	0.00064	0.60	0.00034	0.87
30%	0.00166	0.60	0.00034	1.02	0.00077	0.64	0.00029	0.99
40%	0.00213	0.64	0.00149	0.76	0.00066	0.92	0.00027	1.01
50%	0.00489	0.46	0.00178	0.77	0.00058	1.09	0.00034	0.98
60%	0.00486	0.54	0.00292	0.70	0.00244	0.70	0.00023	1.09
70%	0.00808	0.44	0.00499	0.61	0.00274	0.69	0.00064	0.91
80%	0.00774	0.58	0.01156	0.48	0.00273	0.74	0.00044	1.07
90%	0.02439	0.32	0.02960	0.32	0.00234	0.96	0.00245	0.80
CMORPH-5%	0.00044	0.63	0.00025	0.97	0.00036	0.67	0.00000	1.66
10%	0.00071	0.56	0.00046	0.87	0.00069	0.65	0.00004	1.23
20%	0.00093	0.60	0.00068	0.84	0.00077	0.66	0.00028	0.90
30%	0.00136	0.62	0.00147	0.73	0.00158	0.51	0.00005	1.41
40%	0.00211	0.58	0.00191	0.72	0.00161	0.61	0.00007	1.36
50%	0.00439	0.41	0.00770	0.47	0.00191	0.62	0.00020	1.17
60%	0.00414	0.54	0.00623	0.56	0.00191	0.67	0.00028	1.11
70%	0.00570	0.54	0.01022	0.50	0.00193	0.73	0.00085	0.93
80%	0.01165	0.42	0.07585	0.12	0.00191	0.91	0.00129	0.86
90%	0.02423	0.27	0.12788	0.06	0.00305	0.83	0.00136	0.85

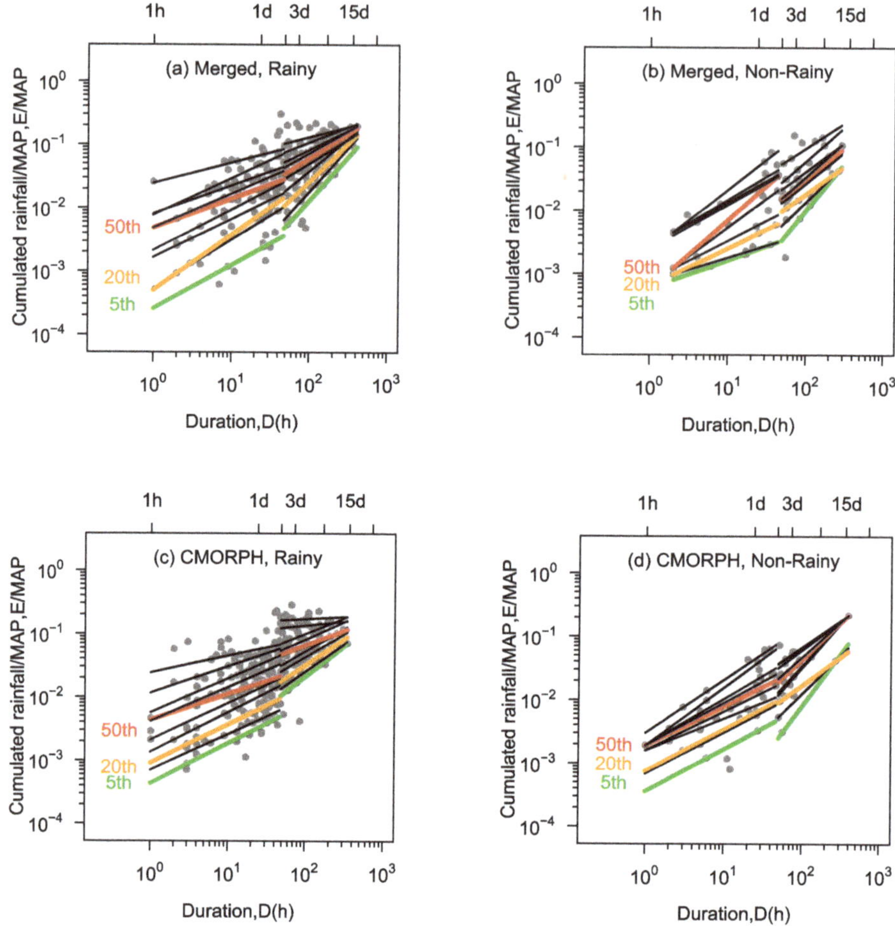

Figure 4. Same as Figure 3, but for normalized E–D thresholds (EMAP–D).

4. Discussions

So far, four groups of rainfall thresholds in China have been established. Each group consists of four sub-thresholds in rainy season and non-rainy season (both are divided into short duration and long duration), respectively. For thresholds below the quantile level of 50% derived from both the merged and CMORPH rainfall products, the slope of the regression line of the rainfall threshold for long durations is higher than that for short durations, which may be attributed to different mechanisms in triggering landslides by long- and short-duration rainfall.

In our previous study, a statistical model for landslide has been established by using the landslide susceptibility and a global rainfall threshold, showing that around 70% of the landslide events occurred in China during 2016–2017 can be warned in advance by the model [11,12]. In addition, when the rainfall threshold exceeds the value at the quantile level of 20%, attention should be paid. Therefore, the rainfall threshold at the quantile level of 30% is compared to other existing thresholds, and it is validated by using the landslides occurred in 2016–2017. Finally, the warning ability is compared with the previous statistical model for landslide.

The E–D thresholds with other global and regional thresholds in the literature are listed in Table 3. The I–D rainfall thresholds are converted to the E–D thresholds, and then all these thresholds are plotted on the log10-log10 graph (Figure 5). For the long duration, the thresholds derived from both the merged and CMORPH rainfall products are higher in the rainy season than those in the non-rainy season. For the short duration, the thresholds derived from the merged rainfall product are higher in the rainy season than those in the non-rainy season, but the thresholds from the CMORPH rainfall product in the rainy season are similar to those in the non-rainy season.

It can be seen that the thresholds established in this study are lower than most of the thresholds in previous studies shown here, especially for the short duration, including the global thresholds [43–45], the regional thresholds in Puerto Rico [46], Taiwan as well as Zhejiang of China [47,48]. This provides important information for the early warning of landslides in China, that is, the landslides occurred in China might be triggered by lower rainfall thresholds than that recognized before, which might result in missing alarms. The long-duration thresholds established in this study are close to the global thresholds defined by Guzzetti et al. (2008) (the line indicated by number 3-2 in Figure 5) [25] and the regional thresholds for Yan'an in China (the line indicated by number 10 in Figure 5) defined by Chen and Wang (2014) [49]. In addition, Li et al. (2017) defined a rainfall threshold (the line indicated by number 6 in Figure 5) for the whole China with 60 landslide events occurred in June to September during 2005–2011. This threshold is higher than the threshold defined in this study in the rainy season (May to September) [15]. It might because more landslide events (60 versus 660 events) were used in this study to obtain the warning thresholds.

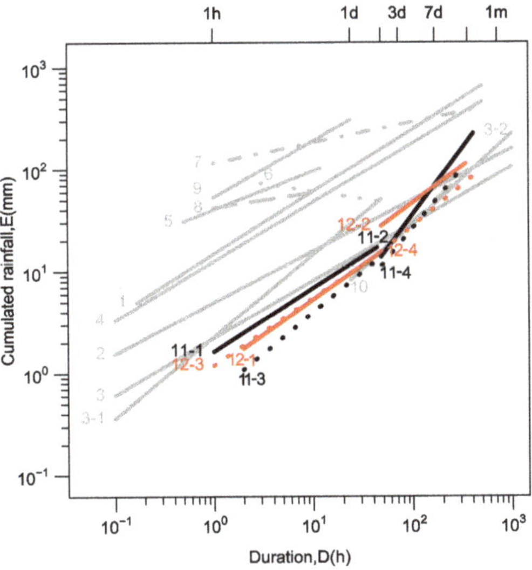

Figure 5. Comparison between E–D thresholds defined in this work and previous published thresholds. The numbers refer to No. in Table 4.

The EMAP–D thresholds are also compared with other thresholds defined in previous studies, as listed in Table 5, and these thresholds are plotted in Figure 6. The thresholds in rainy season calculated by using merged and CMORPH rainfall products are all higher than those in the non-rainy season. For short duration, the thresholds in this study are close to the thresholds calculated by Guzzetti et al. (2008) [25], higher than the Japan threshold established by Saito et al. (2010) [24]. For the long duration, the thresholds are higher than or very close to the thresholds calculated by Guzzetti et al. (2008) and Saito et al. (2010) [24,25].

Table 4. Rainfall thresholds for landslide occurrence in this work compared with those in previous publications.

No.	Reference	Equation	Range(h)	Area
1	(Caine, 1980) [43]	$I = 14.84D^{-0.39}$	$0.167 < D < 240$	world
2	(Innes, 1983) [45]	$E = 4.93D^{0.504}$	$0.1 < D < 1000$	world
3	(Guzzetti et al., 2008) [25]	$I = 2.20D^{-0.44}$	$0.1 < D < 1000$	world
3-1	(Guzzetti et al., 2008) [25]	$I = 2.28D^{-0.2}$	$0.1 < D < 48$	world
3-2	(Guzzetti et al., 2008) [25]	$I = 0.48D^{-0.11}$	$48 \leq D < 1000$	world
4	(Hong et al., 2007) [44]	$I = 12.45D^{-0.42}$	$0.1 < D < 500$	world
5	(Jibson, 1989) [46]	$I = 39.71D^{-0.62}$	$0.5 < D < 12$	Japan
6	(Li et al., 2017) [15]	$I = 85.72D^{-1.15}$	$3 < D < 45$	China
7	(Chien-Yuan et al., 2005) [48]	$I = 115.47D^{-0.8}$	$1 < D < 400$	Taiwan
8	(Jibson, 1989) [46]	$I = 41.83D^{-0.85}$	$1 < D < 12$	Hong Kong
9	(Ma et al., 2015) [47]	$I = 52.86D^{-0.45}$	$1 \leq D \leq 24$	Zhejiang, China
10	(Chen and Wang, 2014) [49]	$I = 0.448D^{-0.08654}$	$24 < D < 336$	Yanan, Shanxi, China
	(Dahal et al., 2008) [50]	$I = 73.90D^{-0.79}$	$5 < D$	Himalaya, Nepal
11-1	This work	$E = 0.53D^{0.7}$	$1 \leq D \leq 44$	China, rainy season, merge rainfall
11-2	This work	$E = 0.032D^{1.33}$	$48 \leq D \leq 412$	China, rainy season, merge rainfall
11-3	This work	$E = 0.45D^{0.68}$	$2 \leq D \leq 44$	China, non-rainy season, merge rainfall
11-4	This work	$E = 0.064D^{1.18}$	$49 \leq D \leq 291$	China, non-rainy season, merge rainfall
12-1	This work	$E = 0.86D^{0.48}$	$1 \leq D \leq 47$	China, rainy season, CMORPH rainfall
12-2	This work	$E = 0.18D^{1.07}$	$48 \leq D \leq 602$	China, rainy season, CMORPH rainfall
12-3	This work	$E = 0.48D^{0.65}$	$1 \leq D \leq 47$	China, non-rainy season, CMORPH rainfall
12-4	This work	$E = 0.59D^{0.84}$	$48 \leq D \leq 248$	China, non-rainy season, CMORPH rainfall

Table 5. EMAP–D rainfall thresholds for landslide occurrence in this study compared with those in previous publications.

No.	Reference	Equation	Range(h)	Area
1	(Guzzetti et al., 2008) [25]	$IMAP = 0.0016D^{-0.4}$	$0.1 < D < 1000$	world
1-1	(Guzzetti et al., 2008) [25]	$IMAP = 0.0017D^{-0.13}$	$0.1 < D < 48$	world
1-2	(Guzzetti et al., 2008) [25]	$IMAP = 0.0005D^{-0.13}$	$48 \leq D < 1000$	world
2	(Saito et al., 2010) [24]	$IMAP = 0.0007D^{-0.21}$	$3 < D < 357$	Japan
3	(Jibson, 1989) [46]	$IMAP = 0.02D^{-0.68}$	$1 < D < 12$	Hong Kong
4-1	this work	$EMAP = 0.00053D^{0.78}$	$1 \leq D \leq 44$	China, rainy season, Merge rainfall
4-2	this work	$EMAP = 0.00005D^{1.28}$	$48 \leq D \leq 412$	China, rainy season, Merge rainfall
4-3	this work	$EMAP = 0.00074D^{0.4}$	$2 \leq D \leq 44$	China, non-rainy season, Merge rainfall
4-4	this work	$EMAP = 0.000054D^{1.2}$	$49 \leq D \leq 291$	China, non-rainy season, Merge rainfall
5-1	this work	$EMAP = 0.00057D^{0.6}$	$1 \leq D \leq 47$	China, rainy season, CMORPH rainfall
5-2	this work	$EMAP = 0.00024D^{1}$	$48 \leq D \leq 602$	China, rainy season, CMORPH rainfall
5-3	this work	$EMAP = 0.00089D^{0.41}$	$1 \leq D \leq 47$	China, non-rainy season, CMORPH rainfall
5-4	this work	$EMAP = 0.00017D^{1.05}$	$48 \leq D \leq 248$	China, non-rainy season, CMORPH rainfall

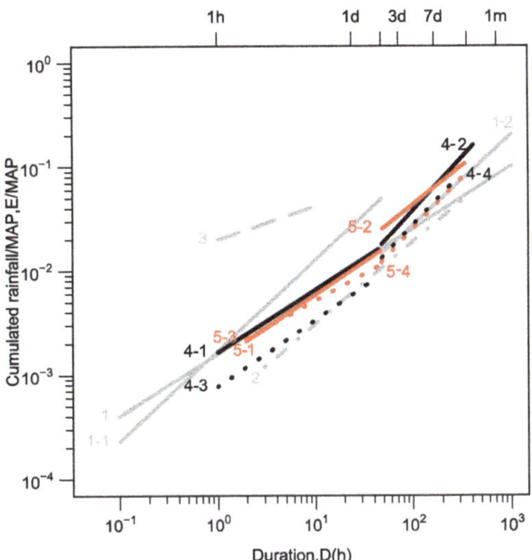

Figure 6. Comparison between the EMAP–D thresholds defined in this study and previous published thresholds. The numbers refer to No. in Table 4.

The comparison with other thresholds indicates that the landslides in China can be triggered by less severe rainfall conditions than previously recognized, which provides important information for the assessment of landslide hazards. In other words, China is highly prone to landslides compared with most other regions in the world. This may be caused by the large areas of mountains and hills, the monsoon climate and active human activities (constructions) of China.

The rainfall thresholds are validated by the 106 landslide and debris flow events occurred during 2016–2017, and the rates of these events, which can be warned in advance by the thresholds in this paper or the original landslide model developed by Wang et al. (2016), is called a "positive rate" here [11]. As shown in Figure 7, it can be seen that the positive rate of the original statistical model for the landslide events in 2016–2017 is 66%, and that of the rainfall thresholds established in this paper is 80–86%. In the rainy season, the positive rate of the original model is 72%, while that of the improved rainfall thresholds is 82%–89%. In the non-rainy season, the positive rate of the original model is only 35%, while that of the improved rainfall thresholds reaches 55%–73%. The positive rates in the whole year, the rainy season and the non-rainy season have increased by 16%–20%, 10%–17% and 20%–38%, respectively. Thus, if these thresholds can be applied in the EWS, more landslides can be warned in advance.

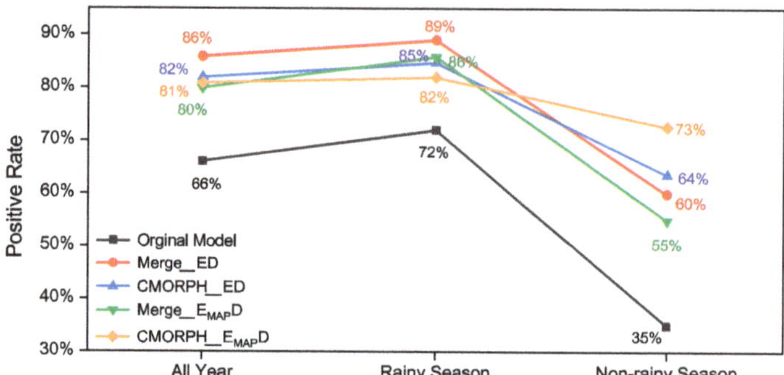

Figure 7. Comparison of positive rates between the improved thresholds and those in the original model [11].

Moreover, the "positive rate" is significantly improved if the thresholds in this study are adopted, however, a higher positive rate does not mean the threshold is better, because higher "positive rate" usually means the false alarm might also be higher, thus in real world application, the threshold defined in this paper should be carefully tested in real-time warning application. Actually, it's impossible to calculate the false alarms because the landslide inventory collected in this study is incomplete. Although the number of events in the catalog is much more than that in previous studies, it's still far from enough. For instance, according to the National Geological Disaster Bulletin, there is an average of 12,126 landslides per year during 2007–2016, but only a total of 771 events in 1998–2017 are collected through reports and the news online. So the false alarms would be inaccurate even if the calculation is conducted on them, which is one of the limitations in this study. In the future work, a more complete dataset of landslides and debris flows will be collected to find the most suitable rainfall thresholds for the warning of landslides and debris flows in the regression threshold lines at different quantile levels.

Furthermore, the differences between thresholds in different durations and seasons suggest that soil moisture plays an important role in the rainfall conditions triggering the landslides, and thus the antecedent rainfall is also important for landslides triggered by long-duration rainfall events. A detailed analysis of these factors, which is related to the physical process of landslide dynamic, is not provided in this study. The rainfall thresholds represent an empirical relationship between the landslides and rainfall that is generally similar to the parameterization of cloud processes in the weather model. Thus, it is still effective for the early warning of landslides. The uncertainty of forecasted precipitation in weather models has an important impact on the early warning of landslides. Recently, the ensemble forecast has been developed to cope with this uncertainty in weather models, which provides a new train of thought for the early warning of landslides. In further studies, based on the rainfall thresholds in this study and others, a landslide ensemble forecasting system (LEFS) is expected to be established to improve the prediction for landslides.

5. Conclusions

In this study, based on the landslide events occurred in China during 1998–2017, four groups of rainfall thresholds for landslide occurrence are defined by using the regression values at different quantile levels of the quantile regression. Based on both the merged rainfall product and the CMORH rainfall product, two kinds of rainfall thresholds—the cumulated event rainfall-rainfall duration thresholds (E–D) and the thresholds normalized by MAP (EMAP–D) are established respectively. Each group consists of four sub-thresholds in rainy season and non-rainy season (both are divided into short duration and long duration).

The slope of thresholds in the events with long durations is higher than that with short durations for thresholds below the quantile level of 50%, which may be due to different mechanisms of landslides triggered by long duration rainfall and short duration rainfall. The evaporation and antecedent rainfall become more important for landslides triggered by long duration rainfall events. Besides, the rainfall thresholds in the non-rainy season are generally lower than those in the rainy season.

The thresholds defined in this study are compared with those in previous studies. The E–D thresholds in this study are generally lower than most of the other thresholds, including the global thresholds, the regional and national thresholds in China, which may be because more landslide events in China were used in this study. Moreover, it also suggests that the combined effects of special geological environment, topographical relief and climate conditions make China suffer landslides more frequently and severely than other regions in the world. The landslide events occurred in 2016–2017 are used to validate the rainfall thresholds. Compared with the previous statistical model for landslides, based on the improved rainfall thresholds, the positive rates for the landslide and debris flow events have increased by 16%–20%, 10%–17% and 20%–38% in the whole year, rainy season and non-rainy season, respectively.

This paper aims to establish rainfall thresholds for its application in landside early warning systems. In addition to the triggering factors, the geological environment is also important, which decides the landslide susceptibility in a specific region. Thus, combining multiple rainfall thresholds with the landslide susceptibility map, and by using the real-time rainfall produced by CMORPH (CMORPH-RAW) and the forecasted rainfall from ensemble numerical weather prediction models, the LEFS will be established in future work. We expect it will contribute to saving lives and mitigating property damages caused by landslide disasters.

Author Contributions: Conceptualization, J.W.; methodology, J.W. and S.H.; software, J.W. and S.H.; validation, S.H. and S.L.; formal analysis, S.H.; investigation, S.H.; resources, J.W.; data curation, S.H.; writing—original draft preparation, S.H.; writing—review and editing, J.W.; visualization, S.H. and S.L.; supervision, J.W.; project administration, J.W.; funding acquisition, J.W. All authors have read and agreed to the published version of the manuscript.

Funding: This work was supported by the National Key Research and Development Program of China (Grant No. 2016YFA0600703) and the National Natural Science Foundation of China (Grant No.41605084).

Acknowledgments: We thank Wu Jia and Gao Xuejie for providing CN05.1 data. And we acknowledge the NOAA CPC for providing the CMORPH data and the Conservation Science Program of World Wildlife Fund for providing DEM. We thank Nanjing Hurricane Translation for reviewing the English language quality of this paper.

Conflicts of Interest: The funders had no role in the design of the study; in the collection, analyses, or interpretation of data; in the writing of the manuscript, or in the decision to publish the results.

References

1. Kirschbaum, D.; Stanley, T.; Zhou, Y. Spatial and temporal analysis of a global landslide catalog. *Geomorphology* **2015**, *249*, 4–15. [CrossRef]
2. Lin, Q.; Wang, Y. Spatial and temporal analysis of a fatal landslide inventory in China from 1950 to 2016. *Landslides* **2018**, *15*, 2357–2372. [CrossRef]
3. Fischer, E.M.; Knutti, R. Anthropogenic contribution to global occurrence of heavy-precipitation and high-temperature extremes. *Nat. Clim. Chang.* **2015**, *5*, 560–564. [CrossRef]
4. Crozier, M.J. Deciphering the effect of climate change on landslide activity: A review. *Geomorphology* **2010**, *124*, 260–267. [CrossRef]
5. Gariano, S.L.; Guzzetti, F. Landslides in a changing climate. *Earth Sci. Rev.* **2016**, *162*, 227–252. [CrossRef]
6. Saez, J.L.; Corona, C.; Stoffel, M.; Berger, F. Climate change increases frequency of shallow spring landslides in the French Alps. *Geology* **2013**, *41*, 619–622. [CrossRef]
7. Campbell, R.H. Soil slopes, debris flows, and rainstorms in the Santa Monica Mountains and vicinity, southern California. U.S. U.S. Government Printing Office: Washington, DC, USA, 1975; Geological Survey professional paper volume 851, pp. 1–51.
8. Baum, R.L.; Godt, J.W. Early warning of rainfall-induced shallow landslides and debris flows in the USA. *Landslides* **2009**, *7*, 259–272. [CrossRef]

9. Devolo, G.; Kleivane, I. *Landslide Early Warning System and Web Tools for Real-Time Scenarios and for Distribution of Warning Messages in Norway. Engineering Geology for Society and Territory*; Graziella, D., Ingeborg, K., Monica, S., Nils-Kristian, O., Ragnar, E., Erik, J., Hervé, C., Eds.; Springer: Berlin, Germany, 2015; Volume 2, pp. 625–629.
10. Osanai, N.; Shimizu, T.; Kuramoto, K.; Kojima, S.; Noro, T. Japanese early-warning for debris flows and slope failures using rainfall indices with Radial Basis Function Network. *Landslides* **2010**, *7*, 325–338. [CrossRef]
11. Wang, J.; Wang, H.J.; Hong, Y. *A Realtime Monitoring and Dynamical Forecasting System for Floods and Landslides in China*; China Meteorological Press: Beijing, China, 2016; pp. 111–152.
12. He, S.S.; Wang, J.; Wang, H.J. Projection of landslides in China during the 21st century under the RCP8.5 Scenario. *J. Meteor. Res.* **2019**, *33*, 138–148. [CrossRef]
13. Huang, J.; Ju, N.P.; Liao, Y.J. Determination of rainfall thresholds for shallow landslides by a probabilistic and empirical method. *Nat. Hazards Earth Syst. Sci.* **2015**, *15*, 2715–2723. [CrossRef]
14. Zhou, W.; Tang, C. Rainfall thresholds for debris flow initiation in the Wenchuan earthquake-stricken area, southwestern China. *Landslides* **2014**, *11*, 877–887. [CrossRef]
15. Li, W.; Liu, C.; Scaioni, M.; Sun, W.; Chen, Y.; Yao, D.; Chen, S.; Hong, Y.; Zhang, K.; Cheng, G. Spatio-temporal analysis and simulation on shallow rainfall-induced landslides in China using landslide susceptibility dynamics and rainfall I-D thresholds. *Sci. China Earth Sci.* **2017**, *60*, 720–732. [CrossRef]
16. Iverson, R.M. Landslide triggering by rain infiltration. *Water Resour. Res.* **2000**, *36*, 1897–1910. [CrossRef]
17. Chung, M.-C.; Tan, C.-H.; Chen, C.-H. Local rainfall thresholds for forecasting landslide occurrence: Taipingshan landslide triggered by Typhoon Saola. *Landslides* **2016**, *14*, 19–33. [CrossRef]
18. Wu, S.-J.; Hsiao, Y.-H.; Yeh, K.-C.; Yang, S.-H. A probabilistic model for evaluating the reliability of rainfall thresholds for shallow landslides based on uncertainties in rainfall characteristics and soil properties. *Nat. Hazards* **2017**, *87*, 469–513. [CrossRef]
19. Posner, A.J.; Georgakakos, K.P. Soil moisture and precipitation thresholds for real-time landslide prediction in El Salvador. *Landslides* **2015**, *12*, 1179–1196. [CrossRef]
20. Guzzetti, F.; Peruccacci, S.; Rossi, M.; Stark, C.P. Rainfall thresholds for the initiation of landslides in central and southern Europe. *Meteorol. Atmos. Phys.* **2007**, *98*, 239–267. [CrossRef]
21. Giannecchini, R.; Yuri, G.; Michele, B. *Rainfall Intensity-Duration Thresholds for Triggering Shallow Landslides in the Eastern Ligurian Riviera (Italy)*; Engineering Geology for Society and Territory; Graziella, D., Ingeborg, K., Monica, S., Nils-Kristian, O., Ragnar, E., Erik, J., Hervé, C., Eds.; Springer: Berlin, Germany, 2015; Volume 2, pp. 1581–1584.
22. Segoni, S.; Rosi, A.; Rossi, G.; Catani, F.; Casagli, N. Analysing the relationship between rainfalls and landslides to define a mosaic of triggering thresholds for regional-scale warning systems. *Nat. Hazards Earth Syst. Sci.* **2014**, *14*, 2637–2648. [CrossRef]
23. Rossi, M.; Luciani, S.; Valigi, D.; Kirschbaum, D.; Brunetti, M.T.; Peruccacci, S.; Guzzetti, F. Statistical approaches for the definition of landslide rainfall thresholds and their uncertainty using rain gauge and satellite data. *Geomorphology* **2017**, *285*, 16–27. [CrossRef]
24. Saito, H.; Nakayama, D.; Matsuyama, H. Relationship between the initiation of a shallow landslide and rainfall intensity—Duration thresholds in Japan. *Geomorphology* **2010**, *118*, 167–175. [CrossRef]
25. Guzzetti, F.; Rossi, M.P.; Stark, C.P. The rainfall intensity-duration control of shallow landslides and debris flows: An update. *Landslides* **2008**, *5*, 3–17. [CrossRef]
26. Lehner, B.; Verdin, K.; Jarvis, A. *HydroSHEDS Technical Documentation*; User's Guide; Lehner, B., Verdin, K., Jarvis, A., Eds.; World Wildlife Fund US: Washington, DC, USA, 2006.
27. Shen, Y.; Pan, Y.; Yu, J.; Zhao, P.; Zhou, Z. Quality assessment of hourly merged precipitation product over China. *Trans. Atmos. Sci.* **2013**, *36*, 37–46.
28. Joyce, R.J.; Janowiak, J.E.; Arkin, P.A.; Xie, P. CMORPH: A method that produces global precipitation estimates from passive microwave and infrared data at high spatial and temporal resolution. *J. Hydrometeorol.* **2004**, *5*, 487–503. [CrossRef]
29. Wu, J.; Gao, X. A gridded daily observation dataset over China region and comparison with the other datasets. *Chin. J. Geophys.* **2013**, *56*, 1102–1111.
30. Segoni, S.; Piciullo, L.; Gariano, S.L. A review of the recent literature on rainfall thresholds for landslide occurrence. *Landslides* **2018**, *15*, 1483–1501. [CrossRef]

31. Chen, S.; Hu, J.J.; Zhang, A.; Min, C.; Huang, C.Y.; Liang, Z.Q. Performance of near real-time Global Satellite Mapping of Precipitation estimates during heavy precipitation events over northern China. *Theor. Appl. Climatol.* **2019**, *135*, 877–891. [CrossRef]
32. Li, C.M.; Tang, G.Q.; Hong, Y. Cross-evaluation of ground-based, multi-satellite and reanalysis precipitation products: Applicability of the Triple Collocation method across Mainland China. *J. Hydrol.* **2018**, *562*, 71–83. [CrossRef]
33. Wei, G.H.; Lu, H.S.; Crow, W.T.; Zhu, Y.H.; Wang, J.Q.; Su, J.B. Comprehensive evaluation of GPM-IMERG, CMORPH, and TMPA precipitation products with gauged rainfall over mainland China. *Adv. Meteorol.* **2018**, *2018*, 1–18. [CrossRef]
34. Xu, Y.; Gao, X.; Shen, Y.; Xu, C.; Shi, Y.; Giorgi, F. A daily temperature dataset over China and its application in validating a RCM simulation. *Adv. Atmos. Sci.* **2009**, *26*, 763–772. [CrossRef]
35. Aleotti, P. A warning system for rainfall-induced shallow failures. *Eng. Geol.* **2004**, *73*, 247–265. [CrossRef]
36. Giannecchini, R. Rainfall triggering soil slips in the southern Apuan Alps (Tuscany, Italy). *Adv. Geosci.* **2005**, *2*, 21–24. [CrossRef]
37. Brunetti, M.T.; Peruccacci, S.; Rossi, M.; Luciani, S.; Valigi, D.; Guzzetti, F. Rainfall thresholds for the possible occurrence of landslides in Italy. *Nat. Hazards Earth Syst. Sci.* **2010**, *10*, 447–458. [CrossRef]
38. Koenker, R.; Hallock, K.F. Quantile regression. *J. Econ. Perspect.* **2001**, *15*, 143–156. [CrossRef]
39. Team, R.C. *R: A Language and Environment for Statistical Computing*; User's Guide; R Core Team: R Foundation for Statistical Computing: Vienna, Austria, 2017.
40. Koenker, R. Quantile Regression in R: A Vignette. In *Quantile Regression (Econometric Society Monographs)*; Cambridge University Press: Cambridge, UK, 2005; pp. 295–316. [CrossRef]
41. Ding, Y.; Wang, Z. A study of rainy seasons in China. *Meteorol. Atmos. Phys.* **2008**, *100*, 121–138.
42. Pradhan, A.M.S.; Lee, S.R.; Kim, Y.T. A shallow slide prediction model combining rainfall threshold warnings and shallow slide susceptibility in Busan, Korea. *Landslides* **2018**, *16*, 647–659. [CrossRef]
43. Caine, N. The rainfall intensity: Duration control of shallow landslides and debris flows. Geografiska annaler. *Ser. A Phys. Geogr.* **1980**, *62*, 23–27.
44. Hong, Y.; Robert, F. Adle; J.G. Evaluation of the potential of NASA multi-satellite precipitation analysis in global landslide hazard assessment. *Geophys. Res. Lett.* **2007**, *33*, L22402. [CrossRef]
45. Innes, J.L. Debris flows. *Prog. Phys. Geogr.* **1983**, *7*, 469–501. [CrossRef]
46. Jibson, R.W. Debris flows in southern Puerto Rico. *Spec. Pap. Geol. Soc. Am.* **1989**, *236*, 29–56.
47. Ma, T.; Li, C.; Lu, Z.; Bao, Q. Rainfall intensity–Duration thresholds for the initiation of landslides in Zhejiang Province, China. *Geomorphology* **2015**, *245*, 193–206. [CrossRef]
48. Chien-Yuan, C.; Tien-Chien, C.; Fan-Chieh, Y.; Wen-Hui, Y.; Chun-Chieh, T. Rainfall duration and debris-Flow initiated studies for real-time monitoring. *Environ. Geol.* **2005**, *47*, 715–724. [CrossRef]
49. Chen, H.; Wang, J. Regression analyses for the minimum intensity-duration conditions of continuous rainfall for mudflows triggering in Yan'an, northern Shaanxi (China). *Bull. Eng. Geol. Environ.* **2014**, *73*, 917–928. [CrossRef]
50. Dahal, R.K.; Hasegawa, S. Representative rainfall thresholds for landslides in the Nepal Himalaya. *Geomorphology* **2008**, *100*, 429–443. [CrossRef]

Article

The Selection of Rain Gauges and Rainfall Parameters in Estimating Intensity-Duration Thresholds for Landslide Occurrence: Case Study from Wayanad (India)

Minu Treesa Abraham [1,*], Neelima Satyam [1], Ascanio Rosi [2], Biswajeet Pradhan [1,3,4] and Samuele Segoni [2]

1 Discipline of Civil Engineering, Indian Institute of Technology Indore, Madhya Pradesh 453552, India; neelima.satyam@iiti.ac.in (N.S.); Biswajeet.Pradhan@uts.edu.au (B.P.)

2 Department of Earth Sciences, University of Florence, Via Giorgio La Pira, 4, 50121 Florence, Italy; ascanio.rosi@unifi.it (A.R.); samuele.segoni@unifi.it (S.S.)

3 Centre for Advanced Modelling and Geospatial Information Systems (CAMGIS), Faculty of Engineering and Information Technology, University of Technology Sydney, Sydney, Broadway, Sydney P.O. Box 123, Australia

4 Department of Energy and Mineral Resources Engineering, Sejong University, Choongmu-gwan, 209 Neungdong-ro, Gwangjin-gu, Seoul 05006, Korea

* Correspondence: minu.abraham@iiti.ac.in

Received: 9 March 2020; Accepted: 31 March 2020; Published: 1 April 2020

Abstract: Recurring landslides in the Western Ghats have become an important concern for authorities, considering the recent disasters that occurred during the 2018 and 2019 monsoons. Wayanad is one of the highly affected districts in Kerala State (India), where landslides have become a threat to lives and properties. Rainfall is the major factor which triggers landslides in this region, and hence, an early warning system could be developed based on empirical rainfall thresholds considering the relationship between rainfall events and their potential to initiate landslides. As an initial step in achieving this goal, a detailed study was conducted to develop a regional scale rainfall threshold for the area using intensity and duration conditions, using the landslides that occurred during the years from 2010 to 2018. Detailed analyses were conducted in order to select the most effective method for choosing a reference rain gauge and rainfall event associated with the occurrence of landslides. The study ponders the effect of the selection of rainfall parameters for this data-sparse region by considering four different approaches. First, a regional scale threshold was defined using the nearest rain gauge. The second approach was achieved by selecting the most extreme rainfall event recorded in the area, irrespective of the location of landslide and rain gauge. Third, the classical definition of intensity was modified from average intensity to peak daily intensity measured by the nearest rain gauge. In the last approach, four different local scale thresholds were defined, exploring the possibility of developing a threshold for a uniform meteo-hydro-geological condition instead of merging the data and developing a regional scale threshold. All developed thresholds were then validated and empirically compared to find the best suited approach for the study area. From the analysis, it was observed that the approach selecting the rain gauge based on the most extreme rainfall parameters performed better than the other approaches. The results are useful in understanding the sensitivity of Intensity–Duration threshold models to some boundary conditions such as rain gauge selection, the intensity definition and the strategy of subdividing the area into independent alert zones. The results were discussed with perspective on a future application in a regional scale Landslide Early Warning System (LEWS) and on further improvements needed for this objective.

Keywords: landslides; thresholds; Wayanad; early warning; GIS; rainfall intensity

1. Introduction

Landslides can be considered as processes that move earth and rock downwards by sliding, falling and flowing in response to the extant conditions [1]. Within a span of seven years from 2004 to 2010, a total of 2620 landslides were recorded globally, which led to the loss of 32,322 lives [2]. In India, most of the highlands are affected by landslides and rainfall is identified as the major triggering factor in the Himalayas and Western Ghats [3–6]. The rise in population demands for the urbanization in high-altitude regions, which are usually susceptible to mass movements; therefore, when such places become densely populated areas, landslides cause severe fatalities. Recent changes in climate are also worsening the situation, with an increase in high intensity rainfalls and the consequent triggering of rapid mass movements [7,8], such as the debris flows, which occurred in Wayanad district in the state of Kerala during 2018 and 2019. The region is affected by a number of debris flows, with run-out distances as long as 3 km. Most of the slope failures that occurred during the 2018 monsoon also occurred during the 2019 monsoon as well. Thus, the increasing vulnerability of the region emphasizes the need for landslide early warning systems (LEWS) to forecast future events. Research has been carried out for establishing LEWS using the relationship between rainfalls and landslides in the Indian Himalayas [9–12], but detailed investigations for the Western Ghats have not been conducted yet. A LEWS should be developed on a regional scale for Wayanad district, incorporating monitoring tools and rainfall thresholds so that warnings can be issued to authorities and the local community. As a first step, this study focuses on establishing intensity–duration thresholds for the study area using statistical analysis.

Rainfall thresholds can be defined as a critical state of rainfall parameters from which an effect or result (landslides) can happen [13]. The minimum quantity of accumulated rainfall parameters which are required to trigger a landslide event will define the rainfall threshold for a region. Empirical and process-based approaches are widely used by researchers for developing rainfall thresholds [8,14–23]. The definition of a process-based threshold is associated with detailed site investigations and precise measurements. This approach is suitable for local scale or site-specific studies where the hydro-meteo-geological parameters can be monitored with required accuracy. Owing to the difficulties in estimating such parameters on a regional scale, this research focuses on an empirical approach to derive the rainfall thresholds using historical data. A rainfall event is usually characterized by three parameters: rainfall event (E), intensity (I), and duration (D). An event rainfall is the total accumulated amount of rainfall during a period of continuous precipitation, dubbed the duration of rainfall. The classical definition of intensity of rainfall is the average rate of precipitation usually expressed in mm/h or mm/day. Intensity–duration thresholds were first established by considering 73 landslides in several parts of the world by Caine in 1980 [24]. The definition of threshold, considering the minimum boundary, was then followed by researchers across the globe for analyzing local, regional and global scale thresholds [25–30]. Moreover, high intensity rainfalls are often associated with landslides in hilly areas [31] and when an empirical approach is pursued in regional scale studies, rainfall is more influential that site and slope characteristics [31–33]. For defining the minimum boundary or threshold, different statistical approaches can be used [34]. In literature, different definitions are adopted to calculate the intensity used in threshold analysis, such as mean intensity of an event [24], peak intensity [25] or most extreme intensity of sub-events [23].

Deducing the rainfall event associated with the occurrence of a landslide is the key factor in determining the threshold conditions. A recent review highlights that rain gauges are the most widely used instrument to collect rainfall data [35]. The selection of rain gauge in data-scarce regions are mostly forced, as the nearest rain gauge is selected based on spatial constraints and minor refinements [5,35,36]. In some studies, when rain gauge density is very small, landslides outside a specific radius from rain gauges are discarded, thus further reducing the amount of available data [37]. Moreover, some scholars observed that the threshold definition can be very sensitive to some boundary conditions, such as the rain gauge selection and characteristics or the delimitation of alert zones [22,38].

This research focuses on the effect of different approaches of analysis, including different definitions of rainfall intensity and different rain gauge configurations by using the database developed for the study area of Kerala (India). The term "approach" is used in this study, denoting the process of identification of a rainfall event which results in a landslide. The objective is to understand how the identification of rainfall parameters can affect the development of the rainfall threshold on both local and regional scales. Being one of the most followed approaches, the reliability of developing rainfall thresholds based on the nearest rain gauge is used as a benchmark and quantitatively compared with other strategies.

2. Description of Study Area

Kerala state is located in the southernmost part of the Indian subcontinent which is characterized by all the three physiographic division: coastal plains, midlands and highlands. The geomorphic features vary from coastal plains below sea level to mountain peaks with an elevation of 2695 metres. In addition, 40% of the state's area is occupied by the Western Ghats, the most significant orographic feature of Indian Peninsula [39]. The Western Ghats is a humid forested area where debris flows, initiated by rainfall, being the primary agents of landscape evolution [40]. Torrential rains during the months of June, July and August 2018 triggered around 341 major landslides across ten districts of the state [41]. Wayanad district was one of the worst affected districts which suffered severe socioeconomic setbacks due to these landslides. About 36.74% of the net cropped area of the district was damaged in the disaster [41]. Most of the locations that were affected in the 2018 landslide disaster were reactivated during the 2019 monsoon also, making the situation critical.

The 2130 km^2 area of Wayanad district lies between 11°30′ N to 12°3′ N latitudes and 75°39′ E to 76°30′ E longitudes, as shown in Figure 1. Geologically, the district can be divided into four sectors: peninsular gneissic complex, migmatite complex, charnockite group and the Wayanad group in north-central, south-central, southern and northern parts respectively [42]. Wayanad group rocks are found on the northern side as bands. Charnockite rocks form the hilly terrains of south and southeast parts of Wayanad, with narrow bands of pyroxene granulite and magnetite quartzite within charnockite. Biotite hornblende gneiss found over large parts of south-central Wayanad represents the migmatite complex.

The altitude of the district ranges up to 2084 m above sea level (Figure 1). The high altitude Western Ghats and the denuded Wayanad Plateau constitutes the physiography of the region. The plateau of Wayanadu is sloping towards the east and is bordered by isolated structural hills in the east. Most of the district is drained by the Kabani River and its tributaries. The river which flows to the east, along with its tributaries, contributed to the major carvings in the landscape of the catchment.

The topography of the region consists of features ranging from rugged high ranges to flood plains [42]. Hill ranges in the west, northwest and south-western parts of Wayanad can be classified as high ranges with rugged topography, occupied with dense forests with steep slopes. The eastern hills of the districts are high ranges with moderately rugged topography, with an elevation ranging from 1000 m to 1400 m. The valleys between high ranges are formed by the process of deposition and erosion. The flood plains of the region form productive aquifers, and the alluvial thickness of more than 10 m are typical in such plains. This topographical diversity increases the chances of landslides in the region.

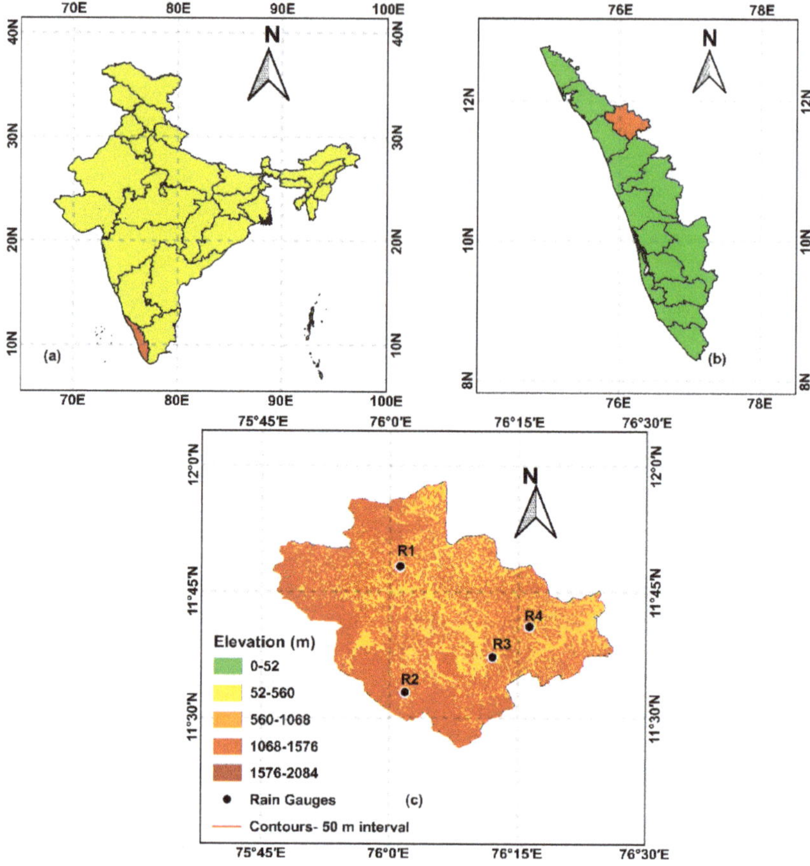

Figure 1. Location and digital elevation model of Wayanad District (**a**) India (**b**) Kerala (**c**) Wayanad (Modified after [43]).

Four major soil types are found common in Wayanad district [42]. Reddish-brown lateritic soil, formed due to the tropical climatic conditions, has its particle size ranging from clay to silt, with minor coarse fractions. Mananthavady, Kalpetta and Sulthan Bathery areas are rich in forest soil, formed by weathering under forest cover. Deep brownish hydromorphic soil is seen in the undulating topography in the district. This type of soil is formed by transportation and sedimentation of hill slope materials. Alluvial soils are found along the riverbanks, consisting of sandy and clayey fractions. During the monsoon, landslides are frequent in the region and they can be considered as the main geomorphological process, shaping the landscape.

2.1. Landslides in Wayanad

The types of landslides observed in the study region are mainly debris flows or slides of rapid to very rapid nature. A general agreement is that such landslides are triggered by high-intensity rainfalls. Hence, intensity and duration are the two parameters based on which rainfall thresholds are derived from in this study. A threshold line in the form of power-law is derived for the region using frequentist approach for the study area. Images of some of the landslide events that occurred during the 2018 monsoon in the study region are shown in Figure 2, pointing to the typology of landslides observed in the study area. In the debris flow that occurred in Pancharakkolli (Figure 2a), 10 acres of land comprising of 4 acres forest and 6 acres agricultural land was lost. A total of nine houses were damaged in the course, out of which five were completely destroyed. In Padinjarethara (Figure 2b), four debris flows were initiated from the forest area and caused a severe loss of agricultural land. A total of 10 acres of land was lost and 24 families were affected by the disaster. Several translational and rotational earth slides also occurred during the 2018 monsoon (Figure 2c). Such events have substantially decreased the stability of existing slopes and have affected the functionality of buildings and roads. The debris flow that occurred in Kurichermala (Figure 2d) was the largest in terms of the run-out, where around 150 acres of land (130 acres tea estate and 20 acres agricultural land) was washed out along with the debris. A total of 17 families lost their homes in the disaster [44]. The high elevation zones in the district is characterized by long and large volume debris flows (Figure 2a,b,d) owing to the high regolith thickness combined with slope steepness. In zones of low elevation, debris/earth slides are observed as riverbank failures and cut slope failures.

Figure 2. Images of landslides that occurred in Wayanad district during the 2018 monsoon.

Concerning landslide data, a variety of methods can be combined to obtain inventories as complete and detailed as possible, including direct surveys, newspapers, internet news and official reports from technical or administration offices [27,35,45]. In this study, the details of landslide events that occurred during the study period (2010–2018) were collected from reports of the Geological Survey of India, District Soil Conservation Office Wayanad and media reports. The spatial distribution of 123 landslide events identified are plotted in Figure 3, along with the drainage map of Wayanad. Most of the landslide locations are close to the streams in the region, at higher elevations. The landslides that occurred in the low-lying regions often took place under anthropogenic influences, along cuts and slopes or along a riverbank, triggered during rainfalls. The database consists of the location, date of initiation and typology of the landslide. The precision of the location and time reported is subject to the availability of data from the report. The dates of occurrences of landslides were collected from the government reports with daily accuracy. When the dates were not mentioned clearly, as in the case of media reports, the event was assumed to occur one day before the reporting date. Locations were deduced with a spatial accuracy of the nearest mentioned place in the reports, i.e., the names of villages in most cases. A general idea about the landslide typology was available from most of the reports, but the details of the mode of failure were not available. Another constraint with the available reports is their bias towards the fatalities. As the high altitude and unstable slopes are mostly within the forest or less occupied areas, any possible slope failure in such areas is not reported by the media or government reports. As the objective of the study is also to aid in LEWS to reduce the impact of disaster on the population, ignoring such events is acceptable.

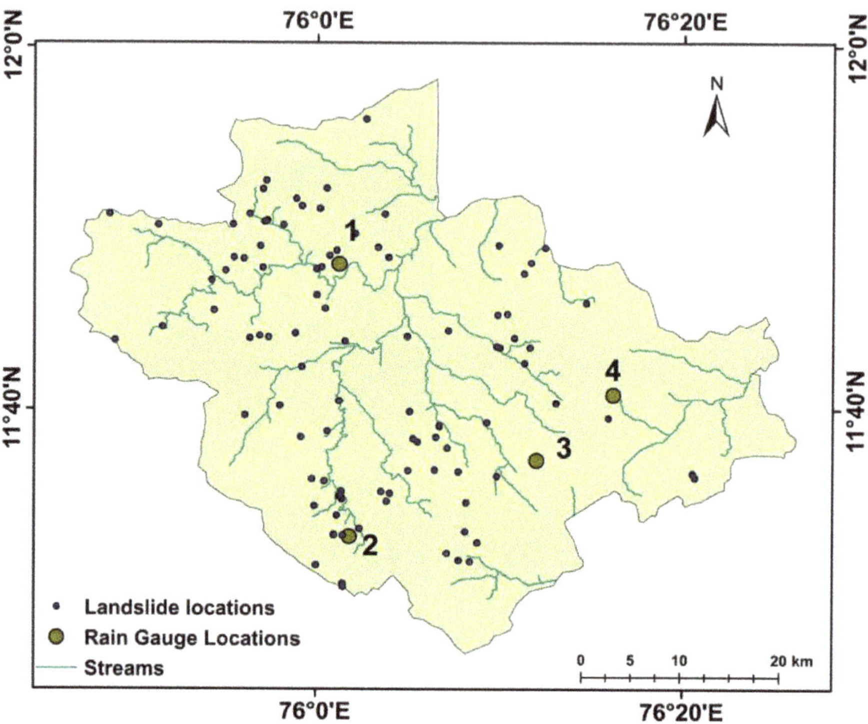

Figure 3. Drainage map of Wayanad overlaid by locations of landslides and rain gauges considered for analysis.

2.2. Rain Gauges and Rainfall in Wayanad

For threshold analyses, the starting point is the collection of rainfall and landslide data [27,35]. Concerning rainfall, despite some recent advances relying on radar measurements, rain gauge is by far the most commonly used method of measurement in threshold analysis [35]. Therefore, rainfall data with the best possible temporal accuracy (daily time steps) were collected from 4 rain gauge stations maintained in the area by the India Meteorological Department (IMD) [46]. The data from 2010 to 2018 are used to carry out the analysis for Wayanad district. The locations of the rain gauge stations are Mananthavady (R1), Vythiri (R2), Ambalavayal (R3) and Kuppadi (R4). The annual cumulative rainfall during the study period is shown in Figure 4. Maximum rainfall was observed in the year of 2018 with a cumulative rainfall of 3832 mm.

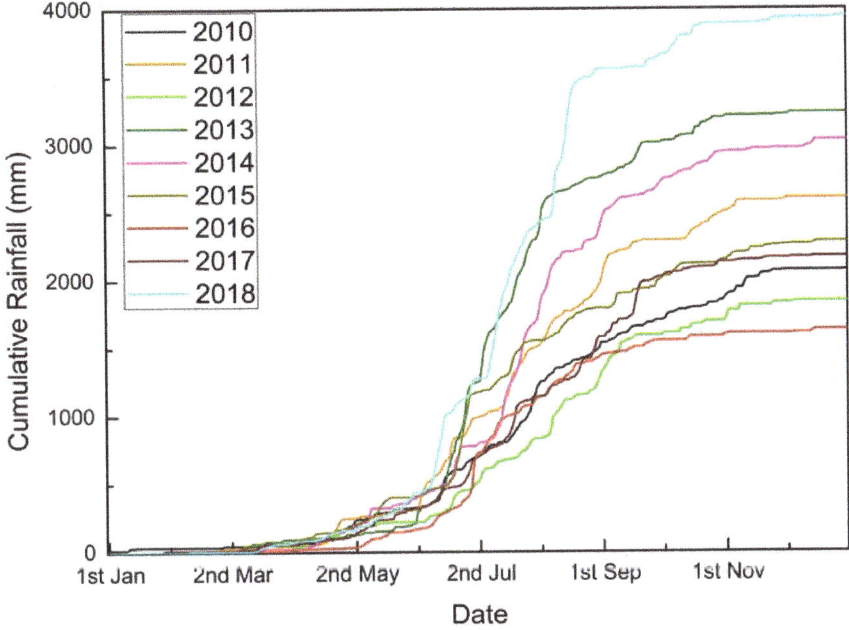

Figure 4. Annual cumulative rainfall during the study period—district average values.

The decision on choosing the rain gauge or source of rainfall data is an important step in threshold definition. The variation of annual rainfall in four rain gauge stations is depicted in Figure 5. During the study period, R3 and R4 recorded minimum annual rainfalls, and in most of the cases, the annual precipitations recorded at these two gauges were comparable. R1 received a higher amount of rainfall than R3 and R4, while R2 recorded the maximum rainfall in all the 9 years considered. The study can be refined with the availability of rainfall data of better spatial and temporal resolution. The variation in annual rainfall as observed in Figure 5 stresses the difference in meteorological conditions of the four rain gauge areas and the need for separate local scale thresholds.

2.3. Approaches Used to Configure the Threshold Analysis

In this study, different approaches were considered for deriving intensity–duration thresholds. The method of choosing rain gauges based on the nearest rain gauge is still one of the simplest and, although criticized, widely used methods [35]. To correlate each landslide to the nearest rain gauge, the district was divided into four zones, based on the spatial distribution of rain gauges, by means of the Thiessen polygons technique, as shown in Figure 6. The method is based on a proximal mapping in which the estimate of rainfall at any point is considered equal to the observation of the nearest sampling point in the area [47]. Based on the location of the landslides, the triggering rainfall was identified using the data from the rain gauge stations, using different approaches. Since the temporal resolution of available rainfall data is one day, multiple landslides within the same polygon that occurred on the same day were counted as a single landslide event for the definition of intensity–duration threshold. The first approach was based on a regional scale, where the nearest rain gauge was chosen as the reference gauge and data from all four rain gauges are merged to establish one single regional scale threshold (merged threshold).

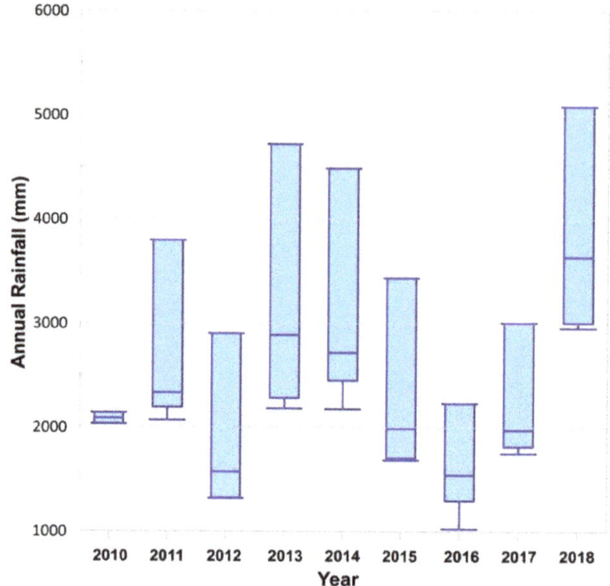

Figure 5. Box and Whisker plot showing the variation of annual rainfall across the four rain gauge stations.

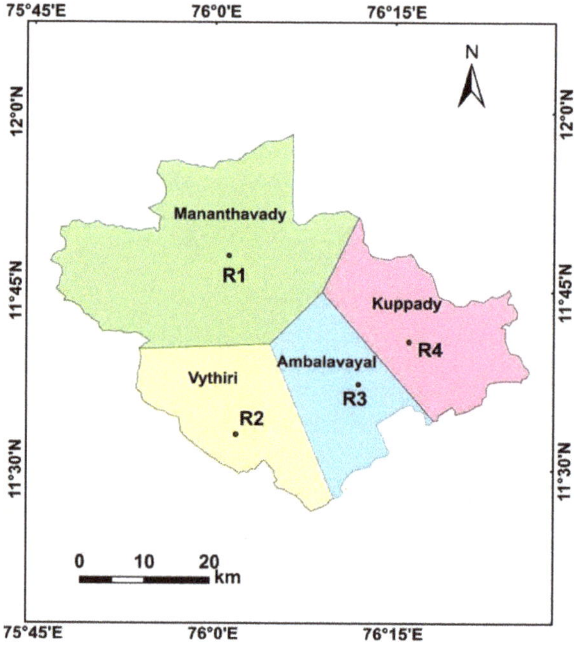

Figure 6. Location of rain gauges and definition of their area of influence based on nearest rain gauge.

Considering the vastness of area and geographical settings, the measurement from the nearest available rain gauge could underestimate the threshold as the localized convective storms might not be recorded at the rain gauge location [35]. Hence, another approach was adopted in which the rain gauge which records the maximum average intensity was chosen as the reference gauge, irrespective of the spatial distribution (Imax threshold).

In the third approach, the peak daily intensity observed from the beginning of landslide event was considered for the analysis instead of the average intensity (Peak I threshold). The reason for a deviation from the conventional average intensity approach is the possible avoidance of the underestimation of thresholds due to the low density of rain gauges.

Lastly, analyses were conducted separately on a local scale, so that four thresholds were derived separately for each polygon (R1, R2, R3 and R4 thresholds).

Separate intensity–duration thresholds were defined for each approach with different exceedance probabilities of 5%, 2.5%, 1% and 0.05% to find the best suited method.

3. Results

3.1. Statistical Analysis

According to Caine [24], intensity–duration thresholds are defined as a power-law in the following form:

$$I = \propto D^{\beta} \tag{1}$$

where I is the mean intensity, expressed in mm h^{-1}, D is the duration in hours and α and β are empirically derived parameters. α is the scaling constant, which defines the intercept, and β defines the slope of the power-law curve. In this study, two rainfall events are considered separate if there is no precipitation for a minimum period of 24 hours (one day) in between, and the frequentist method is adopted to establish a regional scale threshold for Wayanad district.

This approach uses the least square method to find the best fit line [34]. Taking into account the variation of intensity values, the values were first log-transformed to avoid problems in fitting the data. In a log vs. log plot, the data is fitted using a straight line with the equation

$$\text{Log}\, I = \text{Log}\propto + \beta \text{Log} D \tag{2}$$

which is equivalent to the power-law in Equation (1). From Equation (2), the values of α and β can be calculated, as Logα is the intercept and β is the slope of the straight line.

The difference in y coordinates of each event with the best fit line is then calculated and termed as δI, which is obtained by the following equation:

$$\delta I = \text{Log}[I_f D] - \text{Log}[I(D)] \tag{3}$$

where Log $I_f(D)$ is the y co-ordinate on best-fit line, and Log I(D) is the mean intensity associated with each event. The distribution of δI is then fitted using a kernel density function of the form.

$$\delta I = \text{Log}[I_f(D)] - \text{Log}[I(D)] \tag{4}$$

The data are found to follow a distribution similar to the standard Gaussian distribution. The Gaussian fit of the probability density function is shown in Figure 7 as the dotted line. The threshold lines of different exceedance probabilities were calculated using the fitted distribution of δI. The distance Δ between the best-fit line and the T line is used to calculate the intercept of threshold line in the log vs. log plot. An exceedance probability of 5% indicates that the probability of occurrence of landslides below this threshold is less than 5% [34]. The threshold of 5% exceedance is plotted according to the shift Δ.

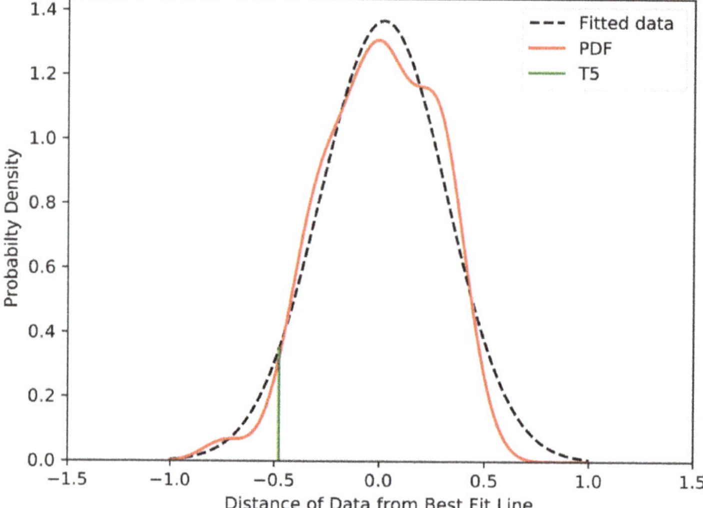

Figure 7. The probability density function of distribution of δI, fitted using a kernel density function—example using merged dataset.

3.2. Empirical Thresholds

Among the 123 landslide events recorded in the study period, 51 events occurred within the first polygon Mananthavady, 36 events in the second polygon Vythiri, 15 events in the third polygon Ambalayavayal and 21 events within the last polygon, Kuppadi. As described earlier, four different approaches were used to define thresholds using the Frequentist method for the study area.

Using the frequentist approach, thresholds of several exceedance probabilities are defined for the study area (Figure 8a). Out of the 123 landslide events considered for the analysis, 5% of events are expected to fall below the T5 line, 2.5% below the T2.5 line and 1% events are below T1 line. No events are expected to fall below the threshold line of 0.05% exceedance probability for such a small dataset, which makes the line well below the possible critical conditions. All defined threshold lines are observed to follow this pattern.

For Wayanad district, new rainfall thresholds were defined for possible landslide initiation, based on the frequentist approach. All three regional scale thresholds are following the pattern as depicted in Figure 8a, in terms of percentage events below each threshold line. The merged thresholds have the highest slope of −0.24 but lesser intercept values than Imax thresholds. For Imax thresholds, the rainfall event with maximum average intensity among the four rain gauge stations was considered for analysis. In most of the cases, it was observed that the nearest rain gauge recorded the maximum average precipitation, and, in some cases, other rain gauges were chosen for analysis. Thus, the intercept of the threshold is slightly higher than that of merged data with a lesser slope. The peak I threshold follows a pattern which is different from the power-law form associated to intensity–duration thresholds as described in the work of Caine [24]: the resulting slope of the threshold is a positive value. The reason for this could be that in the meteo-climatic setting of the study area, the total duration (D) of the main event and the peak intensity registered in one of the sub-events are completely independent and the relationship among them leads to an equation form that does not follow the power-law form discovered by Caine. On the contrary, the longer the main rainfall event, the higher the possibility of more intense bursts of rain, hence the positive exponent of the power-law function reported in Figure 8d.

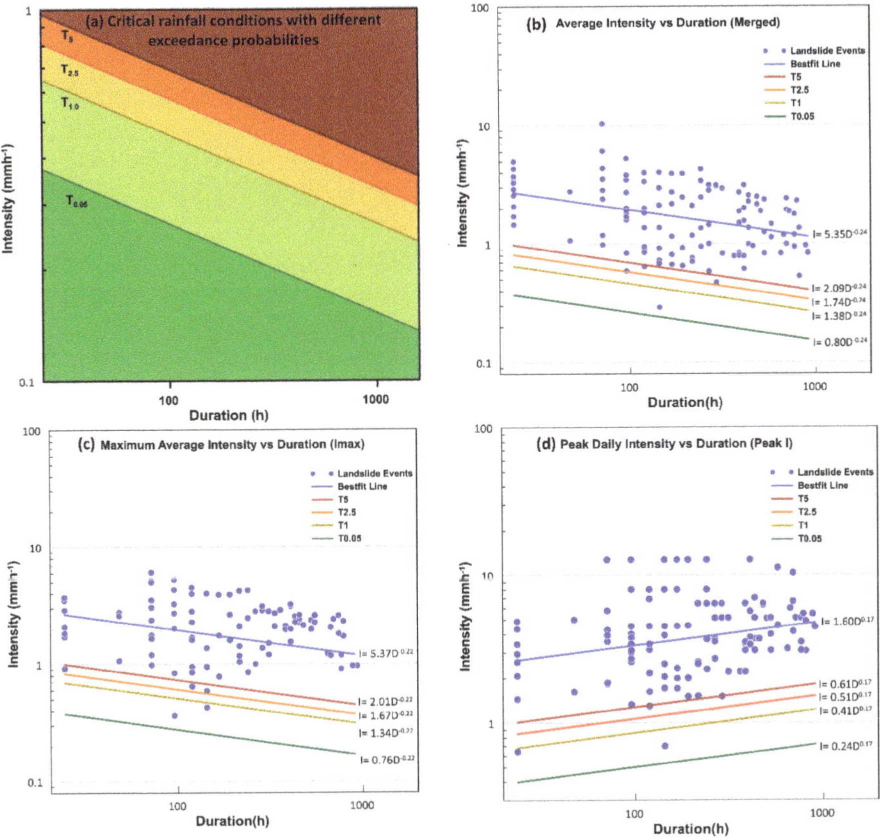

Figure 8. Regional scale intensity–duration thresholds established for Wayanad with different exceedance probabilities. (**a**) Critical rainfall conditions with different exceedance probabilities, (**b**) Average intensity vs duration (Merged), (**c**) Maximum average intensity vs duration (Imax), (**d**) Peak daily intensity vs duration (Peak I).

From Figure 9, it can be inferred that the rainfall conditions that triggered landslides in the four separate polygons are slightly different from each other. Polygon 1 (Manathavady) covers the maximum area and most of the landslide incidences are found to be located within the boundary. The rainfall parameters that triggered landslides in Polygon 1 and Polygon 2 are characterized by relatively higher intensity, and hence the slope of threshold curves is less than that of the merged data. These regions are affected by large flows as the high-altitude regions in the district falls within these polygons. In Polygon 3 (Ambalavayal), the number of events is the least and the observed events were the results of relatively higher intensity rainfalls. In Polygon 3 and Polygon 4 (Kuppadi), most of the incidences recorded were earth slides and cut slope failures. These polygons are at lesser elevations, with moderate to low dissected plateau geomorphological conditions, and the slope failures are induced by anthropogenic activities in the pursuit of infrastructure development. Since the number of events considered in each polygon is lower, the percentage distribution of landslides below each threshold line is slightly different from that shown in Figure 8a. For a better comparison of the threshold pattern, all thresholds were plotted on the same graph for all the four exceedance probabilities as shown in Figure 10. This helps for an easy comparison of the defined thresholds at each level of exceedance.

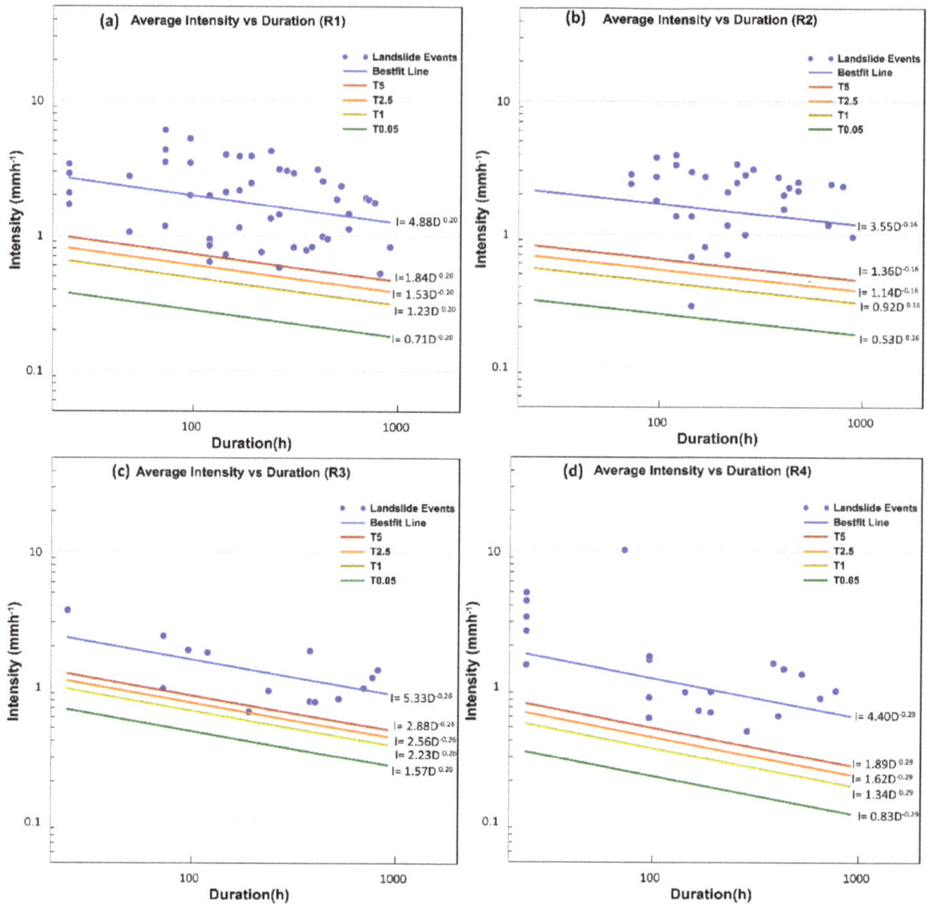

Figure 9. Local scale intensity–duration thresholds established for Wayanad with different exceedance probabilities. (**a**) Average intensity vs duration (R1), (**b**) Average Intensity vs Duration (R2), (**c**) Average Intensity vs Duration (R3), (**d**) Average Intensity vs Duration (R4).

From Figure 10, it can be observed that for all exceedance levels, the relative positions of all the defined thresholds follow a similar pattern. Peak I and R3 thresholds are much higher than all the other thresholds. At lesser durations, Peak I thresholds are observed to be lower than R3 and the reverse is observed during higher durations. R4 thresholds are the lowest in all cases, as the region is characterized by less intensity rainfalls. R2 thresholds are conservative at lesser durations, but as the duration increases, the threshold curve crossed merged, Imax and R1 thresholds. R1 thresholds are observed to be in close similarity with the Imax values as at some exceedance probabilities, R1 is higher than both Imax and merged thresholds and generally at higher durations, the threshold becomes more conservative. The Imax thresholds are always higher than that of the merged thresholds with similar values at low durations. The shift between the two threshold lines increases as the duration increases.

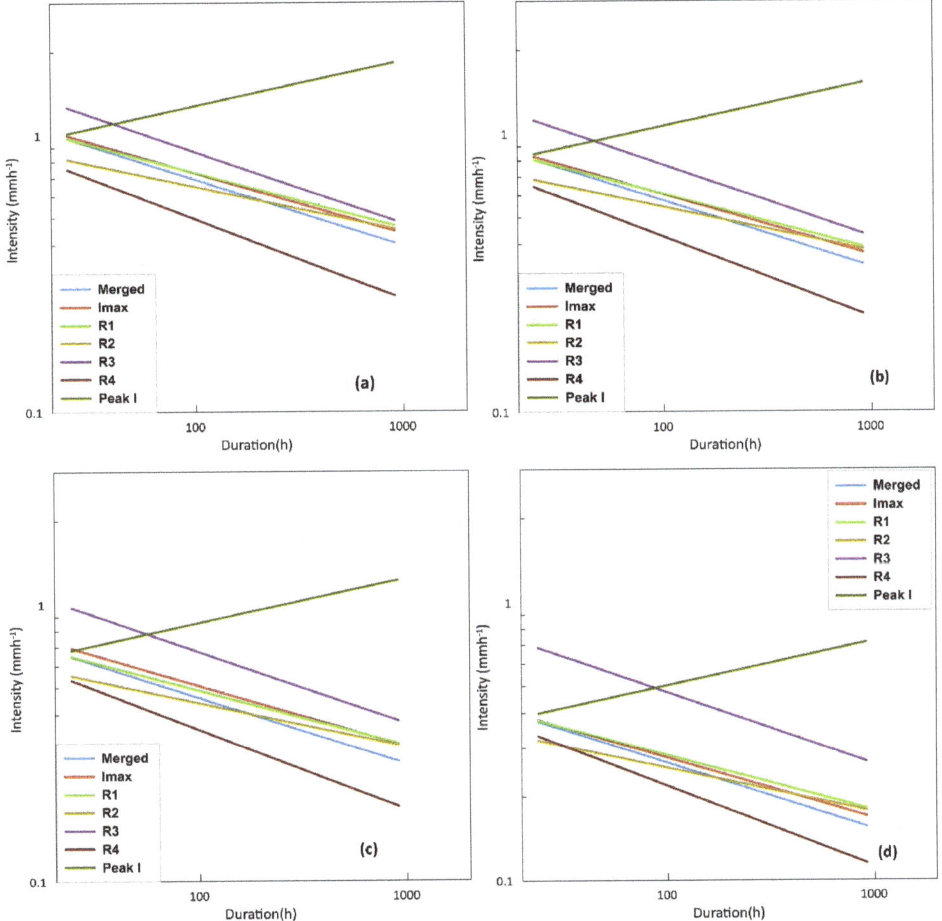

Figure 10. Comparison of different thresholds at (**a**) 5% exceedance probability (**b**) 2.5% exceedance probability (**c**) 1% exceedance probability (**d**) 0.05% exceedance probability.

4. Discussions

Choosing the best suited method from the obtained observations demands a detailed analysis of the effect of thresholds on the study area. Separate analyses were conducted for occurrences of landslides in each polygon using statistics. As pointed out in a recent study, statistical attributes are reliable parameters that can be used to compare different methodologies for the definition of threshold [48]. When the available information about the distribution of rainfall is coarse, the possibilities for underestimating the threshold values are higher. From an operational point of view, this could possibly lead to a number of false alarms. Hence, it is important to complete quantitative analysis using statistical attributes. The attributes are calculated using a confusion matrix, comparing the prediction of each defined threshold and the occurrence of landslides. Everyday prediction of thresholds during the study period (2010–2018) was used for the verification of the thresholds. True positives are counted when the threshold is crossed, and a landslide is reported on the day. If no landslides are reported when a threshold is crossed, it is counted as a false positive. Similarly, if landslides are reported without the crossing of threshold, it is considered as a missed alarm and counted as a false negative. When thresholds are not exceeded and landslides are not reported, true negatives are counted.

A perfect model would predict all the landslides correctly without any false positives or false negatives. The performances of our tests are far from perfection, but the main objective of this work is to discuss the results in a relative way to compare the different approaches used for threshold analysis. The higher the number of true positives and true negatives, the better the model. Here, we use derived parameters like efficiency, sensitivity, specificity and the likelihood ratio for a better understanding of the relative performance of different thresholds. The aforementioned statistics can help optimize the threshold model configuration, identifying a balance between false and missed alarms prediction. The results are summarized in Table 1 below.

The higher number of false positives in all the cases points towards less positive prediction power of the model. The very high number of true negatives in comparison with the order of other parameters increases the efficiency of the model. It can be stated that the thresholds are conservative in nature with much lower false negatives, and the Negative Predictive Power is very close to one in all the cases. As expected, with the decrease in exceedance probability, the number of false positives is increasing, which reduces the efficiency considerably. If the exceedance probability increases by more than 5, the number of missed alarms will increase beyond 5%, which is also not acceptable. Hence, when defining a threshold, a 5% exceedance probability can be considered.

A perfect prediction model should have the sensitivity and specificity values as one. Sensitivity is a key towards the true positive rate of the model, and specificity is an indication towards the true negative rate. In this study, T0.05 thresholds have sensitivity values as one, but this happens at the cost of very low specificity values, which is not acceptable. The likelihood ratio can be considered as the term which considers the effect of both sensitivity at the same time and can be taken as a reliable parameter for comparison of different methods [49]. It can be understood from the analysis that Imax thresholds have the maximum likelihood ratio in three different exceedance probabilities. At the same time, in each polygon, the separate thresholds derived perform well. Hence, this study proposes a regional scale threshold of 5% exceedance probability using the Imax approach and four separate thresholds for each polygon operating on a local scale.

If polygons defined thresholds as lower than the merged dataset (R2 and R4), single regional scale thresholds perform better than the separate polygon-wise threshold due to a lower number of false alarms. In the other two polygons, polygon-wise thresholds (R1 and R3) can be opted over the regional scale thresholds. Separate local scale thresholds have the advantage of more uniform climatic and geological conditions, but the lower number of events used for calibration especially in R3 and R4 is the major constraint in the definition. However, while creating a single dataset for the whole region, the merged approach of considering the nearest rain gauge is less likely to be adopted than the Imax approach. In case of peak I approach, the occurrence of high intensity rainfalls in the beginning of rainfall event will produce a false alarm which will be sustained throughout the event, predicting the possibility of a landslide. Even though the defined thresholds appear to be higher than all the other approaches, this method is not found to be effective in reducing false alarms.

These results are useful to understand the sensitivity of I–D threshold models to some boundary conditions such as the rain gauge selection, the intensity definition and the strategy of subdividing the area into independent alert zones. Unfortunately, the derived thresholds are not ready to be operated into a LEWS, but still the results highlighted the shortcomings that could be addressed with future improvements. For instance, it would be very useful to use rainfall with higher temporal resolutions (e.g., hourly) and to take into account the effect of antecedent rainfall conditions during the monsoon season by using some state-of-the-art approaches like weighted antecedent precipitation indexes [49–51] or soil moisture estimates [21].

Table 1. Statistical Comparison of the derived thresholds. (The maximum likelihood ratio values are highlighted in bold.).

Statistical Attributes	T5				T2.5				T1				T0.05			
	R	Merged	Imax	Peak I	R	Merged	Imax	Peak I	R	Merged	Imax	Peak I	R	Merged	Imax	Peak I
Ture Positives (a)	119	119	118	118	122	121	120	121	122	122	122	121	123	123	123	123
False Positives (b)	2130	2080	1967	2114	2439	2388	2301	2394	2721	2772	2705	2750	3515	3727	3620	3562
False Negatives (c)	4	4	5	5	1	2	3	2	1	1	1	2	0	0	0	0
True Negatives (d)	10895	10945	11058	10911	10586	10637	10724	10634	10304	10253	10320	10275	9510	9298	9405	9463
Efficiency = (a + d)/(a + b + c + d)	0.84	0.84	0.85	0.84	0.81	0.82	0.82	0.82	0.79	0.79	0.79	0.79	0.73	0.72	0.72	0.73
Misclassification rate = (b + c)/(a + b + c + d)	0.16	0.16	0.15	0.16	0.19	0.18	0.18	0.18	0.21	0.21	0.21	0.21	0.27	0.28	0.28	0.27
Odds ratio = (a + d)/(b + c)	5.16	5.31	5.67	5.20	4.39	4.50	4.71	4.49	3.83	3.74	3.86	3.78	2.74	2.53	2.63	2.69
Positive predictive power = a/(a + b)	0.05	0.05	0.06	0.05	0.05	0.05	0.05	0.05	0.04	0.04	0.04	0.04	0.03	0.03	0.03	0.03
Negative predictive power = d/(c + d)	1.00	1.00	1.00	1.00	1.00	1.00	1.00	1.00	1.00	1.00	1.00	1.00	1.00	1.00	1.00	1.00
Sensitivity = a/(a + c)	0.97	0.97	0.96	0.96	0.99	0.98	0.98	0.98	0.99	0.99	0.99	0.98	1.00	1.00	1.00	1.00
Specificity = d/(b + d)	0.84	0.84	0.85	0.84	0.81	0.82	0.82	0.82	0.79	0.79	0.79	0.79	0.73	0.71	0.72	0.73
False positive rate = b/(b + d)	0.16	0.16	0.15	0.16	0.19	0.18	0.18	0.18	0.21	0.21	0.21	0.21	0.27	0.29	0.28	0.27
False negative rate = c/(a + c)	0.03	0.03	0.04	0.04	0.01	0.02	0.02	0.02	0.01	0.01	0.01	0.02	0.00	0.00	0.00	0.00
Likelihood ratio = Sensitivity/(1 − Specificity)	5.92	6.06	**6.35**	5.91	5.30	5.37	**5.52**	5.35	4.75	4.66	**4.78**	4.66	**3.71**	3.49	3.60	3.66

5. Conclusions

A new landslide catalogue was prepared for Wayanad district, Kerala, India, compiling information from different data sources. The catalogue consists of landslide events that happened from 2010 to 2018 in the district. With the available data, rainfall events associated with each landslide was identified using the data from four rain gauges located at different places in the district.

The catalogue was used to determine intensity–duration thresholds on regional and local scales, which is the first attempt of its kind for the study area. Four different approaches were adopted in the study to develop intensity–duration thresholds by varying the selection of rain gauge, area considered and definition of rainfall parameters. After the analysis, it can be concluded that on a regional scale, selecting the rain gauge based on maximum average intensity (Imax) performs better than choosing the nearest rain gauge. Four separate thresholds for each polygon considered are also proposed in this study.

On a regional scale, with 5% exceedance probability, a rainfall of intensity 1mm h^{-1} of a one-day duration is potent enough to trigger landslides in Wayanad district. It is also observed that Mananthavady and Vythiri polygons are more susceptible to landslides than the other two regions. The intensity of rainfall of a one-day duration which can possibly trigger a landslide in the Mananthavady, Vythiri, Ambalavayal and Kuppady polygons are 0.97 mm h^{-1}, 0.82 mm h^{-1}, 1.26 mm h^{-1} and 0.75 mm h^{-1} respectively. The Ambalavayal polygon can be considered as a relatively less vulnerable region with a lower number of landslide events and higher threshold values.

The study emphasizes the importance of the preparation of landslide catalogues and determination of rainfall thresholds for Wayanad region. An effective LEWS is an immediate requirement in the region, and the study has to be further enhanced with state-of-the-art models developed for other parts in the world. The existing model can also be conceptually modified using precise field monitoring techniques as well. Attempts must be made to reduce the false alarms to develop an operational rainfall threshold model to function as a Landslide Early Warning System for the region.

Author Contributions: Conceptualization, M.T.A., S.S. and N.S.; methodology, M.T.A. and S.S.; data curation, M.T.A.; writing—original draft preparation, M.T.A. and A.R.; writing—review and editing, S.S., B.P. and N.S.; supervision, N.S. and B.P; funding acquisition, S.S. All authors have read and agreed to the published version of the manuscript.

Funding: Florence University, in the framework of the project SAMUELESEGONIRICATEN20.

Acknowledgments: The authors express their sincere gratitude to Geological Survey of India, Kerala SU, District Soil Conservation Office Wayanad and Kerala State Disaster Management Authority (KSDMA) for their support throughout the study.

Conflicts of Interest: The authors declare no conflict of interest.

References

1. Griffiths, J.S.; Whitworth, M. Engineering geomorphology of landslides. In *Landslides: Types, Mechanisms and Modeling*; Clague, J.J., Stead, D., Eds.; Cambridge University Press: Cambridge, UK, 2015; pp. 172–186. ISBN 9780511740367.
2. Petley, D. Global patterns of loss of life from landslides. *Geology* **2012**, *40*, 927–930. [CrossRef]
3. Soja, R.; Starkel, L. Extreme rainfalls in Eastern Himalaya and southern slope of Meghalaya Plateau and their geomorphologic impacts. *Geomorphology* **2007**, *84*, 170–180. [CrossRef]
4. Kuriakose, S.L.; Jetten, V.G.; van Westen, C.J.; Sankar, G.; van Beek, L.P.H. Pore water pressure as a trigger of shallow landslides in the Western Ghats of Kerala, India: Some preliminary observations from an experimental catchment. *Phys. Geogr.* **2008**, *29*, 374–386. [CrossRef]
5. Abraham, M.T.; Pothuraju, D.; Satyam, N. Rainfall Thresholds for Prediction of Landslides in Idukki, India: An Empirical Approach. *Water* **2019**, *11*, 2113. [CrossRef]
6. Dikshit, A.; Satyam, D.N. Estimation of rainfall thresholds for landslide occurrences in Kalimpong, India. *Innov. Infrastruct. Solut.* **2018**, *3*, 24. [CrossRef]
7. Gariano, S.L.; Guzzetti, F. Landslides in a changing climate. *Earth Sci. Rev.* **2016**, *162*, 227–252. [CrossRef]

8. Martelloni, G.; Segoni, S.; Fanti, R.; Catani, F. Rainfall thresholds for the forecasting of landslide occurrence at regional scale. *Landslides* **2012**, *9*, 485–495. [CrossRef]
9. Teja, T.S.; Dikshit, A.; Satyam, N. Determination of Rainfall Thresholds for Landslide Prediction Using an Algorithm-Based Approach: Case Study in the Darjeeling Himalayas, India. *Geosciences* **2019**, *9*, 302. [CrossRef]
10. Dikshit, A.; Satyam, N.; Pradhan, B. Estimation of Rainfall—Induced Landslides Using the TRIGRS Model. *Earth Syst. Environ.* **2019**, *3*, 575–584. [CrossRef]
11. Dikshit, A.; Satyam, N. Probabilistic rainfall thresholds in Chibo, India: Estimation and validation using monitoring system. *J. Mt. Sci.* **2019**, *16*, 870–883. [CrossRef]
12. Abraham, M.T.; Satyam, N.; Pradhan, B.; Alamri, A.M. Forecasting of Landslides Using Rainfall Severity and Soil Wetness: A Probabilistic Approach for Darjeeling Himalayas. *Water* **2020**, *12*, 804. [CrossRef]
13. Berti, M.; Martina, M.L.V.; Franceschini, S.; Pignone, S.; Simoni, A.; Pizziolo, M. Probabilistic rainfall thresholds for landslide occurrence using a Bayesian approach. *J. Geophys. Res. Earth Surf.* **2012**, *117*. [CrossRef]
14. Capparelli, G.; Calabria, U.; Tiranti, D.; Calabria, U. Forecasting of landslides induced by rainfall—F. La. IR hydrological model application on Piemonte Region (NW Italy). *Geophys. Res. Abstr.* **2007**, *9*, 02298.
15. Zhou, W.; Tang, C. Rainfall thresholds for debris flow initiation in the Wenchuan earthquake-stricken area, southwestern China. *Landslides* **2014**, *11*, 877–887. [CrossRef]
16. Capparelli, G.; Tiranti, D. Application of the MoniFLaIR early warning system for rainfall-induced landslides in Piedmont region (Italy). *Landslides* **2010**, *7*, 401–410. [CrossRef]
17. Capparelli, G.; Versace, P. FLaIR and SUSHI: Two mathematical models for early warning of landslides induced by rainfall. *Landslides* **2011**, *8*, 67–79. [CrossRef]
18. Lagomarsino, D.; Segoni, S.; Fanti, R.; Catani, F. Updating and tuning a regional-scale landslide early warning system. *Landslides* **2013**, *10*, 91–97. [CrossRef]
19. Rosi, A.; Canavesi, V.; Segoni, S.; Dias Nery, T.; Catani, F.; Casagli, N. Landslides in the Mountain Region of Rio de Janeiro: A Proposal for the Semi-Automated Definition of Multiple Rainfall Thresholds. *Geosciences* **2019**, *9*, 203. [CrossRef]
20. Rosi, A.; Peternel, T.; Jemec-Auflič, M.; Komac, M.; Segoni, S.; Casagli, N. Rainfall thresholds for rainfall-induced landslides in Slovenia. *Landslides* **2016**, *13*, 1571–1577. [CrossRef]
21. Segoni, S.; Rosi, A.; Lagomarsino, D.; Fanti, R.; Casagli, N. Brief communication: Using averaged soil moisture estimates to improve the performances of a regional-scale landslide early warning system. *Nat. Hazards Earth Syst. Sci.* **2018**, *18*, 807–812. [CrossRef]
22. Segoni, S.; Rosi, A.; Fanti, R.; Gallucci, A.; Monni, A.; Casagli, N. A regional-scale landslide warning system based on 20 years of operational experience. *Water* **2018**, *10*, 1297. [CrossRef]
23. Segoni, S.; Rossi, G.; Rosi, A.; Catani, F. Landslides triggered by rainfall: A semi-automated procedure to define consistent intensity-duration thresholds. *Comput. Geosci.* **2014**, *63*, 123–131. [CrossRef]
24. Caine, N. The rainfall intensity-duration control of shallow landslides and debris flows: An update. *Geogr. Ann. Ser. A Phys. Geogr.* **1980**, *62*, 23–27.
25. Aleotti, P. A warning system for rainfall-induced shallow failures. *Eng. Geol.* **2004**, *73*, 247–265. [CrossRef]
26. Crosta, G.B.; Frattini, P. Rainfall thresholds for soil slip and debris flow triggering. In Proceedings of the EGS 2nd Plinius Conference on Mediterranean Storms, Siena, Italy, 16–18 October 2001; pp. 463–487.
27. Guzzetti, F.; Peruccacci, S.; Rossi, M.; Stark, C.P. Rainfall thresholds for the initiation of landslides in central and southern Europe. *Meteorol. Atmos. Phys.* **2007**, *98*, 239–267. [CrossRef]
28. Guzzetti, F.; Peruccacci, S.; Rossi, M.; Stark, C.P. The rainfall intensity-duration control of shallow landslides and debris flows: An update. *Landslides* **2008**, *5*, 3–17. [CrossRef]
29. Ma, T.; Li, C.; Lu, Z.; Bao, Q. Rainfall intensity-duration thresholds for the initiation of landslides in Zhejiang Province, China. *Geomorphology* **2015**, *245*, 193–206. [CrossRef]
30. Guo, X.; Cui, P.; Li, Y.; Ma, L.; Ge, Y.; Mahoney, W.B. Intensity-duration threshold of rainfall-triggered debris flows in the Wenchuan Earthquake affected area, China. *Geomorphology* **2016**, *253*, 208–216. [CrossRef]
31. Lazzari, M.; Piccarreta, M. Landslide disasters triggered by extreme rainfall events: The case of montescaglioso (Basilicata, Southern Italy). *Geosciences* **2018**, *8*, 377. [CrossRef]

32. Marc, O.; Stumpf, A.; Malet, J.P.; Gosset, M.; Uchida, T.; Chiang, S.H. Initial insights from a global database of rainfall-induced landslide inventories: The weak influence of slope and strong influence of total storm rainfall. *Earth Surf. Dyn.* **2018**, *6*, 903–922. [CrossRef]
33. Segoni, S.; Rosi, A.; Rossi, G.; Catani, F.; Casagli, N. Analysing the relationship between rainfalls and landslides to define a mosaic of triggering thresholds for regional scale warning systems. *Nat. Hazards Earth Syst. Sci.* **2014**, *2*, 2185–2213. [CrossRef]
34. Brunetti, M.T.; Peruccacci, S.; Rossi, M.; Luciani, S.; Valigi, D.; Guzzetti, F. Rainfall thresholds for the possible occurrence of landslides in Italy. *Nat. Hazards Earth Syst. Sci.* **2010**, *10*, 447–458. [CrossRef]
35. Segoni, S.; Piciullo, L.; Gariano, S.L. A review of the recent literature on rainfall thresholds for landslide occurrence. *Landslides* **2018**, *15*, 1483–1501. [CrossRef]
36. Palenzuela, J.A.; Jiménez-Perálvarez, J.D.; Chacón, J.; Irigaray, C. Assessing critical rainfall thresholds for landslide triggering by generating additional information from a reduced database: An approach with examples from the Betic Cordillera (Spain). *Nat. Hazards* **2016**, *84*, 185–212. [CrossRef]
37. Winter, M.G.; Dent, J.; Macgregor, F.; Dempsey, P.; Motion, A.; Shackman, L. Debris flow, rainfall and climate change in Scotland. *Q. J. Eng. Geol. Hydrogeol.* **2010**, *43*, 429–446. [CrossRef]
38. Marra, F.; Destro, E.; Nikolopoulos, E.I.; Zoccatelli, D.; Dominique Creutin, J.; Guzzetti, F.; Borga, M. Impact of rainfall spatial aggregation on the identification of debris flow occurrence thresholds. *Hydrol. Earth Syst. Sci.* **2017**, *21*, 4525–4532. [CrossRef]
39. Kuriakose, S.L. Physically-based dynamic modelling of the effect of land use changes on shallow landslide initiation in the Western Ghats of Kerala, India. Ph.D. Thesis, University of Twente, Enschede, The Netherlands, 2010.
40. Iida, T. A stochastic hydro-geomorphological model for shallow landsliding due to rainstorm. *Catena* **1999**, *34*, 293–313. [CrossRef]
41. United Nations Development Programme. *Kerala Post Disaster Needs Assessment Floods and Landslides-August 2018*; United Nations Development Programme: Thiruvananthapuram, India, 2018.
42. Department of Mining and Geology. *District Survey Report of Minor Minerals*; Department of Mining and Geology: Thiruvananthapuram, India, 2016.
43. CartoDEM. Available online: https://bhuvan-app3.nrsc.gov.in/data/download/index.php (accessed on 20 August 2019).
44. District Soil Conservation Office. *Details of Landslip Damages in Agricultural Lands of Different Panchayats of Idukki District during the Monsoon 2018*; District Soil Conservation Office: Wayanad, India, 2018.
45. Battistini, A.; Segoni, S.; Manzo, G.; Catani, F.; Casagli, N. Web data mining for automatic inventory of geohazards at national scale. *Appl. Geogr.* **2013**, *43*, 147–158. [CrossRef]
46. India Meteorological Department (IMD). Available online: http://dsp.imdpune.gov.in/ (accessed on 3 May 2019).
47. Tabios, G.Q.; Salas, J.D. A Comparative Analysis of Techniques for Spatial Interpolation of Precipitation. *JAWRA J. Am. Water Resour. Assoc.* **1985**, *21*, 365–380. [CrossRef]
48. Lagomarsino, D.; Segoni, S.; Rosi, A.; Rossi, G.; Battistini, A.; Catani, F.; Casagli, N. Quantitative comparison between two different methodologies to define rainfall thresholds for landslide forecasting. *Nat. Hazards Earth Syst. Sci.* **2015**, *15*, 2413–2423. [CrossRef]
49. Ponziani, F.; Pandolfo, C.; Stelluti, M.; Berni, N.; Brocca, L.; Moramarco, T. Assessment of rainfall thresholds and soil moisture modeling for operational hydrogeological risk prevention in the Umbria region (central Italy). *Landslides* **2012**, *9*, 229–237. [CrossRef]
50. Glade, T.; Crozier, M.; Smith, P. Applying probability determination to refine landslide-triggering rainfall thresholds using an empirical Antecedent Daily Rainfall Model. *Pure Appl. Geophys.* **2000**, *157*, 1059–1079. [CrossRef]
51. Chen, G.; Zhao, J.; Yuan, L.; Ke, Z.; Gu, M.; Wang, T. Implementation of a Geological Disaster Monitoring and Early Warning System Based on Multi-source Spatial Data: A Case Study of Deqin County, Yunnan Province. *Nat. Hazards Earth Syst. Sci.* **2017**, 1–15. [CrossRef]

Article

Rainfall Threshold Estimation and Landslide Forecasting for Kalimpong, India Using SIGMA Model

Minu Treesa Abraham [1,*], Neelima Satyam [1], Sai Kushal [1], Ascanio Rosi [2], Biswajeet Pradhan [1,3,4] and Samuele Segoni [2]

1 Discipline of Civil Engineering, Indian Institute of Technology Indore, Madhya Pradesh 453552, India; neelima.satyam@iiti.ac.in (N.S.); ce160004023@iiti.ac.in (S.K.); Biswajeet.pradhan@uts.edu.au (B.P.)
2 Department of Earth Sciences, University of Florence, Via Giorgio La Pira, 4, 50121 Florence, Italy; ascanio.rosi@unifi.it (A.R.); samuele.segoni@unifi.it (S.S.)
3 Centre for Advanced Modelling and Geospatial Information Systems (CAMGIS), Faculty of Engineering and Information Technology, University of Technology Sydney, Sydney P.O. Box 123, Australia
4 Department of Energy and Mineral Resources Engineering, Sejong University, Choongmu-gwan, 209 Neungdong-ro, Gwangjin-gu, Seoul 05006, Korea
* Correspondence: minu.abraham@iiti.ac.in

Received: 1 April 2020; Accepted: 19 April 2020; Published: 22 April 2020

Abstract: Rainfall-induced landslides are among the most devastating natural disasters in hilly terrains and the reduction of the related risk has become paramount for public authorities. Between the several possible approaches, one of the most used is the development of early warning systems, so as the population can be rapidly warned, and the loss related to landslide can be reduced. Early warning systems which can forecast such disasters must hence be developed for zones which are susceptible to landslides, and have to be based on reliable scientific bases such as the SIGMA (sistema integrato gestione monitoraggio allerta—integrated system for management, monitoring and alerting) model, which is used in the regional landslide warning system developed for Emilia Romagna in Italy. The model uses statistical distribution of cumulative rainfall values as input and rainfall thresholds are defined as multiples of standard deviation. In this paper, the SIGMA model has been applied to the Kalimpong town in the Darjeeling Himalayas, which is among the regions most affected by landslides. The objectives of the study is twofold: (i) the definition of local rainfall thresholds for landslide occurrences in the Kalimpong region; (ii) testing the applicability of the SIGMA model in a physical setting completely different from one of the areas where it was first conceived and developed. To achieve these purposes, a calibration dataset of daily rainfall and landslides from 2010 to 2015 has been used; the results have then been validated using 2016 and 2017 data, which represent an independent dataset from the calibration one. The validation showed that the model correctly predicted all the reported landslide events in the region. Statistically, the SIGMA model for Kalimpong town is found to have 92% efficiency with a likelihood ratio of 11.28. This performance was deemed satisfactory, thus SIGMA can be integrated with rainfall forecasting and can be used to develop a landslide early warning system.

Keywords: rainfall thresholds; early warning system; optimization; landslides

1. Introduction

In a global database of landslide disasters given by Froude and Petley (2018) [1], three–quarters of all landslide events between 2004 to 2016 occurred in Asian countries, with substantial events in the Himalayas. Indian Himalayas are highly susceptible to landslides which are triggered primarily by rainfall [2] and Sikkim and Darjeeling Himalayas are among the most highly vulnerable landslide zones. Due to rapid urbanization and increase in population in such areas, landslides and associated loss are an increasing concern [3] and early warning systems are regarded as a promising tool for landslide forecasting and risk management [4].

Rainfall being the most common triggering factor for landslides, early warning systems are typically based on empirical rainfall thresholds that describe the interaction between the primary cause (rainfall) and the final effect (landslide). In a few words, a triggering threshold is represented by a mathematical equation describing the critical rainfall condition above which landslides are triggered. The only input data used for the threshold definition are a dataset of rainfall recordings and a catalogue of landslides for which the time and location of occurrence are known with sufficient approximation. This approach completely bypasses the physical mechanism of triggering, thus simplifying the modeling effort, the computational resources required, and the amount of data needed for the analysis [5].

During the last few decades, many attempts were made across the world to define critical rainfall thresholds based on a number of different rainfall parameter, but the most common are intensity and duration (ID thresholds) [6–13], total event rainfall and duration (ED thresholds) [14–18] and antecedent rainfall [19–22]. The selection of the optimal rainfall parameters which are used for defining the threshold depends mainly on the landslide typology and physical characteristics of the region. It is well accepted that shallow landslides and debris flows are triggered by high intensity-short rainfalls and deep-seated landslides occur as a result of less intense rain over a long time [16,23,24]

When an area is prone to both shallow and rapid and deep-seated and slow moving landslides, a threshold model which can accommodate the effect of both the cases should be defined. A model that holds this characteristic is SIGMA (sistema integrato gestione monitoraggio allerta—integrated system for management, monitoring and alerting), which was developed for managing the risk associated with landslides triggered by rainfall in the Emilia-Romagna Region, Italy [23,25,26]. The model takes cumulative rainfall as input, and it considers the long-term and short-term behavior in order to account for shallow and deep landslides, respectively. Another important advantage of this method is the indication of warning level. The model can be calibrated with respect to the severity of landslides and can be used for developing regional site specific thresholds.

An ongoing research continuously produced new upgrades and optimization of the model SIGMA [19,26,27], making it possible to be used across the world for landslide hazard warning. However, to our knowledge, the model has never been applied outside Italy, thus leaving the claimed flexibility of application only theoretical. Hence, this study applies the SIGMA model to Kalimpong town in the Darjeeling Himalayas and thus tests the exportability of the SIGMA model in different climatic and geomorphological settings.

2. Study Area and Input Data

Kalimpong town is a part of the Kalimpong district of West Bengal state, India, as shown in Figure 1. This hilly town belongs to Darjeeling Himalayas, hemmed between rivers Tista in the west and Relli in the east, with an elevation ranging from 355 m to 1646 m above mean sea level. The slopes in the western face of the town are steep, while the eastern slopes are gentle.

The geological setting of the region is associated with the evolution of Darjeeling Himalayan ranges. Precambrian high-grade gneiss and quartzite, calc–silicate and quartzite, high-grade schist phyletic etc. are the dominant rock types found in the region [28]. Upper sedimentary layers of the young folded mountains get eroded during heavy rainfalls. The area consists of several joints and cracks that accelerates the weathering of the rock and the formation of unconsolidated matter [29]. The bedrock throughout the study area is composed of Daling series quartz mica schist of golden to silver colors [30]. The inclination of bed towards the east and northeast varies from 20° near river Tista to about 40° towards town. These slopes in can be morphometrically classified into escarpment category A (>45°), steep slope category B (30°–45°), moderate steep slope category C (20°–30°) and gentle slope category D (10°–20°). Silt to medium grained sand and loam constitutes a major portion of the topsoil of the area. According to GSI, more than 60% of the region comprises colluvium followed by older debris (24%) and young debris (2.5%). This geomorphological setting makes the region very prone to landslides, and rainfall is the main triggering factor.

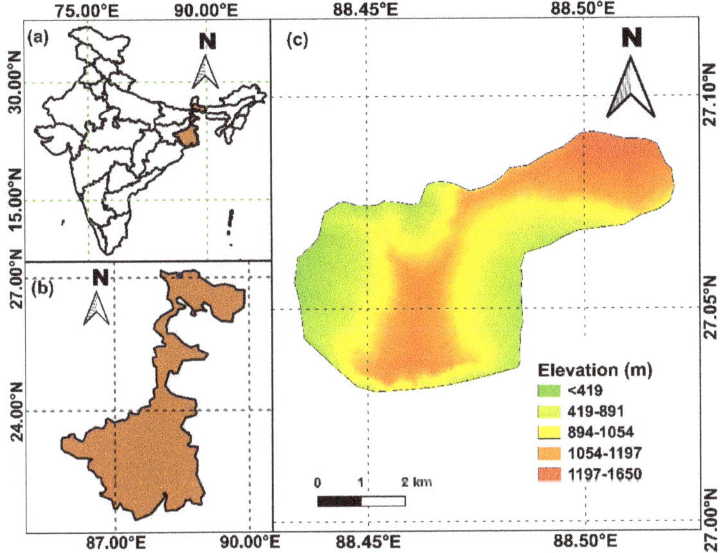

Figure 1. Location details (**a**) India; (**b**) West Bengal; (**c**) digital elevation model of Kalimpong (modified after [31]).

The geology of the area allows rainwater to percolate, increasing the pore pressure, therefore the shear strength of the soil decreases. The change in water content due to intense rainfall leads to the saturation of material and a sudden increase in the unit weight. This mechanism reduces the stability and resistance of parent rocks. The average annual precipitation in this area was observed to be 1872 mm during the study period, and the drainage density of the region is also very high. The area is drained by numerous mountainous natural streams (kholas) and their tributaries (jhoras). The precipitation with daily accuracy was collected for this study from the rain gauge maintained in Tirpai, Kalimpong (Save The Hills). The months from June to September are considered a monsoon period and the monthly rainfall from 2010 to 2017 is given in Table 1.

Table 1. The monthly rainfall (mm) during monsoon seasons in the study area town (2010–2017).

Month	2010	2011	2012	2013	2014	2015	2016	2017
June	317	337	355	248	396	568	327	154
July	666	678	433	424	371	534	870	812
August	425	526	251	401	572	242	263	432
September	268	384	467	113	265	331	367	288

A landslide catalogue was prepared from the reports of the Geological Survey of India, newspapers and field surveys. The database contains the spatial and temporal distribution of rainfall-induced landslide events during 2010–2017. The dataset from 2010 to 2015 was used for model calibration and the dataset from 2016 to 2017 was used for model validation. The major fatal landslides happened in the region were shallow and rapid in nature, but there are some areas which experience continuous sinking because of slow, deep-seated movements, especially near major jhoras [32]. The movements are occurring gradually and are observed as cracks in buildings and roads after each monsoon. Since 2017, these slow movements are monitored using micro-electro-mechanical tilt sensors installed at Chibo [33].

During the validation period (years 2016–2017), ground displacements were observed on 7 days at two locations [2].

The annual cumulative rainfall for these years is plotted in Figure 2a and the temporal distribution of landslides along with the average rainfall is shown in Figure 2b. It is observed that the number of landslide events is maximum in the month of July where the rainfall peak is recorded. From Figure 2b, it is clear that the number of landslides is directly related to the rainfall amount. Landslides are becoming an increasing menace in the region during monsoon season. The havocs related to landslides have multilevel impacts on the livelihood of population. Loss of farm lands and disruption of roads are affecting the income sources of the people. The socioeconomic development of the region is throttled by the disasters and associated setbacks. Hence, it is critical to adopt measures to minimize the impact of landslides in the region. An effective approach is to develop an early warning system using rainfall thresholds to forecast the occurrence of landslides. Since the area is affected by landslides of mixed typology (rapid shallow slides and slow deep seated movements), we took into account the SIGMA model [23], specifically conceived for similar settings, and we customized it for an application in the study area.

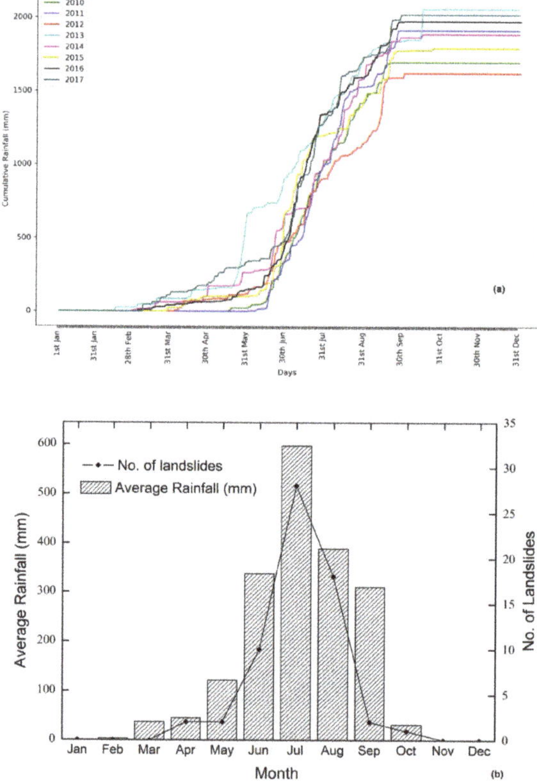

Figure 2. (**a**) Yearly cumulative rainfall; (**b**) Monthly distribution of landslide occurrence and average rainfall, (2010–2017) in mm.

3. The SIGMA Model

The SIGMA model was developed for the Emilia-Romagna region in Italy [23]. This model uses the standard deviation of a statistical distribution as the key parameter for the analysis and defines thresholds as a function of standard deviation, predicting the potential of rainfall to initiate landslide events in the study area. Since the model is based on statistical distribution, it is easily exportable to other regions [23]. Adopting the methodology from Martelloni et al. (2012) [23], a customized SIGMA model for Kalimpong town is derived in this study. The modifications are in accordance with the historical database collected for the study area, thus making SIGMA compatible for a different hydro-meteo-geological setting than the area for which it is originally developed. The daily precipitation data were added at 'n' days, with an 'n' day wide shifting window which moves at everyday time steps throughout rainfall data. The values of 'n' will vary from 1 to 365. To calculate the cumulative probability distribution for each data set, a standard distribution, which is the target function is chosen as a model [34].This transformation relates the cumulative rainfall (z) with the target distribution (y = a,σ) ('σ' is the standard deviation of the series and 'a' is a multiplication constant). For each 'n' day cumulative rainfall series, the values are sorted in ascending order such that

$$z_1 < z_2 < z_3 < \cdots < z_k < \cdots < z_n \tag{1}$$

and a cumulative sample frequency is defined as

$$P_k = \frac{k}{n} - \frac{0.5}{n} = G(y) \tag{2}$$

where, $1 \leq k \leq n$.

The conversion is carried out using the cumulative distribution function of z, termed as F(z). For each value of z_k, $F(z_k)$ defines the probability that the variable z takes a value less than z_k, where k varies between 1 to n.

The original data z transformed to y is obtained as:

$$G^{-1}(F(z)) \rightarrow G^{-1}(P_k) = y \tag{3}$$

After applying the transformation function, from a particular value of standard deviation or its multiple, cumulative sample frequency and precipitation can be calculated. The same procedure is repeated for all values of n from 1 to 365 and precipitation curves (σ curves) are plotted. The probability curves derived are used as the input values in the algorithm. A level of warning is predicted for everyday based on the rainfall thresholds. Rainfall recordings were cumulated with one day time steps for a particular time interval. These values are compared with the precipitation curves, from shorter to longer time frames [23]. In the case of shallow landslides, the analysis should focus on the immediate effect of rainfall: the cumulative rainfall values up to two days before the day of analysis is considered. The decisional algorithm used is given in Equation 4:

$$C_{1\text{-}3} = \left[\sum_{i=1}^{n} P(t+1-i)\right]_{n=1,2,3} \geq [S_n(\Delta)]_{n=1,2,3} \tag{4}$$

where, Δ = a,σ, $C_{1\text{-}3}$ is the vector indicating the cumulated rainfall at time t and $S_n(\Delta)$ are the thresholds relative to number of days n and Δ [23]. In the case of slow movements, the algorithm ponders the effect of cumulative rainfall from 4 days up to 63 days [23]. The condition for crossing the threshold is given by:

$$C_{4-63} = \left[\sum\nolimits_{i=1}^{n+3} P(t-2-i)\right]_{n=1,2,\ldots 60} \geq [S_{n+3}(\Delta)]_{n=1,2,\ldots 60} \tag{5}$$

The definitions of vector C are kept the same and have been used in the study for the analysis. The analysis was carried out in the same method proposed by the developers of the SIGMA model, to define the thresholds for Kalimpong town.

4. Analysis

The rainfall and landslide data for Kalimpong town have been used to develop (2010–2015) and validate (2016 and 2017) rainfall thresholds for the region. The spatial distribution of landslide events during the study period is shown in Figure 3.

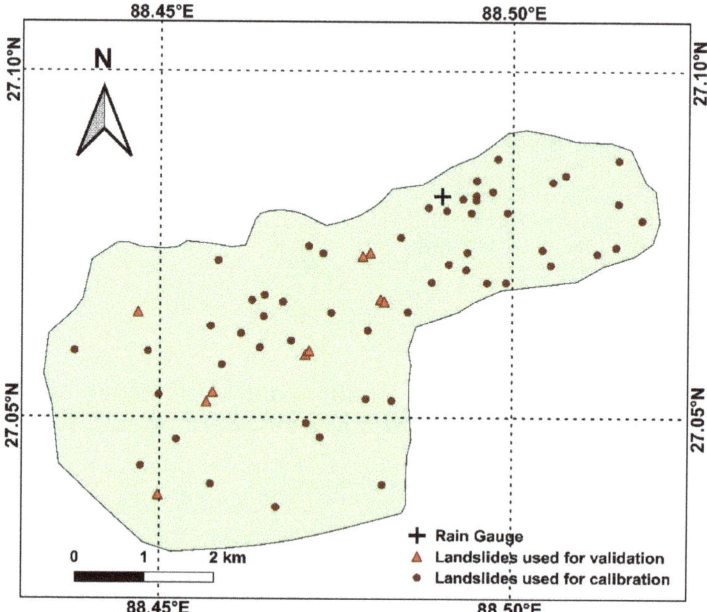

Figure 3. Spatial distribution of rain gauges and landslide events during the study period. (modified after [15]).

For each day, 'n'-day cumulative rainfall values were calculated with n ranging from 1 to 365. Cumulative probability distribution curves were plotted after sorting the values in ascending order. For small values of 'n', the distributions were found to be closer to log-normal, and for higher values of 'n', the distributions tend towards normal. The asymmetric distribution of data sets has been observed by other researchers as well [23]. Choosing Gaussian distribution as a target function, cumulative values corresponding to multiples of SIGMA were calculated by applying the transformation, as shown in Figure 4a.

After applying the transformation, a probability of not overcoming a particular '$a\sigma$' value can be calculated using the reverse procedure. For each value of '$a\sigma$', cumulative values for n-days varying from 1 to 365 were plotted as SIGMA curves. The values of standard curves were initially taken as 1.5σ, 1.75σ, 2σ and 2.5σ and are plotted in Figure 4b.

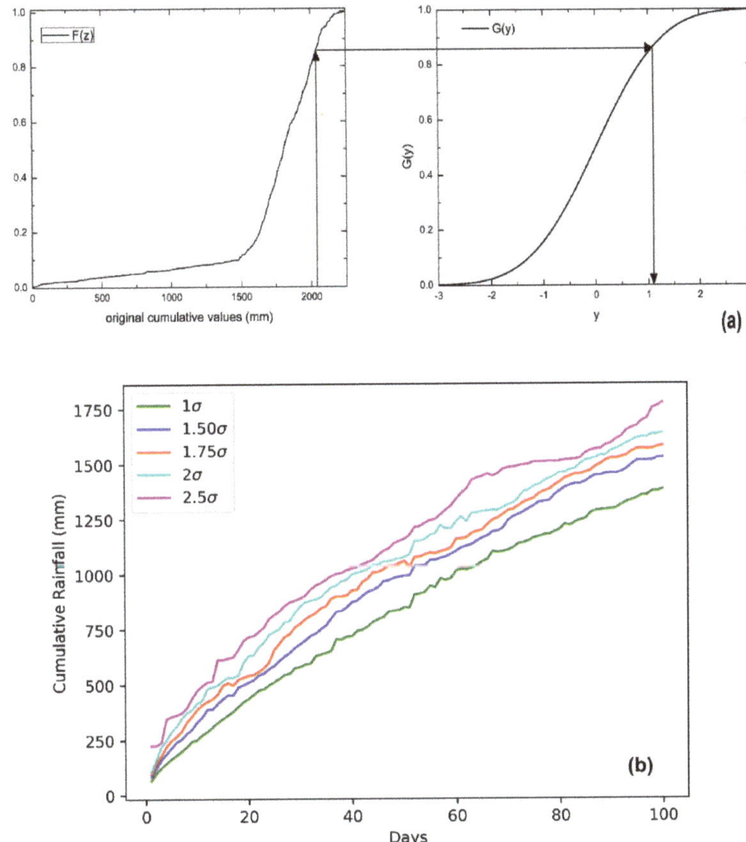

Figure 4. (**a**) Transformation of original cumulative distribution in the target distribution for Kalimpong town; (**b**) An example of SIGMA (sistema integrato gestione monitoraggio allerta—integrated system for management, monitoring and alerting) curves (σ curves) for cumulative periods up to 100 days (2010–2015).

From the probability distribution plots, SIGMA curves have been combined using an algorithm, which is the crucial part of the SIGMA model. The algorithm defines four different levels of alert, such as "red", "orange", "yellow" and "green". These values are used to delineate exceptional rainfall values. The starting algorithm for the proposed model is as shown in Figure 5. It considers the effect of short-term rainfall first, and exceedance of threshold will give high criticality alert for the day. If a red alert case does not exist, first orange alert level and then yellow alert level were checked for each day. If the result is negative in all cases, absent criticality (green color) is defined for the day. Hence, if a landslide happens to continue for a number of days, an effective model should predict the corresponding warning level on each day. The block diagram proposed in Figure 5 has to be considered as a starting point for the work, since it was then calibrated as described in the following paragraphs.

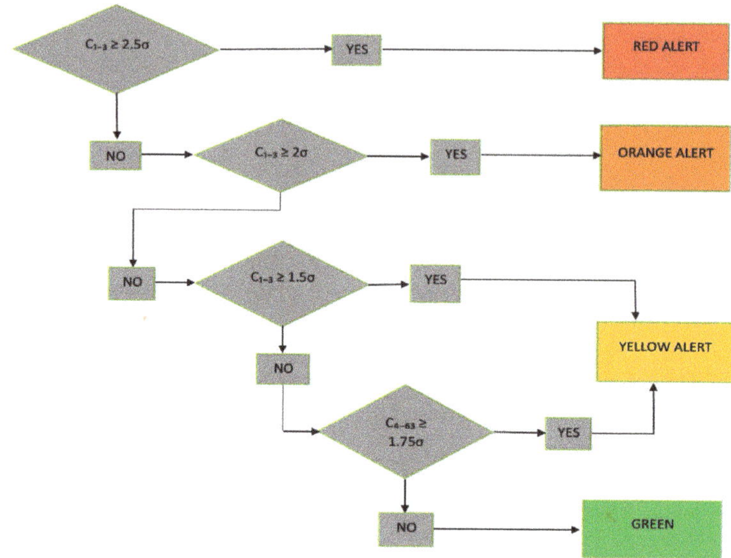

Figure 5. Algorithm used for calibration of the SIGMA model for Kalimpong town.

A threshold is considered to be exceeded if any of the elements in the vector crosses the value. Once a threshold is exceeded, the algorithm defines the level of warning on each day. These outputs were used to calibrate the model (data from 2010 to 2016). A trial and error procedure has been adopted in the optimization module of the algorithm, which relates the daily warning levels with landslide occurrences, as in Martelloni et al. (2012) [23]. The value of threshold is progressively raised so that false alarms are avoided. A visualization of the procedure is shown in Figure 6 where standard SIGMA value of 1.75 was optimized to 1.95. Using the same procedure, other standard values of 1.5 and 2 were optimized to 1.65, and 2.05, respectively. The standard value of 2.5 remained the same after optimization. The thresholds values were increased to minimize false alarms for each event, such that no true alarms are missed. The execution of this module terminates once the algorithm catches an event with an observed warning level conforming to the considered threshold. The standard SIGMA curves remain the same, but the calibration process gives a modified set of SIGMA curves for the region.

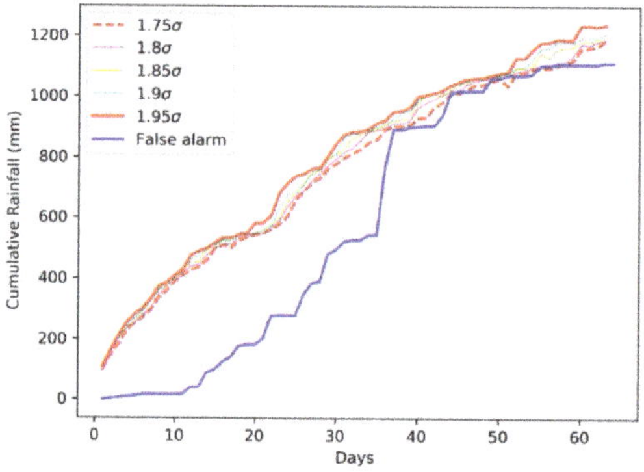

Figure 6. Visualization of calibration algorithm. The threshold value was raised till the cumulative rainfall curve of the event (F) is not crossing the threshold curve (standard threshold of 1.75 is optimized to 1.95).

5. Results

For the validation of results, rainfall and landslide data of 2016 and 2017 have been used. For each day of the validation period, the level of warning predicted by the decisional algorithm was compared with the reported landslide events according to the classical scheme of a confusion matrix, as shown in Figure 7. A confusion matrix is used to describe the performance of a decisional algorithm on a validation dataset for which the true values are known. The performance of an algorithm can be visualized using this matrix.

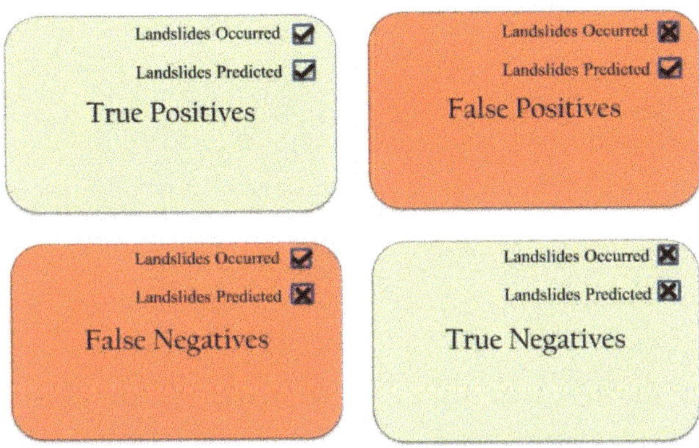

Figure 7. Confusion matrix for quantitative comparison.

Correct predictions can be both true positives (days in which the model forecasted correctly the occurrence of at least a landslide) and true negatives (days in which the model forecasted correctly that no landslides occurred). False negatives are those days in which the algorithm missed the alarms, and false positives are days in which the model issued false alarms. During 2016, eight shallow landslide events are reported in the region. The events were rapid in nature and happened to occur on a single day. Among the eight events, six were correctly predicted by the SIGMA model. Ground displacements were reported at two locations in the Chibo–Pashyor area during seven days in 2017: on 28th–29th July 2017 and 13th–17th August 2017 [2]. In all seven days, ordinary criticality was well-predicted in the present analysis. 55 false alarms were forecasted in a span of two years. An overview of the quantitative validation of the model for Kalimpong is tabulated in Table 2.

Table 2. Validation of the SIGMA model for Kalimpong.

Statistical Attributes	SIGMA Model
T1 = True positives	13
F1 = False positives	55
F2 = False negatives	2
T2 = True negatives	661
Negative predictive power = T2/(F2 + T2)	1.00
Positive predictive power = T1/(T1 + F1)	0.19
Misclassification rate = (F1 + F2)/(T1 + F1 + F2 + T2)	0.08
Efficiency = (T1 + T2)/(T1 + F1 + F2 + T2)	0.92
Odds ratio = (T1 + T2)/(F1 + F2)	11.82
False negative rate = F2/(T1 + F2)	0.13
False positive rate = F1/(F1 + T2)	0.08
Specificity (Sp) = T2/(F1 + T2)	0.92
Sensitivity (Sn) = T1/(T1 + F2)	0.87
Likelihood ratio = Sn/(1 − Sp)	11.28

6. Discussion

The validation of the SIGMA model for the study area gave satisfactory results by predicting all slow movements events correctly and producing two missed alarms. However, this came at the cost of having a relatively high count of false alarms (22).

As pointed out by Lagomarsino et al. (2015) [27], the most complete form of validation for a threshold model is to compare the skill scores to the ones obtained with the application of other models in the same test site. Several rainfall thresholds for landslides in Kalimpong region has already been defined [2,10,15]. Therefore, the validation statistics of SIGMA were compared to those obtained in the same test site by two already published works, which make use of ED and ID thresholds, as shown in Table 3.

Table 3. Comparison of the SIGMA model with other empirical thresholds.

Statistical Attributes	ID Threshold [10,35]	ED Threshold [15]			SIGMA (This Work)
		Threshold	Lower Limit	Upper Limit	
T1 = True positives	8	8	9	8	13
F1 = False positives	98	93	117	75	55
F2 = False negatives	7	7	6	7	2
T2 = True negatives	618	623	599	641	661
Negative predictive power = T2/(F2 + T2)	0.99	0.99	0.99	0.99	1.00
Positive predictive power = T1/(T1 + F1)	0.08	0.08	0.07	0.10	0.19
Misclassification rate = (F1 + F2)/(T1 + F1 + F2 + T2)	0.14	0.14	0.17	0.11	0.08
Efficiency = (T1 + T2)/(T1 + F1 + F2 + T2)	0.86	0.86	0.83	0.89	0.92
Odds ratio = (T1 + T2)/(F1 + F2)	5.96	6.31	4.94	7.91	11.82
False negative rate = F2/(T1 + F2)	0.47	0.47	0.40	0.47	0.13
False positive rate = F1/(F1 + T2)	0.14	0.13	0.16	0.10	0.08
Specificity (Sp) = T2/(F1 + T2)	0.86	0.87	0.84	0.90	0.92
Sensitivity (Sn) = T1/(T1 + F2)	0.53	0.53	0.60	0.53	0.87
Likelihood ratio = Sn/(1 − Sp)	3.90	4.11	3.67	5.09	11.28

It can be seen that during the validation period, SIGMA outperforms the other models. The terms for evaluating the overall performance of model, efficiency and likelihood ratio, are maximum for SIGMA among the models tested. Efficiency being the ratio of true predictions to a total number of predictions, does not show a significant variation amongst different models. This is often reported in LEWS [27] where the number of true negatives are of a higher order than the other three variables. Specificity measures the ratio of correctly predicted days with no landslides to the total number of days without landslides, and sensitivity denotes the ratio of correctly predicted landslides to the total number of landslides. Likelihood ratio is the ratio of sensitivity to 1-specificity, evaluating the effect of both the parameters. The main reason for the better performance of SIGMA seems to be the effectiveness in predicting the slow movements occurred in 2017: a technique based on detecting antecedent rainfall anomalies. SIGMA is more suited than ID and ED thresholds to forecast slope movements with a complex hydrological response [26]. ID and ED thresholds correctly predicted seven out of eight shallow landslides happened in 2016 but failed to forecast the slow movements in 2017 except for one day. It is also observed that the number of false alarms is also less in the SIGMA model, when compared with the other models. However, before having a definitive response on which would be the threshold model more effective to use in an EWS in Kalimpong, further tests are needed and a larger validation dataset needs to be accounted for.

In addition, the validation showed that SIGMA could need to be improved further, especially concerning the high number of false alarms. That was not a surprising outcome since in the calibration, the optimization procedure aimed at minimizing false negatives (missed alarms) instead of searching for a compromise between missed alarms and correct predictions, thus leading to a high number of false positives. A research direction worth exploring is testing different time intervals in the decisional algorithm: the one used in this research are the one resulted optimal for the Emilia-Romagna region (Italy), and they were defined after a long period of adjustments [26]. The different physical settings of Kalimpong allow for a different optimal set of SIGMA values and time intervals to be defined.

7. Conclusions

Forecasting of rainfall-induced landslides in Kalimpong town have been carried out using the SIGMA model and considering the historical rainfall and landslide information. A single parameter, the cumulative rainfall, defines the threshold by means of a set of statistical thresholds.

The algorithm was designed to consider a three day rainfall effect for shallow landslides and more days (up to 63) for slow landslides. The time period and standard SIGMA values were decided by trial and error procedure during calibration, minimizing missed alarms and false alarms. A validation procedure showed satisfactory results and proved that SIGMA performed better than other ED and ID thresholds defined for the same region by previous works. For the study area, where both rapid and slow movements are present, the combined use of short-term and long-term antecedent rainfall is thus a point of strength of the model.

It can be concluded that the SIGMA model is a simple and efficient tool which can be used for landslide early warning on regional scale. The model predicts warning levels associated with each day, which can be directly linked to the severity of landslide events predicted. This study proves that the SIGMA model can be exported in parts of the world other than Italy, where the model was originally conceived, with satisfactory performance.

While applying the SIGMA model for a study area different from Italy, in a different hydro-meteo-geological setting, it was found that the values of $S_n(\Delta)$ of Kalimpong is different from those used for Italy [23]. A simpler algorithm than the one used for Italy was found to provide optimum results, as a smaller area and single rain gauge is considered for the analysis. Being a statistical model, the starting algorithm was decided by trial and error using the meteorological data and was fine-tuned by minimizing false alarms using an optimization procedure. The algorithm correctly predicted warnings on 13 out of 15 days of landslides during the validation period (2016–2017). The events in 2017 were the result of continuous rainfall over a longer time period. It can be concluded that this algorithm-based approach considers the effect of both long-term and short-term rainfall and even slow movements are predicted, providing a performance better than traditional ID and ED thresholds.

The number of false alarms generated has to be reduced either by tuning the SIGMA levels and the time interval lengths, or by improving the model conceptually. As an instance, physical parameters like soil moisture can be considered along with the rainfall data to increase the positive predictive power [18,25]. Also to expand the model for a larger area, spatial variability of meteorological parameters should be considered [36,37]. After further tests and developing standard action plans for each level of warning, this model has the potential to be integrated with rainfall forecasting and to be used as a landslide early warning system on a regional scale.

Author Contributions: Conceptualization, M.T.A. and N.S.; data curation, S.K.; formal analysis, M.T.A. and S.K.; methodology, S.S.; supervision, N.S., B.P. and S.S.; validation, M.T.A.; writing—original draft, M.T.A. and A.R.; writing—review and editing, N.S., B.P. and S.S. All authors have read and agreed to the published version of the manuscript.

Funding: The work has been carried out with the financial support of Department of Science & Technology (DST), New Delhi, for funding the research project Grant No. (NRDMS/02/31/015(G)); Florence University, in the framework of the project SEGONISAMUELERICATEN20.

Acknowledgments: The authors express our sincere gratitude to both the funding agencies. We are also thankful to Praful Rao and Save the Hills organization for their constant support throughout the study.

Conflicts of Interest: The authors declare no conflicts of interest.

References

1. Froude, M.J.; Petley, D.N. Global fatal landslide occurrence from 2004 to 2016. *Nat. Hazards Earth Syst. Sci.* **2018**, *18*, 2161–2181. [CrossRef]
2. Dikshit, A.; Satyam, N. Probabilistic rainfall thresholds in Chibo, India: Estimation and validation using monitoring system. *J. Mt. Sci.* **2019**, *16*, 870–883. [CrossRef]

3. Van Westen, C.J.; Ghosh, S.; Jaiswal, P.; Martha, T.R.; Kuriakose, S.L. From landslide inventories to landslide risk assessment; an attempt to support methodological development in India. In *Landslide Science and Practice*; Springer: Berlin/Heidelberg, Germany, 2013; Volume 1, pp. 3–20.
4. Piciullo, L.; Calvello, M.; Cepeda, J.M. Territorial early warning systems for rainfall-induced landslides. *Earth-Science Rev.* **2018**, *179*, 228–247. [CrossRef]
5. Segoni, S.; Piciullo, L.; Gariano, S.L. A review of the recent literature on rainfall thresholds for landslide occurrence. *Landslides* **2018**, *15*, 1483–1501. [CrossRef]
6. Caine, N. The rainfall intensity-duration control of shallow landslides and debris flows: An update. *Geogr. Ann. Ser. A Phys. Geogr.* **1980**, *62*, 23–27.
7. Aleotti, P. A warning system for rainfall-induced shallow failures. *Eng. Geol.* **2004**, *73*, 247–265. [CrossRef]
8. Guzzetti, F.; Peruccacci, S.; Rossi, M.; Stark, C.P. Rainfall thresholds for the initiation of landslides in central and southern Europe. *Meteorol. Atmos. Phys.* **2007**, *98*, 239–267. [CrossRef]
9. Brunetti, M.T.; Peruccacci, S.; Rossi, M.; Luciani, S.; Valigi, D.; Guzzetti, F. Rainfall thresholds for the possible occurrence of landslides in Italy. *Nat. Hazards Earth Syst. Sci.* **2010**, *10*, 447–458. [CrossRef]
10. Dikshit, A.; Satyam, D.N. Estimation of rainfall thresholds for landslide occurrences in Kalimpong, India. *Innov. Infrastruct. Solut.* **2018**, *3*, 24. [CrossRef]
11. Rosi, A.; Peternel, T.; Jemec-Auflič, M.; Komac, M.; Segoni, S.; Casagli, N. Rainfall thresholds for rainfall-induced landslides in Slovenia. *Landslides* **2016**, *13*, 1571–1577. [CrossRef]
12. Abraham, M.T.; Satyam, N.; Rosi, A.; Pradhan, B.; Segoni, S. The Selection of Rain Gauges and Rainfall Parameters in Estimating Intensity-Duration Thresholds for Landslide Occurrence: Case Study from Wayanad (India). *Water* **2020**, *12*, 1000. [CrossRef]
13. Abraham, M.T.; Pothuraju, D.; Satyam, N. Rainfall Thresholds for Prediction of Landslides in Idukki, India: An Empirical Approach. *Water* **2019**, *11*, 2113. [CrossRef]
14. Zhao, B.; Dai, Q.; Han, D.; Dai, H.; Mao, J.; Zhuo, L. Probabilistic thresholds for landslides warning by integrating soil moisture conditions with rainfall thresholds. *J. Hydrol.* **2019**, *574*, 276–287. [CrossRef]
15. Teja, T.S.; Dikshit, A.; Satyam, N. Determination of Rainfall Thresholds for Landslide Prediction Using an Algorithm-Based Approach: Case Study in the Darjeeling Himalayas, India. *Geosciences* **2019**, *9*, 302. [CrossRef]
16. Melillo, M.; Brunetti, M.T.; Peruccacci, S.; Gariano, S.L.; Roccati, A.; Guzzetti, F. A tool for the automatic calculation of rainfall thresholds for landslide occurrence. *Environ. Model. Softw.* **2018**, *105*, 230–243. [CrossRef]
17. Peruccacci, S.; Brunetti, M.T.; Luciani, S.; Vennari, C.; Guzzetti, F. Lithological and seasonal control on rainfall thresholds for the possible initiation of landslides in central Italy. *Geomorphology* **2012**, *139–140*, 79–90. [CrossRef]
18. Abraham, M.T.; Satyam, N.; Pradhan, B.; Alamri, A.M. Forecasting of Landslides Using Rainfall Severity and Soil Wetness: A Probabilistic Approach for Darjeeling Himalayas. *Water* **2020**, *12*, 804. [CrossRef]
19. Lagomarsino, D.; Segoni, S.; Fanti, R.; Catani, F. Updating and tuning a regional-scale landslide early warning system. *Landslides* **2013**, *10*, 91–97. [CrossRef]
20. Bai, S.; Wang, J.; Thiebes, B.; Cheng, C.; Yang, Y. Analysis of the relationship of landslide occurrence with rainfall: A case study of Wudu County, China. *Arab. J. Geosci.* **2014**, *7*, 1277–1285. [CrossRef]
21. Lee, M.L.; Ng, K.Y.; Huang, Y.F.; Li, W.C. Rainfall-induced landslides in Hulu Kelang area, Malaysia. *Nat. Hazards* **2014**, *70*, 353–375. [CrossRef]
22. Althuwaynee, O.F.; Pradhan, B.; Ahmad, N. Estimation of rainfall threshold and its use in landslide hazard mapping of Kuala Lumpur metropolitan and surrounding areas. *Landslides* **2015**, *12*, 861–875. [CrossRef]
23. Martelloni, G.; Segoni, S.; Fanti, R.; Catani, F. Rainfall thresholds for the forecasting of landslide occurrence at regional scale. *Landslides* **2012**, *9*, 485–495. [CrossRef]
24. Crosta, G. Regionalization of rainfall thresholds: An aid to landslide hazard evaluation. *Environ. Geol.* **1998**, *35*, 131–145. [CrossRef]
25. Segoni, S.; Rosi, A.; Lagomarsino, D.; Fanti, R.; Casagli, N. Brief communication: Using averaged soil moisture estimates to improve the performances of a regional-scale landslide early warning system. *Nat. Hazards Earth Syst. Sci.* **2018**, *18*, 807–812. [CrossRef]
26. Segoni, S.; Rosi, A.; Fanti, R.; Gallucci, A.; Monni, A.; Casagli, N. A regional-scale landslide warning system based on 20 years of operational experience. *Water* **2018**, *10*, 1297. [CrossRef]

27. Lagomarsino, D.; Segoni, S.; Rosi, A.; Rossi, G.; Battistini, A.; Catani, F.; Casagli, N. Quantitative comparison between two different methodologies to define rainfall thresholds for landslide forecasting. *Nat. Hazards Earth Syst. Sci.* **2015**, *15*, 2413–2423. [CrossRef]
28. Wadia, D.N. *Geology of India*; Tata McGrawHill: New Delhi, India, 1975.
29. Sumantra, S. Causes of Landslides in Darjeeling Himalayas during June-July, 2015. *J. Geogr. Nat. Disasters* **2016**, *6*. [CrossRef]
30. Mukherjee, A.; Mitra, A. Geotechnical Study of Mass Movements Along the Kalimpong Approach Road in the Eastern Himalayas. *Indian J. Geol.* **2001**, *73*, 271–279.
31. CartoDEM. Available online: https://bhuvan-app3.nrsc.gov.in/data/download/index.php (accessed on 20 August 2019).
32. Save The Hills Blog. Available online: http://savethehills.blogspot.com/ (accessed on 3 December 2019).
33. Dikshit, A.; Satyam, D.N.; Towhata, I. Early warning system using tilt sensors in Chibo, Kalimpong, Darjeeling Himalayas, India. *Nat. Hazards* **2018**, *94*, 727–741. [CrossRef]
34. Goovaerts, P. *Geostatistics for Natural Resources Evaluation*; Oxford University Press: Oxford, UK, 1997.
35. Dikshit, A.; Satyam, N. Rainfall Thresholds for the prediction of Landslides using Empirical Methods in Kalimpong, Darjeeling, India. In Proceedings of the JTC1 Workshop on Advances in Landslide Understanding, Barcelona, Spain, 24–26 May 2017.
36. Lu, B.; Charlton, M.; Harris, P.; Fotheringham, A.S. Geographically weighted regression with a non-Euclidean distance metric: A case study using hedonic house price data. *Int. J. Geogr. Inf. Sci.* **2014**, *28*, 660–681. [CrossRef]
37. Pasculli, A.; Palermi, S.; Sarra, A.; Piacentini, T.; Miccadei, E. A modelling methodology for the analysis of radon potential based on environmental geology and geographically weighted regression. *Environ. Model. Softw.* **2014**, *54*, 165–181. [CrossRef]

Article

Determination of Empirical Rainfall Thresholds for Shallow Landslides in Slovenia Using an Automatic Tool

Galena Jordanova [1,*], Stefano Luigi Gariano [2], Massimo Melillo [2], Silvia Peruccacci [2], Maria Teresa Brunetti [2] and Mateja Jemec Auflič [3]

1 Faculty of Natural Sciences and Engineering, Department of Geology, University of Ljubljana, 1000 Ljubljana, Slovenia

2 CNR-IRPI—Research Institute for Geo-Hydrological Protection of the Italian National Research Council, 06127 Perugia, Italy; stefano.luigi.gariano@irpi.cnr.it (S.L.G.); massimo.melillo@irpi.cnr.it (M.M.); silvia.peruccacci@irpi.cnr.it (S.P.); maria.teresa.brunetti@irpi.cnr.it (M.T.B.)

3 Geological Survey of Slovenia, 1000 Ljubljana, Slovenia; mateja.jemec-auflic@geo-zs.si

* Correspondence: galena.jordanova@geo.ntf.uni-lj.si; Tel.: +386-40-309-142

Received: 29 April 2020; Accepted: 18 May 2020; Published: 19 May 2020

Abstract: Rainfall-triggered shallow landslides represent a major threat to people and infrastructure worldwide. Predicting the possibility of a landslide occurrence accurately means understanding the trigger mechanisms adequately. Rainfall is the main cause of slope failures in Slovenia, and rainfall thresholds are among the most-used tools to predict the possible occurrence of rainfall-triggered landslides. The recent validation of the prototype landslide early system in Slovenia highlighted the need to define new reliable rainfall thresholds. In this study, several empirical thresholds are determined using an automatic tool. The thresholds are represented by a power law curve that links the cumulated event rainfall (E, in mm) with the duration of the rainfall event (D, in h). By eliminating all subjective criteria thanks to the automated calculation, thresholds at diverse non-exceedance probabilities are defined and validated, and the uncertainties associated with their parameters are estimated. Additional thresholds are also calculated for two different environmental classifications. The first classification is based on mean annual rainfall (MAR) with the national territory divided into three classes. The area with the highest MAR has the highest thresholds, which indicates a likely adaptation of the landscape to higher amounts of rainfall. The second classification is based on four lithological units. Two-thirds of the considered landslides occur in the unit of any type of clastic sedimentary rocks, which proves an influence of the lithology on the occurrence of shallow landslides. Sedimentary rocks that are prone to weathering have the lowest thresholds, while magmatic and metamorphic rocks have the highest thresholds. Thresholds obtained for both classifications are far less reliable due to the low number of empirical points and can only be used as indicators of rainfall conditions for each of the classes. Finally, the new national thresholds for Slovenia are also compared with other regional, national, and global thresholds. The thresholds can be used to define probabilistic schemes aiming at the operative prediction of rainfall-induced shallow landslides in Slovenia, in the framework of the Slovenian prototype early warning system.

Keywords: rainfall thresholds calculation; mean annual rainfall; lithology; Slovenia

1. Introduction

Landslides are one of the most common hazardous natural phenomena in Slovenia and worldwide, threatening the safety of local residents and damaging infrastructure. The main triggering factor of shallow landslides in Slovenia is rainfall, especially short and intense rainstorms, combined with local geological, geomorphological and climatic conditions [1]. In recent decades, intensive rainfall events have become much more frequent. This is also due to global climate change, which leads to a high number of shallow slope failures [2]. Every year dozens to hundreds of new shallow landslides are recorded in Slovenia. Many of them cause damage to infrastructure and properties, including residential buildings and agricultural land. To mitigate possible serious consequences and damage, the use of a landslide early warning system (LEWS) is fundamental. To operate a successful LEWS, it is essential to understand the relationship between rainfall and landslide occurrence. This relationship is commonly defined by means of empirical rainfall thresholds. The calculation of rainfall thresholds for landslide triggering has been a major challenge over the last few decades. Campbell [3] was the first to demonstrate the connection between antecedent rainfall and its infiltration into low-permeable rocks with the triggering of landslides. Nilsen and Turner [4] also proved the impact of rainstorms and antecedent rainfall on the occurrence of slope mass movements and calculated threshold values for the investigated area. Caine [5] proposed a power law equation linking mean rainfall intensity (I) and duration of the rainfall event (D) based on data from different geological, morphological and climatic settings. Since then many different methods and algorithms have been developed for calculating rainfall thresholds [6–8]. LEWSs based on thresholds of different types have been implemented in many countries and regions [9,10], e.g., for the coastal areas of San Francisco [11,12], the metropolitan areas of Rio de Janeiro [13,14] and Vancouver [15], southern Taiwan [16], Italy [17] and regions in Italy such as Emilia-Romagna [18], Piedmont [19], Tuscany [20–22] and Sicily [23].

In Slovenia, Komac [1] calculated rainfall thresholds for individual lithological units on the entire Slovenian territory using the statistical chi-square method. Jemec Auflič and Komac [24] analyzed rainfall patterns for shallow landslides in the Škofjeloško-Cerkljansko hills during six major rainfall events between 1991 and 2010, while Rosi et al. [25] used the MaCumBA (MAssive CUMulative Brisk Analyzer) algorithm by Segoni et al. [26] to determine the first mean intensity-duration *ID* thresholds on a regional scale and for four major river zones in Slovenia. Bezak et al. [27] determined empirical thresholds for flash floods and landslides in Slovenia using a copula-based method. Bezak et al. [28] also worked on the application of hydrological modelling for temporal prediction of shallow landslides, while Jordanova et al. [29] focused on the determination of empirical thresholds for shallow landslides in the Posavsko hills, Eastern Slovenia, with an analysis of antecedent rainfall and the intensity of seven major rainfall events between 2013 and 2017.

With the aim of improving prevention measures, the prototype of a LEWS for Slovenia was developed in 2013 [30]. The system is based on the comparison between the forecasted precipitation for the next 24 h and rainfall thresholds, determined using the chi-square method and 40 years of average rainfall correlated to the lithological unit [1]. Recently, a validation of this LEWS was carried out [29,31] and demonstrated the need for new thresholds.

The definition of empirical rainfall thresholds is often affected by subjective criteria, such as the definition of the rainfall events responsible for landslide triggering, and by uncertainties, such as the quality of rainfall data and the accuracy of the location and timing of landslide occurrences [32,33]. To avoid any subjective bias in the results, Melillo et al. [34,35] proposed an algorithm for the automatic calculation of thresholds for rainfall-induced landslides, which was improved and implemented in a software tool (CTRL-T, Calculation of Thresholds for Rainfall-induced Landslides Tool) [32]. The tool uses objective, standardized criteria for the automatic reconstruction of landslide-triggering rainfall conditions, based on historical rainfall records and landslide occurrence dates. It was applied by several authors in diverse environments in Italy [32,36], India [37] and Bhutan [38].

According to the determination of the amount of rainfall responsible for the landslide occurrence, we propose new rainfall thresholds, calculated using CTRL-T, for the entire national territory of Slovenia and for climatic and geological subdivisions. The new national rainfall thresholds are compared with the global and regional thresholds proposed by Caine [5], Guzzetti et al. [6,7], Rosi et al. [24], Peruccacci et al. [39] and Palladino et al. [40].

2. Study Area

Slovenia (20,273 km^2) lies in Central Europe in the southeastern part of the Alps (Figure 1). The sparse landscape and the diverse geological conditions range from the Pannonian plains and hills and their sediments in the northeast through the Alpine foothills in the Prealpine region (East to Western Slovenia) to the Alpine region in the northwest and the Eocene flysch in the southwestern Mediterranean plateau. The tectonic and structural elements, intersecting the area, have led to unstable rock masses and landslide-prone conditions [41].

The rainfall is unevenly distributed over the country due to its location between the Alps, the Dinarides, the Pannonian Basin and the Adriatic Sea, which contributes to the Mediterranean climate conditions (Figure 1). According to the Slovenian Environment Agency (ARSO), the annual average precipitation between 1981 and 2010 shows that the western part of Slovenia (excluding the southwestern coastal area) and especially the northwestern Alpine region is the rainiest part [42]. The average annual rainfall ranges from 1600 mm to over 3200 mm in the Julian Alps. Rain clouds usually move north and east towards the Alps and Dinarides, which serve as an orographic barrier. Many deep-seated landslides and debris flows (e.g., Stože, Slano Blato, Potoška Planina) are present in the western and northwestern parts of Slovenia, while the east is more prone to shallow landslides. The less rainy area in the northeast accumulates on average almost 1000 mm per year.

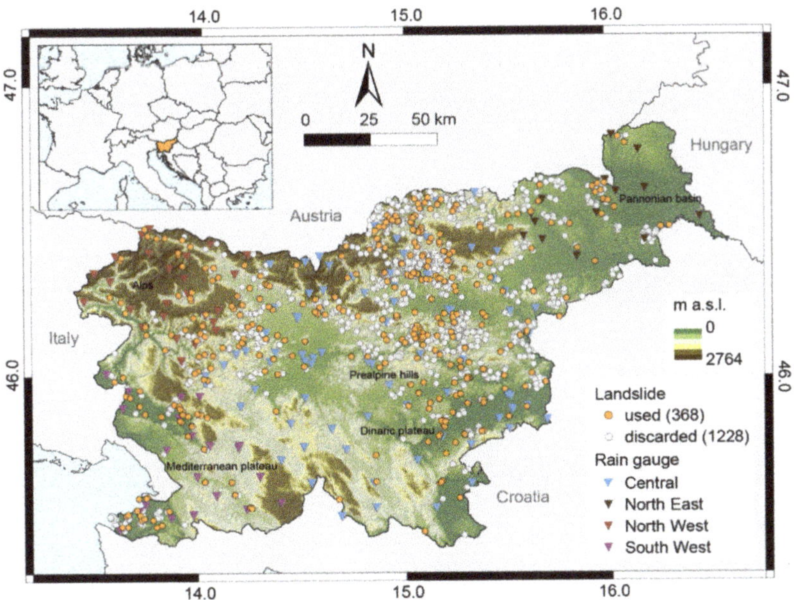

Figure 1. Location of Slovenia, with indication of the rain gauges used to reconstruct the rainfall events responsible for the failures, classified into four geographical areas related to rainfall characteristics, and of the landslides included in the analyzed catalogue. The landslides used for threshold calculations are indicated with orange dots.

3. Methods and Data

3.1. CTRL-T Tool and Threshold Equation

CTRL-T is a tool for the automatic calculation of rainfall thresholds for their use in operative prediction of shallow landslides [32]. The calculation of thresholds is based on continuous sets of hourly rainfall data gathered from rain gauges, and on a landslide database, consisting of known locations (geographic coordinates) and times (accurate dates and, when available, hour) of landslide occurrences. The tool reconstructs rainfall events and determines the events that are more likely to be responsible for the observed slope failures. Two important input parameters were defined prior to the identification of rainfall events, i.e., (i) the maximum permissible distance between the representative rain gauge and the landslide (15 km) and (ii) the maximum acceptable delay between the end of a rainfall event and the occurrence of a landslide (48 h).

The calculations are performed by three separate segments, each of which performs specific tasks [32]. The first segment performs the reconstruction of the individual rainfall events from the continuous rainfall series and calculates the duration (D, in hours) and the cumulated rainfall (E, in mm) of the rainfall events. The separation of consecutive rainfall events is based on climatic and seasonal settings: two "no rain" time intervals are distinguished for a warm/dry and cold/rainy season, respectively. The determination of the two seasons is based on monthly soil–water balance (MSWB) model [43–45]. In more detail, the MSWB model exploits monthly rainfall and temperature data and allows estimating the average monthly potential and real evapotranspiration utilizing a water balance over the mean hydrological year. Furthermore, the aridity index (AI), i.e., the ratio between the average monthly rainfall and the average monthly potential evapotranspiration, is used to define the length of the two seasons for each of the four regions. The warm/dry season has AI < 1, while in the cold/rainy season AI ≥ 1. Once the length of the two seasons in each region has been defined, the ratio between the total amount of real evapotranspiration in the warm and the cold seasons is used to define the ratio between the "no rain" time intervals in each season.

The task of the second segment is to select the nearest rain gauge for each landslide. The maximum allowed distance between a landslide and a rain gauge is within a circular area of a given radius. This task is followed by the selection of single or multiple rainfall conditions (MRC) that are most likely responsible for the slope failures. Each MRC is assigned a weight to select the representative rain gauge and the rainfall conditions associated with the landslide. The weight is equal to the ratio between the cumulated rainfall (E) times the mean rainfall intensity (I) divided by the square of the distance between the rain gauge and the landslide.

The third segment is the calculation of cumulated event rainfall–rainfall duration—*ED*—thresholds at different non-exceedance probabilities (NEPs), and the associated uncertainties, where the MRC with the maximum weight for each failure (MPRC, Maximum Probability Rainfall Condition) are selected. The thresholds are defined using a frequentist approach [46,47] and have a power law form linking E to D:

$$E = (\alpha \pm \Delta\alpha) \cdot D^{(\gamma \pm \Delta\gamma)} \tag{1}$$

where α is the scaling parameter and γ is the shape parameter, i.e., the intercept and slope of the power law curve respectively; $\Delta\alpha$ and $\Delta\gamma$ represent the relative uncertainties of the two parameters [46,47]. A more detailed description of CTRL-T can be found in Melillo et al. [32,34,35].

3.2. Landslide Data

Initially, the landslide database consisted of 2179 landslides that occurred between 18 September 2007 and 5 May 2018. We classified all landslides in the database as shallow landslides. The failures, probably caused by snowmelt during the winter and the first months of spring (i.e., from early December to early April), were discarded. Landslides with unknown dates of occurrence or location and double entries (e.g., two landslides in the same place and time) were also excluded from the analysis. In total, we manually removed 583 landslides from the database, leaving 1596 landslides for further analysis.

The exact time of the failures was not known; therefore, all landslides were recorded as they had occurred at the end of the day. This could introduce uncertainties in the amount of rainfall responsible for the landslides (see e.g., [48]), which however were not evaluated. In particular, all rainfall up to the end of the day of the recorded dates was considered, although the landslides probably occurred earlier and could have consisted of lower amounts of rainfall.

3.3. Rainfall Data

In Slovenia, several different types of gauges measure precipitation data and other climatic variables. For this analysis, ARSO provided rainfall data of all automatic rain gauges. The measures were extracted with a temporal resolution of 30 minutes and aggregated in hourly time steps for the period between 18 September 2007 and 5 May 2018. Of the 144 available rain gauges, only 94 were used for the reconstruction of rainfall conditions that caused the landslides. Average monthly temperature data, useful for the MSWB model, was gathered from all stations.

4. Results and Discussion

4.1. Definition of the Dry and Wet Season

The identification of the warm/dry and cold/rainy seasons in Slovenia was not trivial due to sparse landscapes and different climatic conditions. Four regions were identified: (i) North East, the region with less rainfall; (ii) North West, the Alpine region; (iii) South West, the coastal region with Mediterranean climatic conditions; and (iv) the Central region, the pre-alpine that receives the highest rainfall (more than 3000 mm per year). The rain gauges in each region were classified accordingly (Figure 1). Using the average monthly rainfall and temperature data between 2007 and 2018 and applying the MSWB model, the length of the two seasons in each region was defined. Furthermore, using the aridity index, the ratio among the "no rain" intervals in each season was determined. This ratio resulted equal to 2, corresponding to a "no rain" interval of 48 h (set as minimum) in the warm season and 96 h in the cold season, respectively. Figure 2 shows the results of the analysis for the four considered regions. While the North East, South West and Central regions experience dry periods (AI ≥ 1) in the summer months (with diverse lengths), the North West part (Alpine area) has no dry period. Therefore, in North East, South West and Central regions, rainfall events were reconstructed using a "no-rain" period of 48 h and 96 h in the warm/dry and cold/rainy season, respectively. Conversely, in the North West region, rainfall events were always separated by 96 h.

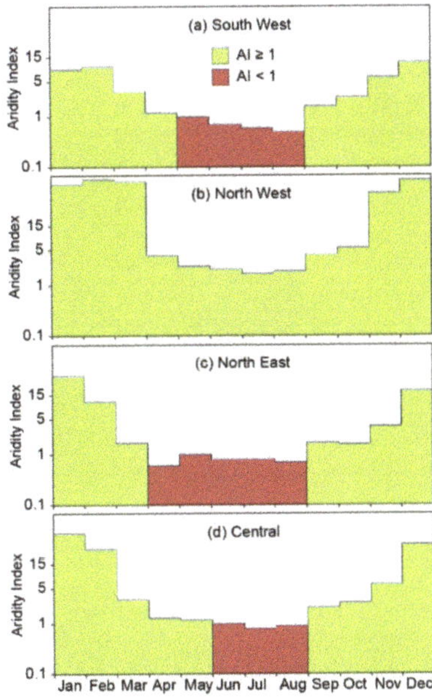

Figure 2. Calculated aridity index for individual regions: (**a**) South West; (**b**) North West; (**c**) North East; (**d**) Central region. Green and red areas indicate the cold/rainy and the warm/dry months, respectively.

4.2. Threshold Calculation

Using CTRL-T we reconstructed 1315 rainfall conditions responsible for the occurrence of landslides in the observed period. For 281 landslides it was not possible to determine the triggering rainfall event for three main reasons: (i) the distance between the landslide and the rain gauges exceeded 15 km (chosen according to the morphology and the rain gauge density of the study area); (ii) the delay between the end of the rainfall condition and the occurrence of the landslide exceeded 48 h; (iii) accurate landslide information or rainfall data were lacking. These landslides were excluded from the calculation. Several landslides occurring on the same day and near the same rain gauges were presumably triggered by the same amount of rainfall. In this case, CTRL-T selected only the rainfall condition corresponding to the first triggered landslide. As a result of the analysis, out of 1315 rainfall events, only 368 survived the selection criteria (Figures 1 and 3).

The values of 15 km and 48 h for maximum distance and delay, respectively, were selected in accordance with previous works [32,33,35,38] and should be considered as conservative upper limits. Most of the landslides (305 out of 368, 83%) were associated with rain gauges located at a maximum distance of 10 km, and half of them within 6 km; in 48 cases the distance was shorter than 2 km. Regarding the delay between the end of the rainfall and the occurrence landslide time, the majority of the landslides (299 out of 368, 81%) that were associated with rainfall conditions ended within a delay of 24 h. Specifically, half of them had a delay of less than 10 h and in 60 cases the delay was null.

Based on the 368 rainfall conditions, the algorithm included in CTRL-T calculated *ED* (cumulated event rainfall—duration) thresholds at different non-exceedance probabilities (Table 1). As a reference with previous works (e.g., [32–35,38,46,47]), Figure 3 shows the rainfall conditions and the threshold at 5% NEP. According to the frequentist method [46,47], the 5% NEP threshold leaves 5% of the empirical *ED* conditions below itself. The relative uncertainties of the parameters of the thresholds were also calculated. The 5% NEP threshold has low relative uncertainties (0.7/6.8 = 10.3%; 0.02/0.4 = 5%), which means a better distribution of the rainfall conditions.

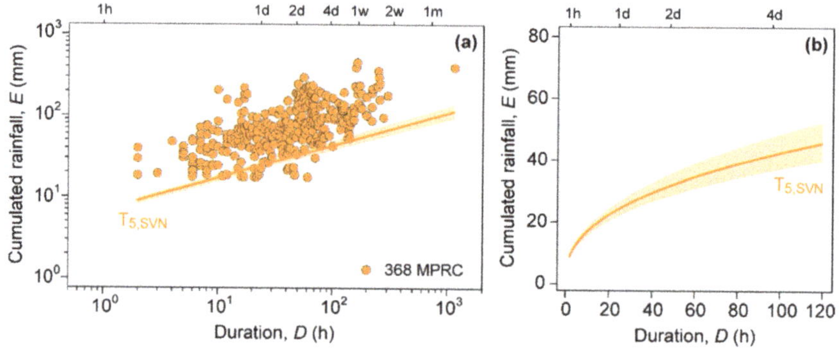

Figure 3. (**a**) Log-log plot with the cumulated event rainfall—duration *ED*, conditions that triggered landslides in Slovenia and the corresponding 5% *ED* threshold ($T_{5,SVN}$). (**b**) $T_{5,SVN}$ threshold in the range 1 h ≤ *D* ≤ 120 h, in linear coordinates. The shaded areas represent the threshold uncertainty.

Table 1. Main characteristics of rainfall thresholds defined in this study.

Name	Region	Area (km^2)	Number of MPRC*	Threshold Equation	Duration Range (h)	$\Delta\alpha/\alpha$ (%)	$\Delta\gamma/\gamma$ (%)
$T_{5,SVN}$	Slovenia	20,273	368	$E = (6.8 \pm 0.7)\cdot D^{(0.40 \pm 0.02)}$	2–1149	10.3	5.0
$T_{1,SVN}$	Slovenia	20,273	368	$E = (4.7 \pm 0.5)\cdot D^{(0.40 \pm 0.02)}$	2–1149	10.6	5.0
$T_{10,SVN}$	Slovenia	20,273	368	$E = (8.2 \pm 0.8)\cdot D^{(0.40 \pm 0.02)}$	2–1149	9.8	5.0
$T_{15,SVN}$	Slovenia	20,273	368	$E = (8.9 \pm 0.9)\cdot D^{(0.40 \pm 0.02)}$	2–1149	10.1	5.0
$T_{20,SVN}$	Slovenia	20,273	368	$E = (10.5 \pm 1.0)\cdot D^{(0.40 \pm 0.02)}$	2–1149	9.5	5.0
$T_{50,SVN}$	Slovenia	20,273	368	$E = (16.5 \pm 1.6)\cdot D^{(0.40 \pm 0.02)}$	2–1149	9.7	5.0
$T_{5,L}$	800 ≤ MAR ≤ 1300 mm	6538	137	$E = (8.3 \pm 1.1)\cdot D^{(0.34 \pm 0.04)}$	2–280	13.2	11.8
$T_{5,M}$	1300 ≤ MAR ≤ 1600 mm	6018	127	$E = (7.3 \pm 1.1)\cdot D^{(0.38 \pm 0.04)}$	2–243	15.0	10.5
$T_{5,H}$	1600 ≤ MAR ≤ 4000 mm	7717	104	$E = (7.2 \pm 1.6)\cdot D^{(0.41 \pm 0.05)}$	5–1149	22.2	11.9
$T_{5,IG}$	Igneous-metamorphic complex	1444	48	$E = (14.8 \pm 3.3)\cdot D^{(0.25 \pm 0.05)}$	2–139	22.3	20.0
$T_{5,LD}$	Limestone and dolomite	8803	72	$E = (8.9 \pm 2.2)\cdot D^{(0.36 \pm 0.06)}$	5–180	24.7	16.7
$T_{5,US}$	Unbound sediments	5601	106	$E = (5.3 \pm 0.9)\cdot D^{(0.47 \pm 0.04)}$	3–1149	17.0	8.5
$T_{5,BS}$	Bound sedimentary rocks	4425	142	$E = (5.9 \pm 0.9)\cdot D^{(0.42 \pm 0.04)}$	2–303	15.2	9.5

* MPRC—Maximum Probability Rainfall Condition.

4.2.1. Thresholds for Different Mean Annual Rainfall Classes

To investigate the role of the rainfall regime for the landslide triggering conditions in Slovenia, we used data on mean annual rainfall (MAR) provided by ARSO [42], which were divided into three classes. Figure 4 shows that the eastern part of Slovenia (32% of the total national territory) is characterized by low values of MAR (800 ≤ MAR ≤ 1300 mm), the central part (30%) by medium values (1300 < MAR ≤ 1600 mm) and the western part (38%) by high values (1600 < MAR ≤ 4000 mm). The number of landslides in the region characterized by a low, medium and high MAR class is 137, 127 and 104, respectively. The lowest density of landslides (one landslide every 74 km^2) is found in the area with high MAR values, while the other two areas are characterized by a similarly higher value of landslide density (one every 47 km^2).

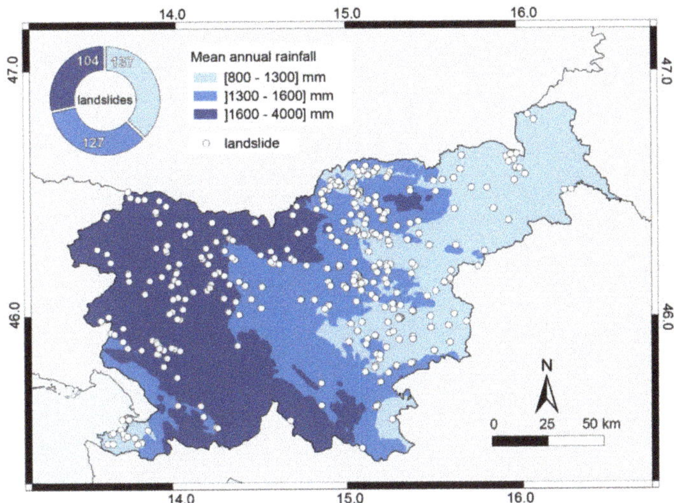

Figure 4. Subdivision of Slovenia based on different mean annual rainfall (MAR) between 1981 and 2010 [42] into three classes, with indication of the landslides used in the analysis. The donut chart shows the number of landslides in each class.

Figure 5a shows the MPRCs classified into three MAR classes, with the corresponding 5% *ED* thresholds, $T_{5,L}$, $T_{5,M}$ and $T_{5,H}$ (Table 1). The three thresholds are also shown in Figure 5b, in linear coordinates and in the range of duration $1 \leq D \leq 120$ h, with the shaded areas representing the uncertainty associated to each threshold. Inspection of Figure 5a and Table 1 reveals that the three point-clouds have different distributions and the subsets have diverse duration ranges, and the resulting thresholds have different parameters. In particular, α increases from 7.2 to 8.3, and γ decreases from 0.41 to 0.34 moving from $T_{5,H}$ to $T_{5,L}$. Therefore, the curves become higher and steeper with an increasing MAR (Table 1), ranging from $\alpha = 8.3 \pm 1.1$ and $\gamma = 0.34 \pm 0.02$ for the low MAR region to $\alpha = 7.1 \pm 1.6$ and $\gamma = 0.41 \pm 0.05$ for the high MAR region. This behavior is in accordance with the findings of Peruccacci et al. [39] in the nearby Italian territory: the rainfall required to trigger landslides increases with the MAR, which proves a sort of adaptation of the landscape to the average rainfall conditions. The relative uncertainty of α increases as the MAR class increases, while $\Delta\gamma/\gamma$ remains stable.

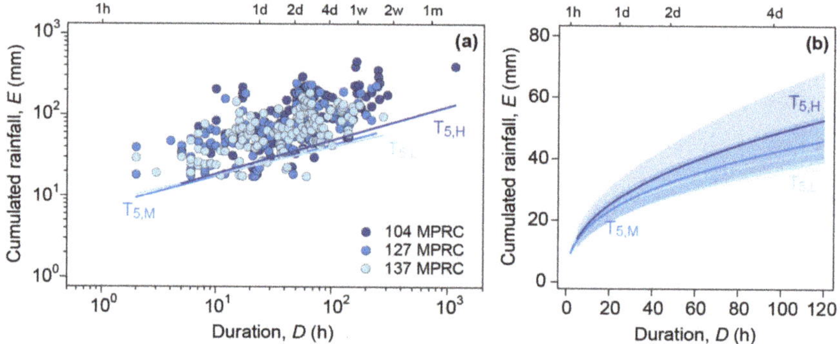

Figure 5. (**a**) Log-log plot with the *ED* (cumulated event rainfall—duration) conditions that triggered landslides in Slovenia classified according to three classes of mean annual rainfall (MAR) and corresponding 5% *ED* thresholds ($T_{5,H}$, $T_{5,M}$, $T_{5,L}$). (**b**) Same thresholds and related uncertainties (shaded areas) in the range $1\ \text{h} \leq D \leq 120\ \text{h}$, in linear coordinates. Legend: L, 800 mm ≤ MAR ≤ 1300 mm; M, 1300 mm < MAR ≤ 1600 mm; H, 1600 mm < MAR ≤ 4000 mm.

4.2.2. Thresholds for Lithological Classes

For the purpose of studying the role of lithology in the triggering of landslides in Slovenia, we used the Slovenian engineering geological map in scale 1:1,000,000 by Ribičič et al. [49] We have reclassified the 29 rock units into four classes (Figure 6): the igneous and metamorphic complex (IG class), limestone and dolomite (LD class), unbound sediments or sedimentary rocks (US class) and bound sedimentary rocks (BS class). Each class represents a unit of similar rock types that occur in Slovenia. The IG class includes diabase, andesites, granites and all types of volcanic sedimentary rocks; the LD class includes all types and forms of these occurring rocks; US class includes all the unconsolidated clastic sediments such as clay, marl, silt, sand, gravel and similar sediments, and the BS class represents all the occurring cemented fine-grained and coarse-grained clastic rocks.

Figure 6 shows the landslides considered for each class, where the class IG has the lowest number (48 of 368), class LD with 72, class US with 106, and class BS with 142 conditions. Overall, 67% of the considered landslides occurred in the areas of sedimentary rocks (bound and unbound), which take roughly half of the total territory of Slovenia (10,026 km^2, 49.5%), while landslides in the area of limestones and dolomites (8803 km^2, 43.4%) account for only 20% of the considered landslides. This proves the impact of lithology on landslide triggering conditions, as reported, e.g., by Jordanova et al. [29], Peruccacci et al. [39], Palladino et al. [40], Vennari et al. [50] and Gariano et al. [51] Sedimentary rocks are relatively unstable masses that are very susceptible to weathering and consequently accumulate thick eluvium, which is the main source of material for shallow landslides.

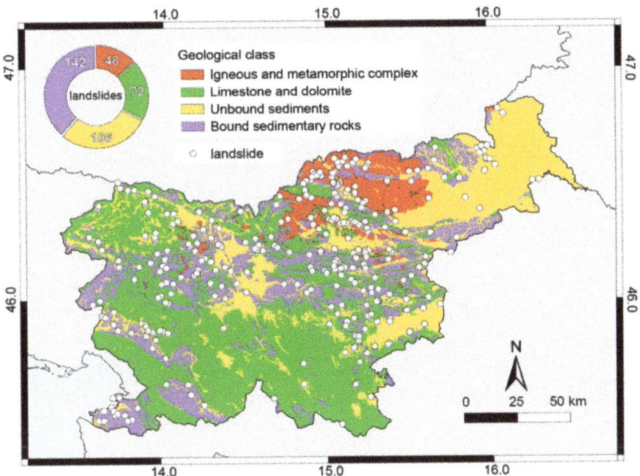

Figure 6. Subdivision of Slovenia into four main lithological classes based on the engineering geological map by Ribičič et al. [49], with indication of the landslides used in the analysis. The donut chart shows the number of landslides in each class.

Figure 7a shows the *ED* conditions in each lithology class in log–log coordinates with the corresponding 5% NEP thresholds, $T_{5,IG}$, $T_{5,LD}$, $T_{5,US}$ and $T_{5,BS}$ (Table 1). The same thresholds are shown in linear coordinates in Figure 7b, with the range of duration (*D*) varying from 1 to 120 h. Due to the small number of conditions for classes IG and LD, the uncertainties are too high, and the thresholds cannot be considered significant for these classes [39,47]. Nevertheless, they indicate the differences in the triggering conditions, with the IG class having the highest threshold and bound sedimentary rocks having the lowest. For obtaining more reliable thresholds, more empirical points are needed. However, a clear distinction in the minimum triggering conditions between landslides that occurred in sedimentary rocks and those that occurred in dolomite, limestone, igneous and metamorphic complexes can be currently observed.

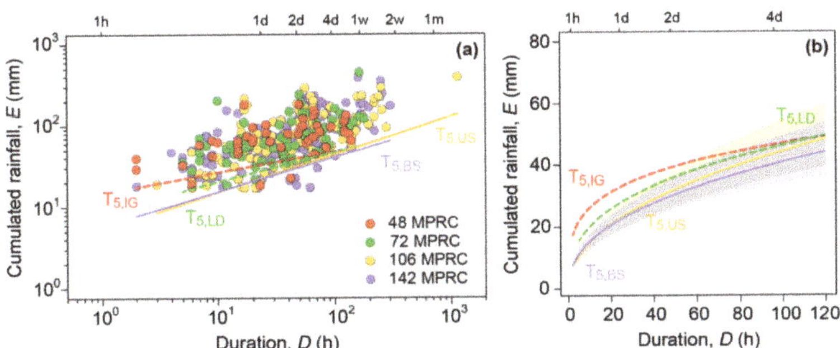

Figure 7. (**a**) Log–log plot with the *ED* (cumulated event rainfall—duration) conditions that triggered landslides in Slovenia classified in four geological classes and corresponding 5% thresholds ($T_{5,IG}$, $T_{5,LD}$, $T_{5,US}$, $T_{5,BS}$). (**b**) Same thresholds and related uncertainties (shaded areas) in the range $1 \text{ h} \leq D \leq 120 \text{ h}$ in linear coordinates. The thresholds with not-acceptable uncertainties are indicated with dotted lines. Legend: IG, Igneous and metamorphic complex; LD, Limestone and dolomite; US, Unbound sediments or sedimentary rocks; BS, Bound sedimentary rocks.

4.3. Threshold Validation

The validation of the national thresholds was based on two subsets of data: (i) a calibration set containing 70% of all reconstructed rainfall conditions (258), and (ii) a validation set containing the remaining 30% (110). The subsets were randomly selected 100 times. In addition, all those rainfall conditions that (presumably) did not cause landslides in the considered period were also reconstructed. The validation was performed 100 times, each time resulting differently; the number of conditions was always the same. The thresholds at different NEPs, calculated using the MRPC in the calibration set, are compared with the MPRC in the validation set and the rainfall conditions that did not trigger landslides. Therefore, 100 contingency tables were determined [33,51], reporting true positives (TP, i.e., landslide-triggering rainfall conditions predicted by the thresholds), false positive (FP, i.e., rainfall conditions not resulting in landslides incorrectly classified as landslide-triggering), true negatives (TN, i.e., rainfall conditions not resulting in landslides not predicted by the thresholds) and false negatives (FN, i.e., landslide-triggering rainfall conditions located below the threshold). Furthermore, three skill scores could be calculated: the true positive rate, i.e., TPR = TP/(TP + FN); the false positive rate, i.e., FPR = FP/(FP + TN); and the true skill statistics, i.e., TSS = TPR − FPR. Moreover, the FPR and TPR values were used to draw the receiver operating characteristic (ROC) curve (Figure 8). The best prediction is achieved when TPR = 1 (all observed landslides correctly detected) and FPR = 0 (no false positives) and is represented by the upper left green point in Figure 8 (best prediction point). The threshold which results closest to the best prediction point is assumed to be optimal.

Table 2 reports the mean values of the performing indexes for calculated thresholds at different NEPs, for the 100 validation runs. As the non-exceedance probability increases, the number of false negatives rises, and that of the true positives decreases. Conversely, lowering the thresholds causes an increase in the number of false positives and a decrease in the number of true negatives. In such cases, if the thresholds are used in a LEWS, false positives lead to false alarms and false negatives lead to missed alarms. It can be noted that the number of false positives can be greatly overestimated due to a lack of landslide information, i.e., many landslides may have occurred, but were not recorded. Likewise, even the true negatives can be overestimated. It has been observed that even a slight underestimation of the number of landslide occurrences can lead to an increase in uncertainty about prediction (and consequently system) performance [51].

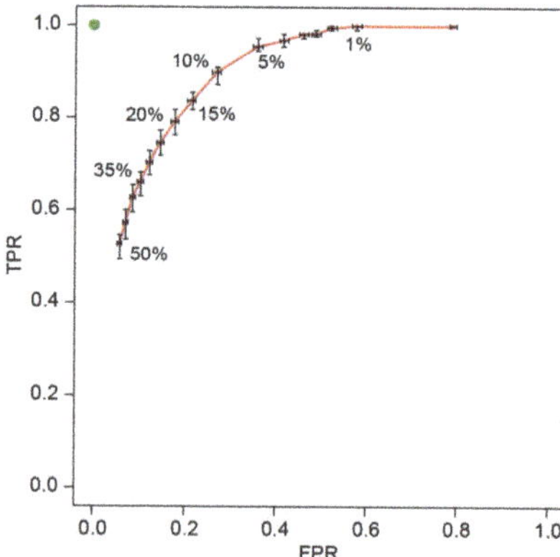

Figure 8. Classification of thresholds at different non-exceeding probabilities (black points) in the ROC space. The threshold closest to the best prediction point (green point) is the optimal threshold. Horizontal and vertical bars represent the range of variation of TPR and FPR for the 100 runs in which the MPRCs are randomly selected.

Table 2. Mean values of the performing indexes for calculated thresholds at different non-exceedance probability. The 15% threshold has the highest scoring indexes. NEP, non-exceeding probability; TP, true positive; FN, false negative; FP, false positive; TN, true negative; TPR, true positive rate; FPR, false positive rate; TSS, true skill statistics; δ, distance from perfect classification point. The optimal value for TPR and TSS is 1, while for FPR and δ is 0.

NEP	TP	FN	FP	TN	TPR	FPR	TSS	δ
1	109	1	5475	5009	0.99	0.52	0.47	0.52
5	105	5	3761	6723	0.96	0.36	0.60	0.36
10	98	12	2815	7669	0.89	0.27	0.63	0.29
15	92	18	2235	8249	0.84	0.21	0.63	0.27
20	86	23	1842	8642	0.79	0.18	0.61	0.28
35	72	37	1062	9423	0.66	0.10	0.94	0.36
50	57	52	590	9894	0.52	0.06	0.46	0.48

The validation showed that the best-performing threshold is that at 15% NEP, which has the shortest distance δ from the best prediction point, and also the highest mean value TSS in the 100 validation runs (Table 2; Figure 8). This threshold is represented by the equation:

$$E = (8.9 \pm 1.0)D^{(0.42 \pm 0.03)} \tag{2}$$

The relative uncertainties of these parameters are slightly higher ($\Delta\alpha/\alpha$ = 11.2%; $\Delta\gamma/\gamma$ = 7%) than the ones reported in Table 1. The reason behind this is in the lower number of rainfall conditions available (258 out of 368).

4.4. Comparison with Other Thresholds

Comparing the proposed new thresholds with the existing ones, in particular with the Slovenian threshold calculated by Rosi et al. [25] (4 in Figure 9), a large difference in the intercept of the thresholds and a small difference in the slope of the functions is noticeable: the new thresholds $T_{5,SVN}$, and $T_{15,SVN}$ are much lower than the previously calculated Slovenian thresholds [25]. Nevertheless, the thresholds defined in this work are higher than those defined for Central and Southern Europe (an area that includes Slovenia) by Guzzetti et al. [6] (2 in Figure 9) and lower, in particular at short durations, than the global thresholds by Caine [5] and Guzzetti et al. [7] (1 and 3 in Figure 9, respectively). In addition, these differences can be ascribed to the use of different sets of input data, such as the number of landslides and time period, as well as on the available rainfall data. Rosi et al. [25] used landslides that occurred between 2007 and 2014 and a limited rainfall dataset (1 rain gauge per 460 km^2). On the other hand, $T_{5,SVN}$ was defined with the same method and has the same resolution of rainfall data (hourly) and the same non-exceeding probability as the thresholds for Italy [39] (5 in Figure 9) and for the Italian Alpine area [40] (6 in Figure 9).

Interestingly, $T_{5,SVN}$ is very similar to the Italian threshold, while it has a slope that is different to the Alpine threshold. Comparing the 5% threshold defined for Slovenia with that defined with the same approach for Italy, some differences are observed. The Slovenian threshold has a similar slope and a lower intersection than the Italian one. Furthermore, the relative uncertainties for the Slovenian case study are higher. This is due to the lower number of empirical data points (368 compared to the 2309 in the Italian case) and also to a different distribution of points in the *ED* graph. In fact, the percentage of MPRC with $D \leq 6$ h is 5.4% in the Slovenian case and 12% in the Italian case. This is due to the coarser (daily vs. hourly) temporal resolution of the landslide data in Slovenia.

The difference between the new Slovenian thresholds and the threshold for the Alpine chain can be ascribed at the same cause. One could have expected that the Slovenian threshold would be similar to the Alpine one, given the similar environment and latitude. However, this difference is again related to the diverse temporal resolution of the two landslide catalogs: daily for Slovenia and hourly for the

Alps. Working with daily information for landslides can result in missing several very short (<6 h) rainfall events that can drive the slope of the threshold.

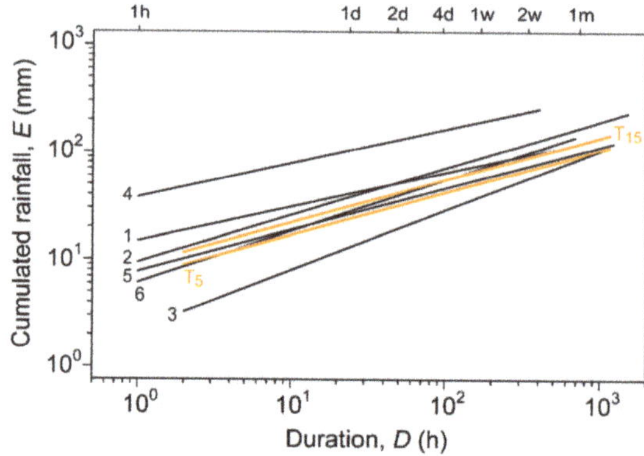

Figure 9. Comparison between the 5% and 15% thresholds for Slovenia and other global (1 and 3), regional (2 and 6) and national (4 and 5) thresholds. Source, numbered chronologically: 1, global threshold by Caine [5]; 2, threshold for Central and Southern Europe by Guzzetti et al. [6]; 3, global threshold by Guzzetti et al. [7]; 4, national threshold for Slovenia by Rosi et al. [25]; 5, national threshold for Italy by Peruccacci et al. [39]; 6, threshold for Alps by Palladino et al. [40].

5. Conclusions

In this paper, new *ED* thresholds for Slovenia were calculated using the automatic tool proposed by Melillo et al. [32] The main objective was to reconstruct the cumulated event rainfall and the duration of the rainfall conditions responsible for landslide occurrences in order to obtain reliable thresholds that could be implemented in a LEWS. Based on the presented results the following conclusions can be drawn.

The proposed *ED* thresholds were determined for the entire Slovenian territory, given that the current national landslide prediction system provides alerts at a national scale. Compared to other thresholds obtained with empirical approaches, the current curves are slightly lower. It should be noted, however, that the threshold used for the comparison is at 5% NEP. The main strength of the frequentist approach lies in the possibility of calculating thresholds at different non-exceeding probabilities, which could be used in probabilistic schemes to produce rising alert levels for landslide occurrence [17].

Thresholds for different MAR classes were also calculated. Due to the high relative uncertainties of the threshold parameters, not all the calculated thresholds can currently be implemented in a LEWS. However, they provide an idea of the landslide triggering conditions in the study area. The higher the mean annual rainfall in an area, the more rainfall is needed to trigger a landslide.

Thresholds for different geological classes were also determined. Due to the high relative uncertainties, not all the defined thresholds are reliable enough to be implemented in a LEWS. They can be considered only as an indicator of the rainfall conditions responsible for landslide occurrences in different lithological units. The sedimentary rocks are more subject to weathering and thus have the lowest thresholds and are by far more susceptible to landslide occurrences than those areas with limestone and magmatic bedrocks.

The main advantages of the tool—whose code is freely downloadable at http://geomorphology.irpi.cnr.it/tools/rainfall-events-and-landslides-thresholds/ctrl-t-algorithm/ctrl-code/ctrl_t_code.r/view—are (i) the fast processing of a large amount of data, which provides results in a short time, (ii) the reduction of subjectivity in the whole process of reconstructing rainfall conditions responsible for the failures and (iii) the definition and validation of rainfall thresholds.

The proposed 15% NEP threshold might be further tested using the national prototype LEWS and critically assessed on the basis of case studies, reviewing the landslide database and ensuring accurate information on the location and the occurrence date/time of the landslide.

Author Contributions: Conceptualization, G.J., S.L.G., M.M. and M.J.A.; Data curation, G.J. and M.J.A.; Formal analysis, G.J., S.L.G., M.M. and M.J.A.; Methodology, S.L.G., M.M., S.P. and M.T.B.; Software, S.L.G., M.M., S.P. and M.T.B.; Visualization, G.J., S.L.G., M.M. and S.P.; Writing—original draft, G.J., S.L.G., M.M., S.P. and M.T.B.; Writing—review and editing, all authors. All authors have approved the submitted version of the manuscript.

Funding: This research received no external funding.

Acknowledgments: This work was supported by the Slovenian Research Agency (Young researcher number 53536 and the research program P1-0195: "Geoenvironment and Geomaterials"). The authors would like to thank the Slovenian Administration for Civil Protection and Disaster Relief and the Ministry for Defense for financing the project Masprem, and the Slovenian Environment Agency (ARSO) for providing rainfall and rain gauge data. The authors would also like to thank the two anonymous reviewers and the editor for their constructive comments that helped to improve the manuscript. The code of CTRL-T was written using the R open-source software and can be freely downloaded at: http://geomorphology.irpi.cnr.it/tools/rainfall-events-and-landslides-thresholds/ctrl-t-algorithm/ctrl-code/ctrl_t_code.r/view. An example of the input files required by the algorithm can be freely downloaded at: http://geomorphology.irpi.cnr.it/tools/rainfall-events-and-landslides-thresholds/ctrl-t-algorithm/input-demo/INPUT.zip/view.

Conflicts of Interest: The authors declare no conflict of interest.

References

1. Komac, M. Intenzivne padavine kot sprožilni dejavnik pri pojavljanju plazov v Sloveniji=Rainstorms as a landslide-triggering factor in Slovenia. *Geologija* **2005**, *48*, 263–279. [CrossRef]
2. Gariano, S.L.; Guzzetti, F. Landslides in a changing climate. *Earth-Sci. Rev.* **2016**, *162*, 227–252. [CrossRef]
3. Campbell, R.H. *Soil Slips, Debris Flows, and Rainstorms in the Santa Monica Mountains and Vicinity, Southern California*; US Government Printing Office: Washington, DC, USA, 1975; Volume 851, p. 51.
4. Nilsen, T.H.; Turner, B.L. *Influence of Rainfall and Ancient Landslide Deposits on Recent Landslides (1950-71) in Urban Areas of Contra Costa County, California*; Government Printing Office: Washington, DC, USA, 1975; Volume 1388, p. 18.
5. Caine, N. The rainfall intensity: Duration control of shallow landslides and debris flows. *Geogr. Ann.* **1980**, *62*, 23–27. [CrossRef]
6. Guzzetti, F.; Peruccacci, S.; Rossi, M.; Stark, C.P. Rainfall thresholds for the initiation of landslides in central and southern Europe. *Meteorol. Atmos. Phys.* **2007**, *98*, 239–267. [CrossRef]
7. Guzzetti, F.; Peruccacci, S.; Rossi, M.; Stark, C.P. The rainfall intensity-duration control of shallow landslides and debris flows: An update. *Landslides* **2008**, *5*, 3–17. [CrossRef]
8. Segoni, S.; Piciullo, L.; Gariano, S.L. A review of the recent literature on rainfall thresholds for landslide occurrence. *Landslides* **2018**, *15*, 1483–1501. [CrossRef]
9. Piciullo, L.; Calvello, M.; Cepeda, J.M. Territorial early warning systems for rainfall-induced landslides. *Earth-Sci. Rev.* **2018**, *179*, 228–247. [CrossRef]
10. Guzzetti, F.; Gariano, S.L.; Peruccacci, S.; Brunetti, M.T.; Marchesini, I.; Rossi, M.; Melillo, M. Geographical landslide early warning systems. *Earth-Sci. Rev.* **2020**, *200*, 102973. [CrossRef]
11. Cannon, S.H.; Ellen, S.D. Rainfall conditions for abundant debris avalanches, San Francisco Bay region, California. *Calif. Geol. Surv.* **1985**, *38*, 267–272.
12. Wieczorek, G.F. Effect if rainfall intensity and duration on debris flows in central Santa Cruz Mountains. In *Debris Flow/Avalanches: Process, Recognition, and Mitigation*; Costa, J.E., Wieczorek, G.F., Eds.; Geological Society of America: Reviews in Engineering Geology: Boulder, CO, USA, 1987; Volume 7, pp. 93–104.
13. Ortigao, B.; Justi, M.G. Rio-watch: The Rio de Janeiro Landslide Alarm System. *Geotech. News* **2004**, *22*, 28–31.
14. Calvello, M.; d'Orsi, R.N.; Piciullo, L.; Paes, N.; Magalhaes, M.; Lacerda, W.A. The Rio de Janeiro early warning system for rainfall-induced landslides: Analysis of performance for the years 2010–2013. *Int. J. Disaster Risk Reduct.* **2015**, *12*, 3–15. [CrossRef]
15. Jakob, M.; Weatherly, H. A hydroclimatic threshold for landslide initiation on the North Shore Mountains of Vancouver, British Columbia. *Geomorphology* **2003**, *54*, 137–156. [CrossRef]

16. Wei, L.-W.; Huang, C.-M.; Chen, H.; Lee, C.-T.; Chi, C.-C.; Chiu, C.-L. Adopting the I3-R24 rainfall index and landslide susceptibility for the establishment of an early warning model for rainfall-induced shallow landslides. *Nat. Hazards Earth Syst.* **2018**, *18*, 1717–1733. [CrossRef]
17. Rossi, M.; Marchesini, I.; Tonelli, G.; Peruccacci, S.; Brunetti, M.T.; Luciani, S.; Ardizzone, F.; Balducci, V.; Bianchi, C.; Cardinali, M.; et al. TXT-tool 2.039-1.1 Italian national early warning system. In *Landslide Dynamics: ISDR-ICL Landslide Interactive Teaching Tools*; Sassa, K., Guzzetti, F., Yamagishi, H., Arbanas, Ž., Casagli, N., McSaveney, M., Dang, K., Eds.; Springer: Cham, Switzerland, 2018; pp. 341–349. [CrossRef]
18. Segoni, S.; Rosi, A.; Fanti, R.; Gallucci, A.; Monni, A.; Casagli, N. A Regional-Scale Landslide Warning System Based on 20 Years of Operational Experience. *Water* **2018**, *10*, 1297. [CrossRef]
19. Aleotti, P. A warning system for rainfall-induced shallow failures. *Eng. Geol.* **2004**, *73*, 247–265. [CrossRef]
20. Rosi, A.; Segoni, S.; Catani, F.; Casagli, N. Statistical and environmental analyses for the definition of a regional rainfall threshold system for landslide triggering in Tuscany (Italy). *J. Geogr. Sci.* **2012**, *22*, 617–629. [CrossRef]
21. Segoni, S.; Rosi, A.; Rossi, G.; Catani, F.; Casagli, N. Analysing the relationship between rainfalls and landslides to define a mosaic of triggering thresholds for regional scale warning systems. *Nat. Hazards Earth Syst.* **2014**, *14*, 2637–2648. [CrossRef]
22. Rosi, A.; Lagomarsino, D.; Rossi, G.; Segoni, S.; Battistini, A.; Casagli, N. Updating EWS rainfall thresholds for the triggering of landslides. *Nat. Hazards* **2015**, *78*, 297–308. [CrossRef]
23. Brigandì, G.; Aronica, G.T.; Bonaccorso, B.; Gueli, R.; Basile, G. Flood and landslide warning based on rainfall thresholds and soil moisture indexes: The HEWS (Hydrohazards EarlyWarning System) for Sicily. *ADGEO* **2017**, *44*, 79–88. [CrossRef]
24. Jemec Auflič, M.; Komac, M. Rainfall patterns for shallow landsliding in perialpine Slovenia. *Nat. Hazards* **2011**, *67*, 1011–1023. [CrossRef]
25. Rosi, A.; Peternel, T.; Jemec Auflič, M.; Komac, M.; Casagli, N. Rainfall thresholds for rainfall-induced landslides in Slovenia. *Landslides* **2016**, *13*, 1571–1577. [CrossRef]
26. Segoni, S.; Rossi, G.; Rosi, A.; Catani, F. Landslides triggered by rainfall: A semi-automated procedure to define consistent intensity-duration thresholds. *Comput. Geosci.* **2014**, *63*, 123–131. [CrossRef]
27. Bezak, N.; Šraj, M.; Mikoš, M. Copula-based IDF curves and empirical rainfall thresholds for flash floods and rainfall-induced landslides. *J. Hydrol.* **2016**, *541*, 272–284. [CrossRef]
28. Bezak, N.; Jemec Auflič, M.; Mikoš, M. Application of hydrological modelling for temporal prediction of rainfall-induced shallow landslides. *Landslides* **2019**, *16*, 1273–1283. [CrossRef]
29. Jordanova, G.; Verbovšek, T.; Jemec Auflič, M. Validation and proposal of new rainfall thresholds for shallow landslide prediction in Posavsko hills, Eastern Slovenia. In Proceedings of the 4th Regional Symposium on Landslides in the Adriatic-Balkan Region, Sarajevo, Bosnia and Herzegovina, 23–25 October 2019; Uljarević, M., Zekan, S., Salković, S., Ibrahimović, D., Eds.; Geotechnical Society of Bosnia and Herzegovina: Sarajevo, Bosnia and Herzegovina, 2019; pp. 37–42.
30. Jemec Auflič, M.; Šinigoj, J.; Krivic, M.; Podboj, M.; Peternel, T.; Komac, M. Landslide prediction system for rainfall induced landslides in Slovenia (Masprem). *Geologija* **2016**, *59*, 259–271. [CrossRef]
31. Jemec Auflič, M.; Šinigoj, J. Validation of the Slovenian national landslide forecast system using contingency matrices. *Geophys. Res. Abstr.* **2019**, *21*, 1. Available online: https://meetingorganizer.copernicus.org/EGU2019/EGU2019-13338.pdf (accessed on 30 April 2020).
32. Melillo, M.; Brunetti, M.T.; Peruccacci, S.; Gariano, S.L.; Roccati, A.; Guzzetti, F. A tool for the automatic calculation of rainfall thresholds for landslide occurrence. *Environ. Model. Softw.* **2018**, *105*, 230–243. [CrossRef]
33. Gariano, S.L.; Melillo, M.; Peruccacci, S.; Brunetti, M.T. How much does the rainfall temporal resolution affect rainfall thresholds for landslide triggering? *Nat. Hazards* **2020**, *100*, 655–670. [CrossRef]
34. Melillo, M.; Brunetti, M.T.; Peruccacci, S.; Gariano, S.L.; Guzzetti, F. An algorithm for the objective reconstruction of rainfall events responsible for landslides. *Landslides* **2015**, *12*, 311–320. [CrossRef]
35. Melillo, M.; Brunetti, M.T.; Peruccacci, S.; Gariano, S.L.; Guzzetti, F. Rainfall thresholds for the possible landslide occurrence in Sicily (Southern Italy) based on the automatic reconstruction of rainfall events. *Landslides* **2016**, *13*, 165–172. [CrossRef]

36. Bordoni, M.; Corradini, B.; Lucchelli, L.; Valentino, R.; Bittelli, M.; Vivaldi, V.; Meisina, C. Empirical and Physically Based Thresholds for the Occurrence of Shallow Landslides in a Prone Area of Northern Italian Apennines. *Water* **2019**, *11*, 2653. [CrossRef]
37. Teja, T.S.; Dikshit, A.; Satyam, N. Determination of Rainfall Thresholds for Landslide Prediction Using an Algorithm-Based Approach: Case Study in the Darjeeling Himalayas, India. *Geosciences* **2019**, *9*, 302. [CrossRef]
38. Gariano, S.L.; Sarkar, R.; Dikshit, A.; Dorji, K.; Brunetti, M.T.; Peruccacci, S.; Melillo, M. Automatic calculation of rainfall thresholds for landslide occurrence in Chukha Dzongkhag, Bhutan. *Bull. Eng. Geol. Environ.* **2019**, *78*, 4325–4332. [CrossRef]
39. Peruccacci, S.; Brunetti, M.T.; Gariano, S.L.; Melillo, M.; Rossi, M.; Guzzetti, F. Rainfall thresholds for possible landslide occurrence in Italy. *Geomorphology* **2017**, *290*, 39–57. [CrossRef]
40. Palladino, M.R.; Viero, A.; Turconi, L.; Brunetti, M.T.; Peruccacci, S.; Melillo, M.; Deganutti, A.M.; Guzzetti, F. Rainfall thresholds for the activation of shallow landslides in the Italian Alps: The role of environmental conditioning factors. *Geomorphology* **2018**, *303*, 53–67. [CrossRef]
41. Komac, M.; Ribičič, M. Landslide susceptibility map of Slovenia at scale 1:250,000. *Geologija* **2006**, *49*, 295–309. [CrossRef]
42. ARSO. Mean Annual Measured Precipitation between 1981 and 2010. Ministry for Environment and Spatial Planning. Environmental Agency of the Republic of Slovenia. 2016. Available online: http://www.meteo.si/uploads/probase/www/climate/image/sl/by_variable/precipitation/mean-annual-measured-precipitation_81-10.png (accessed on 18 December 2019).
43. Thornthwaite, C.W. An approach toward a rational classification of climate. *Geogr. Rev.* **1948**, *38*, 55–94. [CrossRef]
44. Thornthwaite, C.W.; Mather, J.R. Instructions and tables for computing potential evapotranspiration and the water balance. *Publ. Climatol. Lab. Climatol. Drexel Inst. Technol.* **1957**, *10*, 185–311.
45. Dragoni, W.; Cambi, C.; Di Matteo, L.; Giontella, C.; Melillo, M.; Valigi, D. Possible response of two water systems in central Italy to climatic changes. In *Advances in Watershed Hydrology*; Moramarco, T., Barbetta, S., Brocca, L., Eds.; Water Resources Publications, LLC: Denver, CO, USA, 2015; pp. 397–424. ISBN 13 978-1-887-20185-8.
46. Brunetti, M.T.; Peruccacci, S.; Rossi, M.; Luciani, S.; Valigi, D.; Guzzetti, F. Rainfall thresholds for the possible occurrence of landslides in Italy. *Nat. Hazards Earth Syst.* **2010**, *10*, 447–458. [CrossRef]
47. Peruccacci, S.; Brunetti, M.T.; Luciani, S.; Vennari, C.; Guzzetti, F. Lithological and seasonal control of rainfall thresholds for the possible initiation of landslides in central Italy. *Geomorphology* **2012**, *139*, 79–90. [CrossRef]
48. Peres, D.J.; Cancelliere, A.; Greco, R.; Bogaard, T.A. Influence of uncertain identification of triggering rainfall on the assessment of landslide early warning thresholds. *Nat. Hazards Earth Syst.* **2018**, *18*, 633–646. [CrossRef]
49. Ribičič, M.; Komac, M.; Kumelj, Š.; Novak, M. Splošna Inženirsko-Geološka Karta Slovenije (General Engineering Geology Map of Slovenia). Geološki Zavod Slovenije—Geological Survey of Slovenia, 2008. Scale 1:1.000.000. Available online: http://www.egeologija.si/geonetwork/srv/eng/catalog.search#/metadata/d179cbc6-75fa-4a07-9096-814b07ff95a3 (accessed on 30 April 2020).
50. Vennari, C.; Gariano, S.L.; Antronico, L.; Brunetti, M.T.; Iovine, G.; Peruccacci, S.; Terranova, O.; Guzzetti, F. Rainfall thresholds for shallow landslide occurrence in Calabria, southern Italy. *Nat. Hazards Earth Syst.* **2014**, *14*, 317–330. [CrossRef]
51. Gariano, S.L.; Brunetti, M.T.; Iovine, G.; Melillo, M.; Peruccacci, S.; Terranova, O.; Vennari, C.; Guzzetti, F. Calibration and validation of rainfall thresholds for shallow landslide forecasting in Sicily, southern Italy. *Geomorphology* **2015**, *228*, 653–665. [CrossRef]

Article

Numerical Runout Modeling Analysis of the Loess Landslide at Yining, Xinjiang, China

Longwei Yang [1,2,*], Yunjie Wei [2], Wenpei Wang [2] and Sainan Zhu [2]

1 School of Geological Engineering and Geomatics, Chang'an University, Xi'an 710054, China
2 China Institute of Geo-Environment Monitoring, CGS, Beijing 100081, China
* Correspondence: yang0504@chd.edu.cn; Tel.: +86-134-7412-8085

Received: 12 May 2019; Accepted: 20 June 2019; Published: 26 June 2019

Abstract: The Panjinbulake loess landslide is located in the western part of the Loess Plateau, in Yining County, Xinjiang, China. It is characterized by its long runout and rapid speed. Based on a field geological survey and laboratory test data, we used the DAN-W dynamic numerical simulation software (Dynamic Analysis Of Landslides, Release 10, O. Hungr Geotechnical Research Inc., West Vancouver, BC, Canada) and multiple sets of rheological models to simulate the whole process of landslide movement. The best rheological groups of the features of the loess landslide process were obtained by applying the Voellmy rheological model in the debris flow area and applying the Frictional rheological model in the sliding source area and accumulation area. We calculated motion features indicating that the landslide movement duration was 22 s, the maximum movement speed was 20.5 m/s, and the average thickness of the accumulation body reached 5.5 m. The total accumulation volume, the initial slide volume and the long runout distance were consistent with the actual situation. In addition, the potential secondary disaster was evaluated. The results show that the DAN-W software and related model parameters can accurately simulate and predict the dynamic hazardous effects of high-speed and long runout landslides. Together, these predictions could help local authorities make the best hazard reduction measures and to promote local development.

Keywords: loess landslide; DAN-W; numerical simulation; dynamic analysis

---

1. Introduction

Landslides are among the most destructive geological disasters with features of rapid speed, long runout distance, and entrainment effect [1–3]. Catastrophic landslide events are often triggered by heavy rainfall, earthquake, and engineering activities [4–6]. According to the spatial characteristics and trajectory of the sliding body, the entire process of landslide movement is mainly divided into three stages: The starting stage at the slide source area, the propagating stage, and the deposition stage [7–10]. In addition, entrainment, base liquefaction, and air cushioning occur during the landslide movement [11–13]. To reduce the landslide hazard loss, risk assessment is often requested [14–16]. The landslide runout analysis is a very effective method to assess a landslide hazard [17]. Landslide runout analysis involves two aspects: The simulation of previous landslides and the prediction of potential landslides [18]. Runout analysis could be used to design remedial engineering measures, such as barricades and berms [17]. The maximum runout distance, propagation velocity, and the deposit thickness, and provision of the basis for the design of remedial engineering measures, are obtained by landslide runout analysis.

Landslide runout analysis methods mainly include empirical-statistical methods and numerical models [19,20]. Empirical methods establish the geometric relationship between the landslide volume, height difference, and angle of reach (i.e., the angle of the line between the highest point of the rear edge and the farthest point of the sliding distance) to predict the sliding distance. The empirical methods could not precisely predict the runout distance in the different complex geological environments, including entrainment, friction resistance, and impaction. Compared with the empirical method, the numerical models could give more information about the dynamic features of the sliding mass under different geological environments, such as the scraping depth, the thickness of the accumulation area, velocity, and the scope of dangerous area. The numerical simulation methods include the discrete element methods and the continuum methods. The discrete element method is based on Newton's second law and is used to analyze the interaction of particles constituting a landslide. It is suitable for landslides with debris flow patterns, such as the MatDEM (i.e., Fast GPU Matrix computing of Discrete Element Method, Nanjing University) and PFC (i.e., Particle Flow Code, Itasca) [21,22]. The continuum method, based on the momentum and mass equations incorporating the earth pressure theory, simulates the motion characteristics of the slip mass to obtain the velocity, position, and thickness of the slip mass [23,24]. The continuum methods have been successfully used to simulate previous hazards and predict potential hazards, such as debris flow, landslides, landslide bam, and avalanches [25–27]. The methods evolved into models such as the GeoFlow-SPH [28], LS-RAPID [29], Flow-2D [30], Kinematic model [31] and DAN model [23]. Based on the fluid continuity equation and motion equation, Hungr proposed the landslide dynamics model software DAN-W, which regards the sliding body as an equivalent fluid and can accurately calculate sliding motion characteristics [23]. These models provide a good method for the risk assessment of the geologic hazard.

The loess geological hazards frequently occurred in Tajik and Kazakhstan Tian shan area, which have also become a focus [32]. The loess has quite a widespread distribution in the world and it occupies approximately 10% of the total global land area. China is the country with most widely distributed loess area in the world. Loess is mainly distributed in the northwestern part of China, on the Loess Plateau, which covers an area of nearly 630,000 km^2, accounting for 4.4% of China's land area [33,34] (Figure 1a). Due to its special geological structure, loess has high water sensitivity (i.e., loess undergoes a structural collapse when wetted) and is prone to geological hazards. The main types of hazards are loess landslides, such as the Heifangtai Landslide Group and Jingyang Landslide Group [35–39]. Among them, high-speed and long runout loess landslides have caused considerable losses in terms of human lives and property and have become an important research topic. Due to the porosity, weak cementation and water sensitivity of loess as the water content increases, the shear strength of loess declines sharply, and the loess structure is destroyed [40–43]. In addition, the strength loss in loess might also be a chemo-mechanical problem that involves volume and stress changes of the finest component due to changes in pore water salinity. Consequently, a loess slope loses stability and slides. Furthermore, the pore water pressure rises during the sliding process, and the phenomenon of motion liquefaction occurs, which readily forms a high-speed and long runout landslide [44–47].

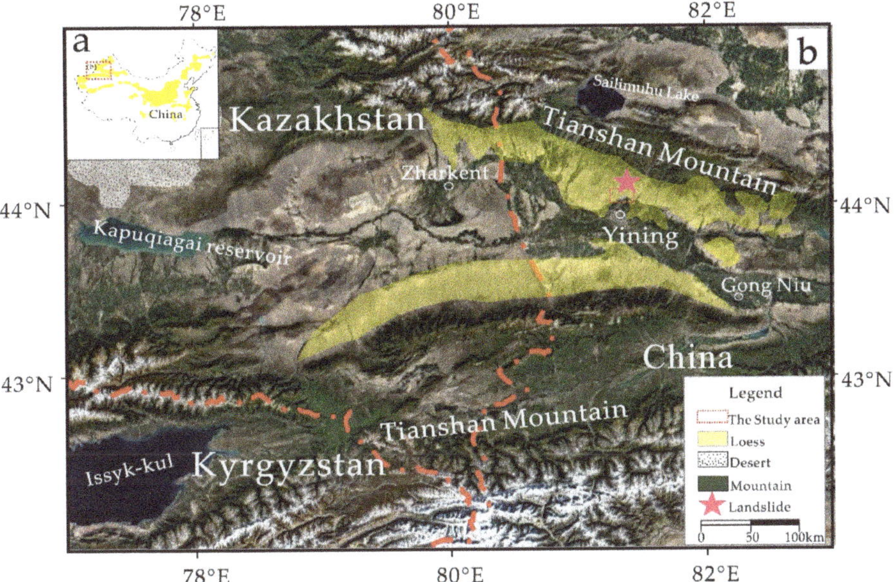

Figure 1. Location map of the Panjinbulake loess landslide (**a** is modified from [33]; **b** is modified from [48]). (**a**) The distribution of loess in China. (**b**) The distribution of loess in the study area.

In the study, we studied the characteristics of the Panjinbulake loess landslide through a field geological survey and aerial image analysis using drones. We used the landslide dynamics model DAN-W and multiple sets of rheological models to calculate the dynamic characteristics of this landslide. By using the different rheological models to simulate the different stages of the loess landslide (i.e., triggering in the sliding source area, propagation in the debris flow channel area, and deposition in the accumulation area), the best rheological model groups and parameters were obtained to improve the accuracy to analyze the loess dynamic characteristic. The potential secondary failure of the landslide was evaluated. This study could offer a basis to predict the potential landslide runout distance and define the hazard area, make necessary measures to prevent landslide induced damages (e.g., engineering measures, landslide early warning systems, and emergency response), and to favour local development.

2. Site Overview

The Panjinbulake loess landslide is located in the Karayagaq Township, Yining County, Xinjiang. The coordinates of the central point of the landslide are 81°30′32″ E and 44°11′48″ N, 53 km from downtown Yining City, and 43 km from Sailimu Lake (Figure 1b).

Among the landslide geological hazards in the Piliqinghe basin, most are high speed and long runout loess landslides, which pose a huge hazard to local agricultural and livestock production. Among these landslides, the Kezileisai landslide buried 402 cattle and 5 sheep, and the direct economic losses reached 490,600 Yuan (¥). There were some similar landslides in the Piliqinghe Basin (Figure 2b). The Panjinbulake loess landslide is a typical high speed and long runout loess landslide that occurred recently in the region. Through analysis of the induced factors and dynamic effects of the landslide, this study provides a reference to the landslide dynamic hazards in the loess area.

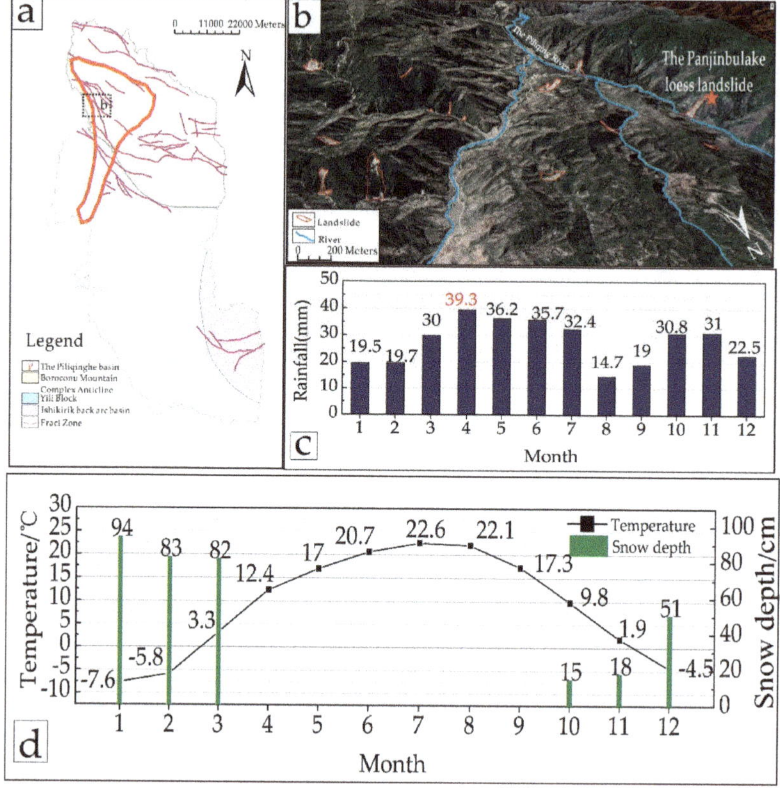

Figure 2. (**a**) The geological structure map in Yining Country. (**b**) The loess landslide distribution in the Piliqinghe Basin. (**c**) The monthly average precipitation of Yining County over time (2011–2016). (**d**) The monthly average temperature–snow depth map of Yining County over time (2011–2016).

3. Post-Failure Behavior and Landslide Influential Factors

3.1. Post-Failure Behavior

The triggering of landslides was mainly from snow infiltration, which turn into long runout and rapid landslides, constitutes a typical disaster model in the loess area. The Panjinbulake landslide belongs to this typical geologic hazard model. Following the instability failure of the landslide, the front loess main body slipped due to the river flushing action. Pore-water pressure increased and soil saturation during the sliding process because of the snow infiltration. The landslide was transformed into debris flow, showing a flow state. Then, the landslide volume increased gradually with entrainment effects and flowed into the Pliqinghe Gully. Finally, the landslide struck the opposite side of the mountain and stopped.

3.2. Landslide Influential Factors

According to the field investigation, the occurrence of the Panjinbulake landslide was caused by a combination of factors, including geological structure and formation lithology, topography and hydrogeological condition.

3.2.1. Geological Structure and Formation Lithology

The landslide research area belongs to the western part of the Yili Valley. The terrain is generally high in the north and low in the south. It is gradually inclined from the northeast to the southwest. The elevation of the area is between 620 and 3700 m. It is a block-like eroded and uplifted mountain, covered with gravel and loess layers, showing a low mountain grassland landscape. The landslide area is located on the southwestern side of the West Tianshan Youdi trough fold belt in the southwestern Tianshan fold of the Tianshan-Xing'an trough fold area. It belongs to the junction of the Boroconu Mountain Complex Anticline and Yili Block and is located 2.7 km south of the Nalati deep fault zone (Figure 2a). The rocks in the study area are mainly Ishikirik group tuff, tuff lava, and gray-green coarse sandstone in the Carboniferous system and basalt in the Dahala Junshan Formation of the Carboniferous system. The surface layer is the Quaternary Holocene loess, with well-developed joints (Figure 2b). The structure was relatively loose, and the wormhole, large void structure, belonging to low-plastic silt. The silt (0.075–0.005 mm) content of the loess in the Piliqinghe area is high, reaching 69.8–86.0%; the fine sand (0.25–0.075 mm) content is 3.7–18.0%; and the clay (<0.005 mm) content is 10.3–12.2%. As a result, the loess expands and collapses after encountering water, and it is prone to motion liquefaction under certain static or dynamic water conditions, which provide good source conditions for landslides.

3.2.2. Topography

The Panjinbulake loess landslide was located on the south bank of the river and had long been subjected to the lateral erosion of the river, resulting in good conditions for the landslide front to be in the air. The hillslope was steep, with a slope of 40° (Figure 3a). Corresponding tensile stress condition occurred near the top of the mountain, and the cracks at the trailing edge of the landslide were gradually enlarged (Figure 3b). According to the Google Earth remote sensing image map from 18 May 2013, the front edge of the landslide had slipped. The sliding volume was about 9000 m^3 (Figure 3a), and several tensile cracks appeared on the trailing edge. Consequently, a steep ridge of up to 1 m was formed (Figure 3c,d), which provided good topographic conditions for loess deformation and stress relief.

3.2.3. Hydrogeological Condition

The landslide area belongs to the temperate continental semi-arid climate. The average annual precipitation (for the period of 2011–2016) is 330.6 mm. Precipitation is highest from March to July, during which the monthly rainfall exceeds 30 mm, accounting for 52.5% of the annual rainfall (Figure 2c). Snowfall mainly occurs from October to the following March. The snowfall thickness can reach 94 mm per month. The fissure water inside the slope is frozen, and the vertical joints and cracks become enlarged due to the frost heaving action. From mid-March, the temperature rises, the snow that covered the surface begins to melt, infiltrating the cracks and joints and forming a certain transient water pressure and transient saturation zone in the surface layer of the slope. This results in a decrease in the anti-sliding force of the slope, thus inducing landslides. The rising water level of the river also causes the hydraulic gradient inside the slope to drop. Ice and snow meltwater can also be stored in the mountain for a long time and can continue to increase the slope sliding force and accelerate the formation of the potential slip surface. In addition, fissures are relatively developed and accumulated in the bedrock, which is exposed in the landslide. There is a large amount of ice and snow meltwater, which readily forms a "pipeline" channel that is in contact with the surface of the Quaternary aeolian loess and is discharged outward in the form of a spring. The flow volume of a spring was measured to be 5×10^{-5} m^3/s. The mineralization of water is less than 1.0 g/L. Based on the soil test, the saturation of the soil is 85.7–91.2%, which shows that it is a very wet sliding body. The natural moisture content of the soil was 17.2–20.4%, the plastic limit was 15.8%, and the liquid limit was 27.6%. According to the measured data of groundwater level, the groundwater in the hill is shallow and buried in the range of 0–15 m. These provide good hydrological conditions for inducing landslides and also provide good groundwater conditions for the rapid conversion of loess landslides into high speed and long runout sliding landslides.

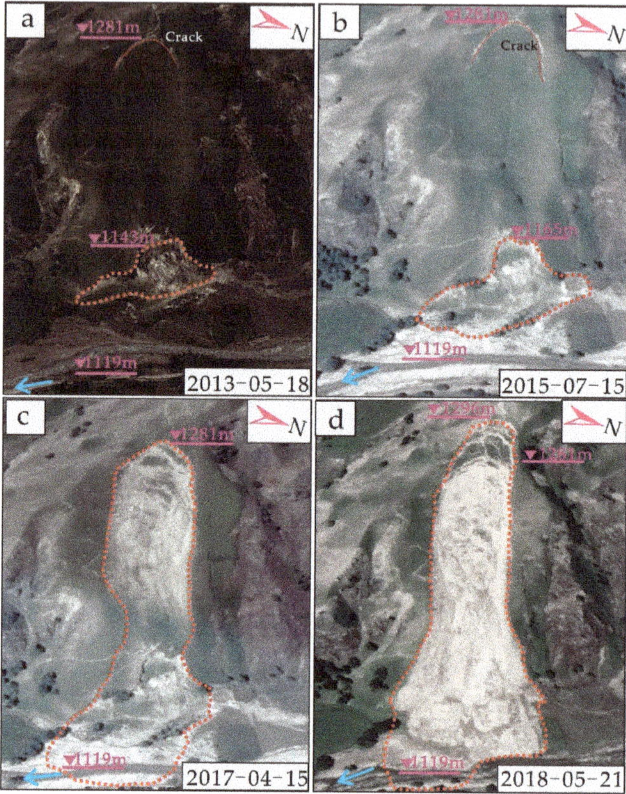

Figure 3. The multi-temporal sensing images of the landslide.

Qualitative analysis of the disaster-formation mechanism of the Panjinbulake landslide was conducted based on a field geological survey and remote sensing satellite images, but this was far from adequate for geological disaster prevention and control. Instead, quantitative analysis methods are required to fully investigate landslide movement and predict the secondary disaster, which can be explored using the landslide dynamic analysis software DAN-W (see Section 5).

4. Basic Characteristics of Landslides and Hazard Zoning

The landslide had a long-strip shape (Figure 4a). The slope before the landslide was close to 40°, and the main slip direction was N 69° E. The elevation of the trailing edge of the landslide was about 1280 m, the elevation of the landslide shearing edge was about 1190 m, and the horizontal distance reached 375 m (Figure 4b). Due to rainfall and snow infiltration, pore water pressure and soil saturation increased during the sliding process. The landslide was transformed into debris flow, showing a flow state. Based on the information obtained from the unmanned aerial vehicle data and remote sensing, geological field surveys, the landslide can be divided into the sliding source area, debris flow area, and accumulation area (Figure 5).

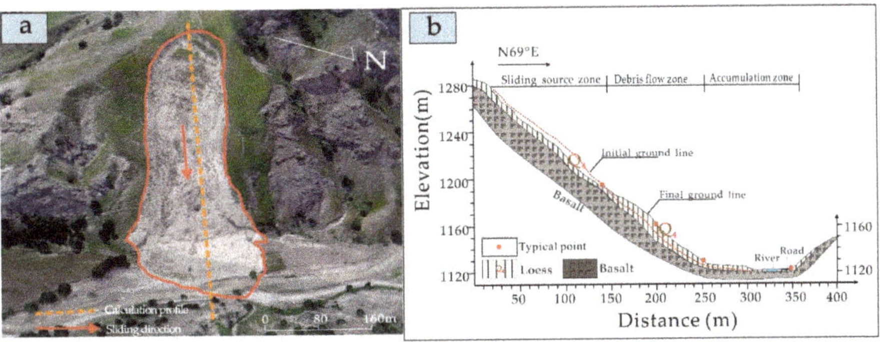

Figure 4. (**a**) Overview of the Panjinbulake landslide. (**b**) The engineering geological section.

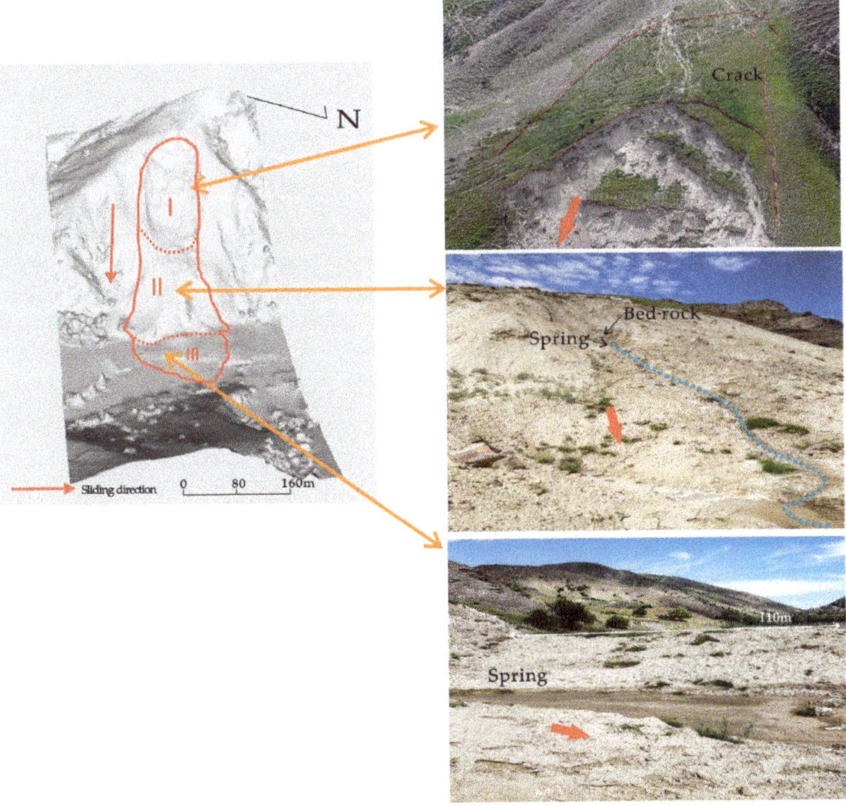

Figure 5. Three-dimensional Digital Elevation Model diagram of the landslide and typical pictures.

4.1. Sliding Source Area (Area I)

The sliding body was mainly Quaternary loess. There were many cracks in the upper part of the trailing edge, the crack width ranged from 23 to 54 cm, and there was an unstable body in the upper part of the trailing edge. Its volume reached 1.2×10^4 m^3. The average width of the sliding source area was 73 m. The area was about 1.2×10^4 m^2, the average thickness of the sliding body was 4–7 m, and the volume was about 5.0×10^4 m^3.

4.2. Debris Flow Area (Area II)

The debris flow area presented a long and narrow shape, and bedrock appeared in the upper part of the debris flow area. The bedrock surface had signs of scratch, and there was an exposed water head. There was also a water head on the western side of the sliding source area. This might suggest that there was sufficient groundwater in the area that promoted the sliding of the landslide. In this area, the volume of the sliding body increased due to the entrainment. The area was 6.8×10^3 m^2, the average thickness of the sliding body was 3–6 m, and the volume was about 25.0×10^4 m^3.

4.3. Accumulation Area (Area III)

The accumulation area had a fan shape, with a length of up to 110 m along the sliding direction and a maximum width of 100 m in the vertical sliding direction. The lithology of the accumulation area is dominated by Quaternary loess, which contains moderately weathered coarse sandstone scraped off the opposite mountain. The landslide struck the opposite side of the mountain and accumulated in the Piliqinghe Gully. The area was 1.46×10^4 m^2, the thickness of the sliding body was 3–6 m, and the volume was about 6×10^4 m^3.

5. Dynamic Analysis

5.1. Theoretical Basis

DAN-W is numerical simulation software developed by Hungr to simulate the whole process of landslide movement and to study the dynamics of landslides [23]. The 3D numerical model was set up according to the two-dimensional simulation conditions provided by the calculation profile in Figure 3b. Based on the aerial views, the path widths of landslide were confirmed. In DAN-W, the Lagrangian analytical solution of the Saint–Venant equation is mainly used to treat the sliding body with the rheological features that are formed by a combination of several blocks with certain materials (Figure 6). In the curve coordinates, the corresponding physical equations and equilibrium equations are established for each block (Figure 6), as in Equations (1)–(7) [23].

$$F = \gamma H_i B_i ds \sin\alpha + P - T \tag{1}$$

Here, F is the sliding force (N); γ is the unit weight (KN/m^3); H is the block height (m); B is the block width (m); ds is the nominal length of the block(m); α is the slope foot (°); P is the internal tangential pressure (N); and T is the base resistance (N); i is the block index.

$$V_i = v_i' + \frac{g(F\Delta t - M)}{\gamma H_i B_i ds} \tag{2}$$

Here, *V* is the new speed when sliding body movement. The new velocity at the end of a time step is obtained from the old velocity, *v* (m/s); g is the gravitational acceleration(m/s^2); Δt is the time step interval(s); M is momentum flux; and the other parameters are the same as in Equation (1).

$$h_j = \frac{2v_j}{(S_{i+1} - S_i)(B_{i+1} + B_i)} \tag{3}$$

Here, h is the average depth of the slip mass; j is block boundary index; i is block index; S is the curve displacement (m); and the other parameters are the same as in Equation (1).

$$V = V_R + \sum V_{point} + \sum_{i=1}^{n} Y_i L_i \tag{4}$$

Here, V is the entire volume of the loess landslide deposits(m^3); V_R is the volume of the initial landslide (m^3); V_{point} is the volume of the unstable body (m^3); Y is the yield rate; L is the length of the i block; and i is the block index.

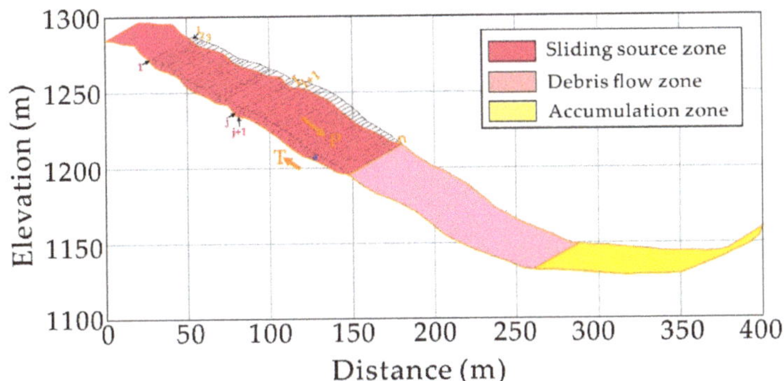

Figure 6. The 3D numerical DAN-W model of the landslide. The parameters are the same as in Equations (1)–(7) (It is modified from [23]).

Momentum and mass during the entrainment of the path material could influence landslide kinematics. To describe the entrainment process quantitatively, an entrainment ratio (ER) could be offered to calculate the increase of the landslide volume for a specific entrainment zone in the DAN model [49].

$$ER = \frac{V_{Entrained}}{V_{Fragmented}} = \frac{V_E}{V_R(1+F_F)} \tag{5}$$

where V_E (i.e., $V_{Entrained}$) is the volume of the entrained path material (m^3); $V_{Fragmented}$ is the volume of the fragmented material in the sliding source area(m^3). V_R is the volume of the initial loess landslide (m^3); and F_F is the fractional amount of volume expansion due to fragmentation (0.25). The entire volume of the loess landslide deposits is equal to $V_R(1+F_F)+V_E$ [40]. In this study, V_R equals 5.0×10^4 m^3 and V_E equals 25.0×10^4 m^3. The length of the entrainment area was approximately 240 m. To simulate the phenomenon of entrainment, an ER equal to 4.0 was used in the DAN model of the loess landslide. According to the pore-water pressure increased and soil saturation during the sliding process because of the snow infiltration. The landslide was transformed into debris flow, showing a flow state, so the scraping volume was huge.

The movement speed of the sliding body and the thickness of the landslide accumulation body are calculated using Equations (1)–(5). In addition, the amount of resistance encountered during the movement of the sliding body is determined by different types of rheological models. In the DAN-W software, the resistance is mainly controlled by different base rheological models. DAN-W provides a range of rheological models. According to the existing research results and the trial-error method [50,51], the Voellmy model (V) and the Frictional model (F) are more suitable for landslide dynamic hazard research. The Frictional model is mainly used for landslides when the particle sizes of the residual body are large. The Frictional model is also used for mountains with open hillside cracks where the turbulent flow is not developed. The Voellmy model is suitable for the simulation of a landslide with fractured particles where there is a visible liquidized layer in the sliding mass. From Equation, it is evident that the rheological model is proportional to the velocity of the sliding body, so it could simulate the energy damage of the turbulent flow. This was caused by the liquefied material that has high moisture content, including the loose soil covering the flow path and a spring appearing in the path. This opinion has been accepted by Geotechnical Engineering Office (GEO) of Hong Kong [14,52].

Voellmy model: The expression of base resistance is as follows

$$\tau = f\sigma + \gamma\frac{v^2}{\xi}, \tag{6}$$

where f is the friction coefficient of the sliding body, σ is total stress perpendicular to the direction of the sliding path, γ is the material unit weight, v is the moving speed of the sliding body, ξ is the turbulence coefficient, and τ is the resistance at the bottom of the sliding body. The constant friction coefficient (f) is a parameter that should be determined using the Voellmy model. The friction coefficient was modified by the pore pressure and could reach much smaller values when the path material shows wet features.

Frictional model: Assume that the flow of the sliding body is controlled by the effective normal stress acting on each block. The expression of resistance τ is as follows

$$\tau = \sigma\left(1 - \gamma_{\mu}\right)\tan\varphi \tag{7}$$

where γ_{μ} is the pore pressure coefficient (specifically, the ratio of pore pressure to total stress); φ is the internal friction angle; σ is the total stress perpendicular to the direction of the sliding path; and τ is the resistance at the bottom of the sliding body.

5.2. Model Selection

In the DAN-W software, the accuracy of the calculation result depends on three important factors: Sliding body motion trajectory, rheological model, and parameter selection. First, based on multi-period remote sensing images, aerial imagery of drones, and field geological surveys, topographic lines before and after landslides were determined (Figure 6). Second, the Panjinbulake loess landslide was divided into the sliding source area, debris flow area, and accumulation area (Figure 5). According to the hazard characteristics of different regions, it is critical to select suitable rheological models for different regions. Since the sliding source area started from the shearing exit, there are signs of scratch on the exposed area of the bedrock. According to the existing research results [23,49–51], the Frictional model(F) was suitable for the sliding source area. The Frictional model and the Voellmy model(V) were used in the debris flow area and the accumulation area, respectively. According to the landslide path sequence, four sets of the rheological model combinations, Frictional–Frictional–Frictional, Frictional–Frictional–Voellmy, Frictional–Voellmy–Voellmy, and Frictional–Voellmy–Frictioanl, were used to simulate the dynamic hazard effects of the Panjinbulake loess landslide, so as to select the most suitable rheological model combination to simulate the movement process of the loess landslide (see Table 1). The simulation results calculated from a combination of the four rheological models were compared to the features of the actual loess landslide. According to the results, we could find that the Frictional–Voellmy–Frictional model fit the above characteristics. From Equation, the Voellmy model is proportional to the velocity of the sliding body, and it could simulate the huge scraping force which removed the surface loess soil. Thus, the Voellmy rheology model had a better fit than the Frictional model in the debris flow area. Finally, we intended to simulate the movement characteristics of the Panjinbulake loess landslide using the Frictional–Voellmy–Frictional rheological model.

Table 1. Hydrodynamic model of the Panjinbulake landslide.

Model	Sliding Source Zone	Debris Flow Zone	Accumulation Zone
FFF	Frictional	Frictional	Frictional
FFV	Frictional	Frictional	Voellmy
FVV	Frictional	Voellmy	Voellmy
FVF	Frictional	Voellmy	Frictional

5.3. Parameter Selection

The dynamics of high speed and long runout landslides have been studied by researchers. The analysis of landslide dynamics depends, to a large extent, on the choice of parameters and the knowledge level of the author [23]. In this paper, the simulation parameters of the Panjinbulake loess landslide were mainly obtained by the field survey data and the existing research results [23,49–51]. For the Voellmy model, the main parameters were ξ = 400 m/s^2 (the software provides a range of 200–500 m/s^2) and f = 0.05. When rain and snow melt water infiltrates, the groundwater level rises, and pore water pressure rises. The sliding body is close to the flow state, so the friction coefficient decreases gradually. The influence of underground groundwater on sliding body motion is realized by changing the friction coefficient in DAN-W (see Section 7.1). For the frictional model, the dynamic friction angle φ_b was set to 19° according the literature [23,49–51]. Due to the infiltration of ice and snow meltwater, the excess pore water pressure increases, and Ru was set to 0.7. Finally, according to the indoor geotechnical test and the engineering analogy method, the typical strength testing index φi was set to 20° and the unit weight (γ) was set to 18 KN/m^3. As shown in Table 2, based on the trial and error method and the existing research results, these rheological model combinations and parameters were used to simulate the dynamic hazard effects of the Panjinbulake loess landslide.

Table 2. Parameters of the Frictional–Voellmy model used for the Panjinbulake landslide.

Model	Unit Weight, γ (KN/m^3)	Internal Frictional Angle, φ_i (°)	Friction Angle, φ_b (°)	Pore Pressure Coefficient, γ_μ	Friction Coefficient, f	Turbulivity, ξ (m/s^2)
Frictional	18	20	19	0.7	-	-
Voellmy	18	20	-	-	0.05	400

6. Results and Analysis

6.1. Speed Analysis

Using the DAN-W software and the Frictional–Voellmy–Frictional model, the total time of the Panjinbulake loess landslide movement was 22 s. It is assumed that the speed of the landslide was 0 m/s when starting in the sliding source area. The sliding body started from the shear exit (i.e., the leading edge of landslide). Due to the steep slope of the landslide, the speed was very fast, reaching 5 m/s very quickly (Figure 7). Under the action of gravitational potential energy, the speed of the sliding body increased rapidly and reached 21.5 m/s at X = 200 m. This process lasted 7 s, and the average acceleration reached 2.3 m/s^2. The sliding body moved from X = 200 m to X = 320 m in debris flow area. This area has exposed spring water (Figure 5), which provides suitable hydrogeological conditions for high speed movement of the sliding body. Thus, the sliding body was always in a high speed state. This phase lasted 8 s. The average speed reached 20 m/s and the sliding body entered the river terrace. Due to the friction of sandy gravel and gradual decrease of the slope, the moving speed of the sliding body dropped sharply, and the speed relative to the opposite mountain was 17.5 m/s. Due to the blocking of the opposite mountain, the sliding body stopped at a horizontal distance of X = 370 m, and the average acceleration in the deceleration phase reached −3.6 m/s^2, which lasted for 7 s.

6.2. Thickness Analysis of the Accumulation Body

As shown in Figure 8, when the landslide sheared and started at X = 140 m, the initial volume of the landslide was 5.62 × 10^4 m^3, and the average thickness of the sliding body in the sliding source area was 4.5–5 m. After sliding for 7 s, it reached X = 200 m. Due to the scraping effect, the landslide volume reached 13.3 × 10^4 m^3, and the thickness of the accumulation body reached 5 m. At 10 s, the sliding body moved forward to the slope toe, X = 250 m, and the volume of the sliding body reached 23.4 × 10^4 m^3. At 15.82 s, the sliding body moved to the opposite side of the slope, at X = 350 m, and the volume of the sliding body reached 27.7 × 10^4 m^3. Finally, at 22 s, the sliding body stopped moving, and the final total volume reached 32×10^4 m^3. The maximum thickness of the accumulation body at X = 255 m reached 6 m, and the average thickness of the accumulation body reached 0.2–1 m in the sliding source area, 1–2 m in the debris flow area, and 5–6 m in the accumulation area. The simulated results were less than the actual measured results because the dynamic model stretched a smooth two dimensional plane into three dimensions (Figure 6).

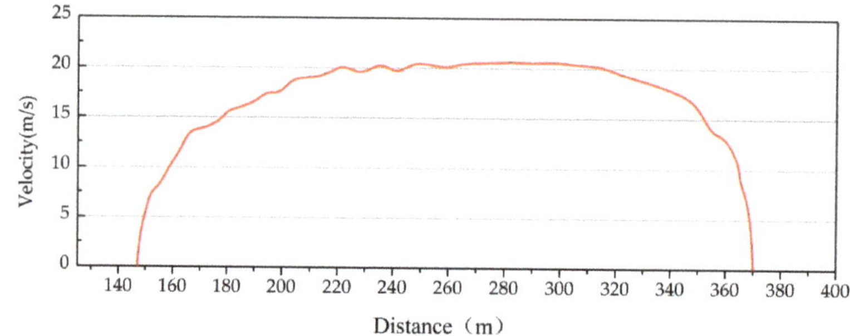

Figure 7. Variation of speed versus sliding range.

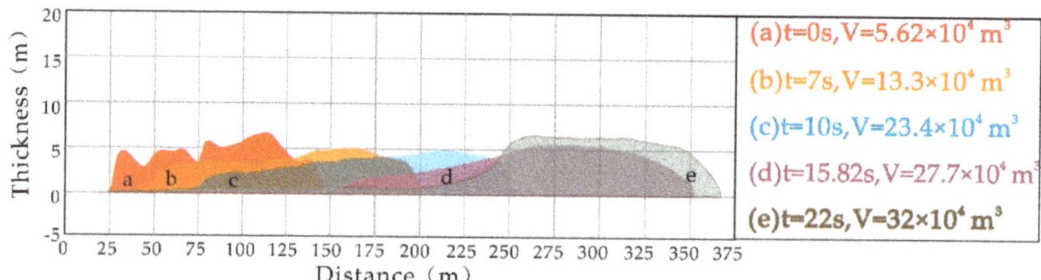

Figure 8. Variation of the thickness of the accumulation body at different times.

6.3. Typical Point Velocity Analysis and Accumulation Body Thickness Analysis

Combined with the actual situation of the field geological survey, the four points with horizontal distances X = 140, 200, 250, and 350 m were selected as the typical points for analysis. The calculation results based on the DAN-W software and F–V–F rheological model are as follows.

(1) The point X = 140 m is located at the landslide shear outlet (Figure 4b). Within 0–10 s, the speed at this point increased linearly from 0 to 17.5 m/s (Figure 9a), which indicates that the acceleration of the sliding body during the starting process increased, reaching 1.75 m/s^2. Also, typical negative terrain was present, and the lower part of the raised bedrock was exposed. This made it so that, after 12 s, the residual sliding body moved to the point where it was hindered by negative terrain with trough shape, the speed gradually attenuated, and the final thickness of the accumulation body at this point was 0.5 m (Figure 9b).

(2) The point X = 200 m is located in the debris flow area (Figure 4b). After the landslide slid for 7 s, it reached this point. Within 7–17.5 s, the speed at this location was always relatively faster, and the average speed reached 17.5 m/s (Figure 9c). In this area, the sliding body had a rapid speed and exposure to spring water, which provides conditions for the occurrence of a high speed and long runout landslide. At 8 s, the thickness of the accumulation body reached 4.53 m. After 17.5 s, the speed at this point gradually decreased, and the final thickness of the accumulation body reached 0.5 m (Figure 9d).

(3) The point X = 250 m is located at the foot of the slope (Figure 4b). After the sliding body slid for 9 s, it reached the foot of the slope. Due to the steep slope, the peak speed reached 19 m/s (Figure 9e), and the thickness of the accumulation body reached 5 m (Figure 9f). Scattering phenomenon was appeared at the foot of the slope where in the interval t = 20–24 s, and there was an increase of debris

flow depth reaching a depth higher to the front passage. In addition, this was also a turning point for the speed of landslide movement. From this point on, the speed of the sliding body began to decrease due to the sudden slowing of the slope and the friction of the sliding body against the sandy gravel of the river terrace.

(4) The point X = 350 m is located at the foot of the opposite slope (Figure 4b). After the sliding body crossed the river terrace, it reached this point at 16 s; then, the speed dropped to 11 m/s (Figure 9g), and the thickness of the accumulation body reached 1.5 m (Figure 9h). From this point on, the sliding body began to hit the opposite side of the mountain. Because of this and even though a considerable amount of energy had been consumed, the sliding body continued to climb 20 m before stopping, indicating that a high-speed and long runout landslide has incredible energy and is severely devastating.

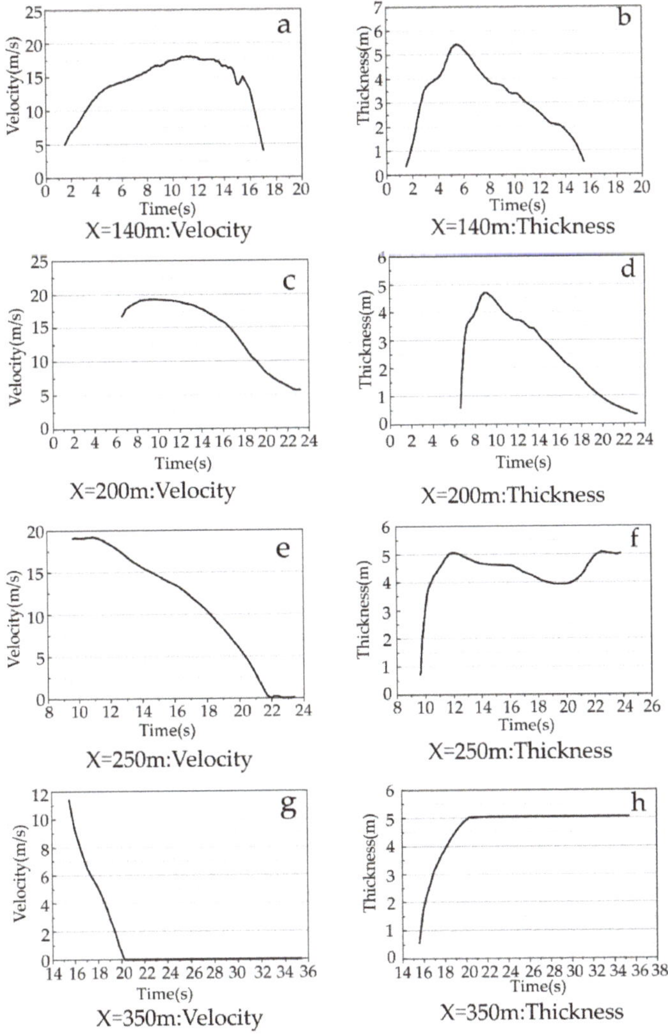

Figure 9. Typical point velocity and variation trend of the accumulation body.

7. Discussion

7.1. Sensitivity Analysis

The DAN-W dynamic model was preferably used to predict the dynamic characteristics of the landslide accumulation area in the Loess Plateau. However, based on a field survey and laboratory tests, some parameters, e.g., the friction coefficient, were obtained by trial and error and the existing research results [23,49–51]. There are many factors affecting the friction coefficient, including the rate of water content, terrain, and ground temperature, with the water content the primary factor. Therefore, it is difficult to provide the friction coefficient in a more efficient way. When using the DAN-W dynamic model to predict the risk assessment, the variability range of the parameters should be considered. In this paper, the influence of friction on the landslide is simulated using multiple sets of working conditions. According to the numerical values recommended by the kinematic model software and existing research results, the friction coefficient is divided into three groups. The first group has a low friction coefficient (i.e., 0.05, 0.1, and 0.2) which means high water content, the second group a moderate friction coefficient (i.e., 0.2, 0.3, and 0.4) which means moderate water content, and the third group a high friction coefficient (i.e., 0.5, 0.55, and 0.6) which means low water content. The calculation results are as follows (Figure 10).

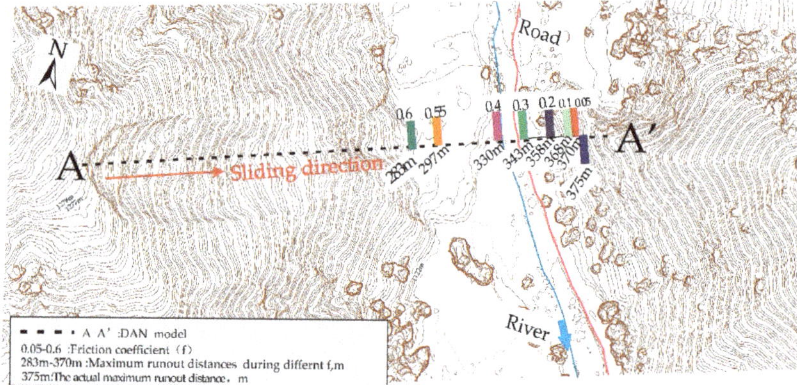

Figure 10. Variation trend of the moving speed versus the sliding rage of the sliding body under the action of different friction coefficients.

According to the numerical simulation results, it can be seen from Figure 10 that, if different friction coefficients are selected, the calculated moving distances are significantly different. For example, when a low friction coefficient (0.05–0.2) is selected, the moving distance of the sliding body is 350–375 m. When the moderate friction coefficient (0.2–0.4) is selected, the moving distance of the sliding body is 320–350 m. When the high friction coefficient (0.4–0.6) is selected, the moving distance of the sliding body is 280–300 m. According to the current geological survey in the field and the aerial view of the drone, the moving distance of the landslide reached a maximum of 366–375 m, which is in line with the calculation results based on the low friction coefficient. The low friction coefficient appears to be due to the saturation degree of its material, and there was sufficient groundwater in the slope area that promoted the sliding of the landslide.

7.2. Empirical-Statistical Model

The empirical formula, mainly based on simple geometric relations of landslides (Figure 11), is simple and effective in the prediction of long runout distance. Figure 11 reveals the geometric relationship between the landslide's apparent friction angle (i.e., the angle between the trailing edge of the landslide and the farthest point of landslide movement), height difference, and motion distance [19]. Based on the concept of the apparent friction angle, Scheidegger proposed an empirical formula called the sled model to calculate the velocity of the sliding body [19]. The specific formula is as follows

$$V = \sqrt{2g(H - t \times L)} \tag{8}$$

where V is the velocity of the estimating point (m/s); g is the acceleration of gravity (m/s^2); and t is the rake ratio between the highest point of the rear edge and the estimating point of the sliding distance (dimensionless); H is the height difference from the highest point of the rear edge and the calculated velocity point of the sliding distance(m); L is the horizontal distance between the landslide trailing edge and the calculated velocity point (m).

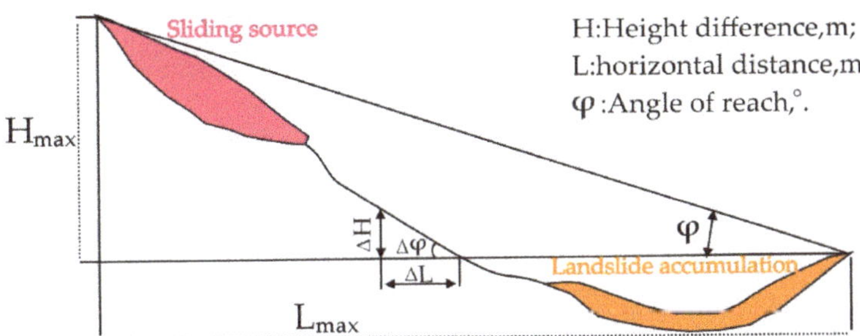

Figure 11. Sketch of the Empirical-statistical model (It is modified from [19]).

According to the calculation of the sled model (Figure 12), the maximum speed of the sliding body was 28 m/s, which occurred near the horizontal distance of 250 m. Similar to the calculation results of the DAN model, the sled model showed that the slide velocity increased sharply after the landslide occurred, and the speed decreased significantly when the slide moved to the front of the road and eventually struck the opposite side of the mountain. The maximum speed obtained by the sled model is far greater than that of the DAN-W dynamic model. Instead of taking the dynamic characteristic of the landslide into account, such as erosion and entrainment, the sled model only gives a preliminary description of the process of landslide movement variation. It could be seen that the calculation results of the DAN model are more accurate. While the concept of the sled model and the apparent friction angle tends to be conservative for landslide hazard prediction, they still comprise a qualitative and effective way to predict disasters.

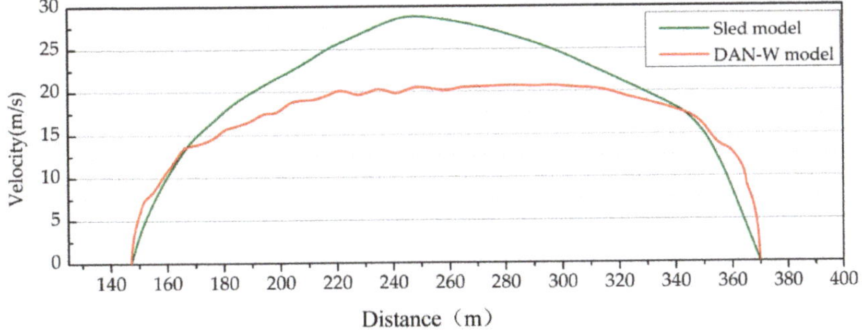

Figure 12. Speed contrast diagram of two models of the Panjinbulake landslide.

The landslide under study is located in Piliqinghe Basin, located in the western part of the Loess Plateau and is part of the "Belt and Road" area. The location has many potential loess landslides, all of which pose a threat to agricultural production. It is especially important to research the prediction of potential landslide disasters, which is of great benefit to disaster prevention and devising mitigation measures. During the period 2017–2018, our team carried out field geological survey work in the area and counted 12 loess landslides that occurred. At the same time, the team measured the basic parameters of the landslide, and calculated the apparent friction angle of each landslide (Table 3, Figure 11). The statistical results indicate that the apparent friction angle of the loess landslide in this area is approximately 25°. Based on the concept of the apparent friction angle, the farthest running distance of the landslide (i.e., L_{max}) can thus be calculated by the formula which is as follows

$$L_{max} = \frac{H_{max}}{\tan 25^{\circ}} = 2.15H_{max} \tag{9}$$

where H_{max} is the height difference from the highest point of the rear edge and the farthest point of the sliding distance (m).

Table 3. Basic geometry of the loess landslides in the Piliqinghe basin.

Number	Loess Landslide	V (m^3)	L_{max} (m)	H_{max} (m)	H_{max}/L_{max}	φ (°)
1	KS1#	5160	130	60	0.46	24.77
2	KS2#	108,000	137	83	0.605	31.21
3	AX1#	109,000	130	85	0.65	33.02
4	AX2#	10,000	75	37	0.49	26.10
5	KZ1#	12,000	335	145	0.43	23.27
6	KZ2#	17500	341	131	0.384	21.01
7	KZ3#	30,000	175	40	0.23	12.95
8	KZ4#	10,000	47	25	0.53	27.92
9	PL1#	66,000	175	85	0.49	26.10
10	PL2#	7000	74	40	0.54	28.36
11	PL3#	3440	136	55	0.41	22.30
12	Panjinbulake	300,000	375	160	0.43	23.26
		Average			0.46	25

The empirical formula and the DAN model can be used to predict and analyze the moving distance of potential landslides from qualitative and quantitative aspects separately.

7.3. Evaluation of Landslide Residual Risk

Based on the field survey and the aerial images from unmanned aerial vehicles, multiple tensile cracks appeared in the upper part of the trailing edge of the landslide and formed an independent unstable block. In the case of rainfall and ice and of snow melt infiltration, the unstable block would be extremely easy to slide, which poses a threat to agricultural production and road operation. To avoid secondary harm, in this paper, we use the DAN model to predict and analyze the movement trend of the unstable body. According to the results of the field geological survey, the unstable block area ranges from 3 to 6 m, with a volume of nearly 1.2×10^4 m^3.

The selected models and parameters are the same as in the previous landslide dynamic hazard analysis. Figure 13 shows the running velocity of the unstable body, the thickness of the deposit, and the predicted disaster threat zone. Since the slope of the unstable body surface is 30°, which is relatively gentle, most of the sliding body is deposited in the slide-source area after the unstable body slides. The average thickness of the sliding body in the sliding source area reaches 5 m, the longest distance reaches 300 m, and the maximum moving speed reaches 18.5 m/s. According to the sled model, the farthest distance of the unstable body motion is 362 m. Combined with the calculation results of the DAN-W model and sled model, it can be concluded that, if the unstable landslide body starts, it will be pose a threaten to road operation. Owing to its rapid movement speed, the landside mass also threatens the safe production of grazing herds of animals. Setting up a warning sign around the landslide to warn herders to locate grazing far from the area is recommended. The local should set engineering measures (such as garbion) around the road to ensure traffic safety.

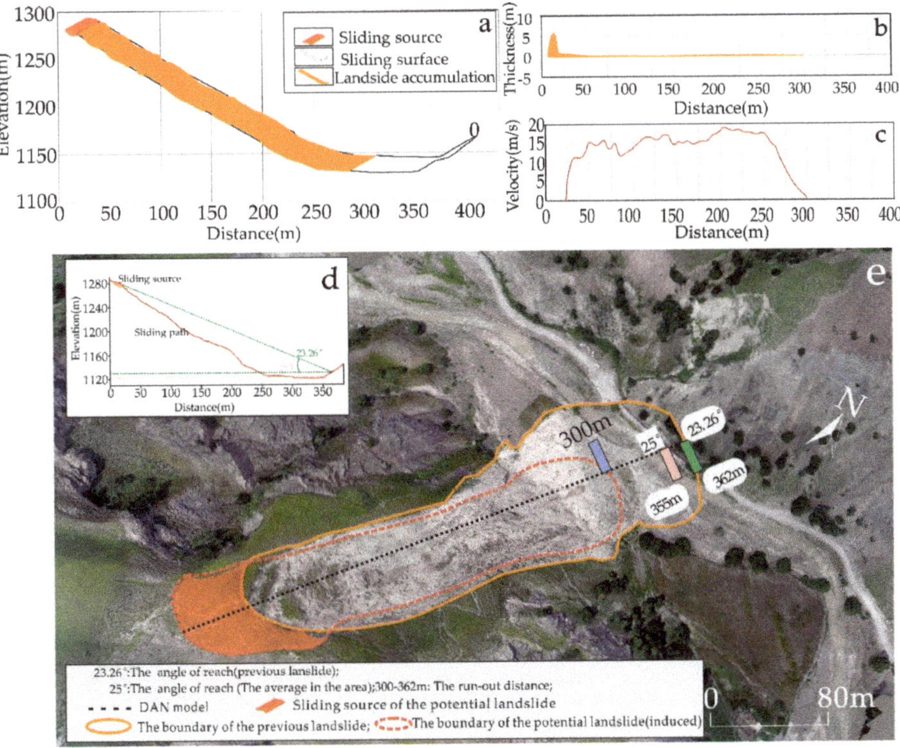

Figure 13. (**a**) The runout distance in the DAN-W. (**b**) Variation of the thickness of the unstable slope in the DAN-W. (**c**) Variation of the velocity of the unstable slope in the DAN-W. (**d**)The sled model for the two different reach angles. (**e**) Maximum extent of the unstable slope runout at different methods.

8. Conclusions

Based on the geological survey in the field, multi-period historical remote sensing images and aerial images of the drone, combined with the geological conditions of the study area, we analyzed the inducing factors and runout process of the Panjinbulake loess landslide and predicted the secondary disaster. Furthermore, the DAN-W dynamic model and a set of combined basal rheological models (Frictional–Voellmy–Frictional models) can suitably simulate the dynamic hazard effects of the Panjinbulake loess landslide. We analyzed the influence of the landslide movement speed, typical point velocity, accumulation body thickness, and friction coefficient. The simulation results showed that the duration of the Panjinbulake loess landslide was 22 s, the maximum speed was 20.5 m/s, and the maximum thickness of the accumulation body was 5.5 m, which is in line with the actual situation based on the field investigation. The basal rheological model combination and parameters obtained through trial and error can be used to simulate and predict the long runout distance of loess landslides, and it is necessary in strengthening the early identification and prevention of loess landslide hazards using multi-precision observation technology and numerical techniques.

Author Contributions: L.Y. analyses, writing and dealing with data; Y.W. analyses the field geological phenomena; W.W. offers the technology of DAN-W; S.Z. analyses the field geological phenomena.

Funding: This study was supported by the National Key Research and Development Program of China [No.2018YFC1505404], China Geological Survey (DD20190647, DD20179609, DD20190637).

Acknowledgments: The authors are grateful to O.Hungr for supplying a copy of the DAN-W software. We thank MDPI (www.mdpi.com) for its linguistic assistance during the preparation of this manuscript. We also thank Gang Liu, Chinese Academy of Sciences, Feihang Qu, Northwest University and Jie Luo, University of Science and Technology of China, for their kind assistance. Finally, we thank editors and reviewers for their thoughtful review and valuable comments to the manuscript.

Conflicts of Interest: The authors declare no conflicts of interest.

References

1. Hungr, O.; Leroueil, S.; Picarelli, L. The Varnes classification of landslide types, an update. *Landslides* **2014**, *11*, 167–194. [CrossRef]
2. Zhang, M.; McSaveney, M.J. Rock-avalanche deposits store quantitative evidence on internal shear during runout. *Geophys. Res. Lett.* **2017**, *44*, 8814–8821. [CrossRef]
3. Cruden, D.M.; Varnes, D.J. Landslides: Investigation and mitigation. Chapter 3-landslide types and processes. In *Transportation Research Board Special Report*; National Research Council, Transportation Research Board: Washington, DC, USA, 1996; p. 247.
4. Segoni, S.; Piciullo, L.; Gariano, S.L. A review of the recent literature on rainfall thresholds for landslide occurrence. *Landslides* **2018**, *15*, 1483–1501. [CrossRef]
5. Yin, Y.; Wang, F.; Sun, P. Landslide hazards triggered by the 2008 Wenchuan earthquake, Sichuan, China. *Landslides* **2009**, *6*, 139–152. [CrossRef]
6. Peng, J.; Zhuang, J.; Wang, G.; Dai, F.; Zhang, F.; Huang, W.; Xu, Q. Liquefaction of loess landslides as a consequence of irrigation. *Q. J. Eng. Geol. Hydrogeol.* **2018**, *51*, 330–337. [CrossRef]
7. Iverson, R.M.; Denlinger, R.P. Flow of variably fluidized granular masses across three-dimensional terrain: 1. Coulomb mixture theory. *J. Geophys. Res.* **2001**, *106*, 537–552. [CrossRef]
8. Stancanelli, L.M.; Lanzoni, S.; Foti, E. Propagation and deposition of stony debris flows at channel confluences. *Water Resour. Res.* **2015**, 51. [CrossRef]
9. Lanzoni, S.; Gregoretti, C.; Stancanelli, L.M. Coarse-grained debris flow dynamics on erodible beds. *J. Geophys. Res. Earth Surf.* **2017**, 122. [CrossRef]
10. Iverson, R.M. The physics of debris flows. *Rev. Geophys.* **1997**, *35*, 245–296. [CrossRef]
11. Hungr, O.; McDougall, S.; Bovis, M. Entrainment of material by debris flows. In *Debris Flow Hazards and Related Phenomena*; Jakob, H., Ed.; Springer: Heidelberg, Germany, 2005; pp. 135–158.
12. Sassa, K.; Nagai, O.; Solidum, R.; Yamazaki, Y.; Ohta, H. An integrated model simulating the initiation and motion of earthquake and rain induced rapid landslides and its application to the 2006 Leyte landslide. *Landslides* **2010**, *7*, 219–236. [CrossRef]
13. Shreve, R.L. Leakage and Fluidization in Air-Layer Lubricated Avalanches. *Geol. Soc. Am. Bull.* **1968**, *79*, 653. [CrossRef]
14. Geotechnical Engineering Office (GEO). *Guidelines on Assessment of Debris Mobility for Open Hillslope Failures: GEO Technical Guidance Note No. 34 (TGN 34)*; Geotechnical Engineering Office, Civil Engineering and Development Department, Hong Kong Government: Hong Kong, China, 2012.
15. Segoni, S.; Piciullo, L.; Gariano, S.L. Preface: Landslide early warning systems: Monitoring systems, rainfall thresholds, warning models, performance evaluation and risk perception. *Nat. Hazards Earth Syst. Sci.* **2018**, *18*, 3179–3186. [CrossRef]
16. Segoni, S.; Tofani, V.; Rosi, A.; Catani, F.; Casagli, N. Combination of Rainfall Thresholds and Susceptibility Maps for Dynamic Landslide Hazard Assessment at Regional Scale. *Front. Earth Sci.* **2018**, *6*, 85. [CrossRef]
17. Mancarella, D.; Hungr, O. Analysis of run-up of granular avalanches against steep, adverse slopes and protective barriers. *Can. Geotech. J.* **2010**, *47*, 827–841. [CrossRef]
18. Loew, S.; Gschwind, S.; Gischig, V.; Keller-Signer, A.; Valenti, G. Monitoring and early warning of the 2012 Preonzo catastrophic rockslope failure. *Landslides* **2016**. [CrossRef]
19. Scheidegger, A.E. On the prediction of the reach and velocity of catastrophic landslides. *Rock Mech.* **1973**, *5*, 231–236. [CrossRef]

20. Pastor, M.; Herreros, I.; Merodo, J.A.F.; Mira, P.; Haddad, B.; Quecedo, M.; Gonzalez, E.; Alvarez-Cedron, C.; Drempetic, V. Modelling of fast catastrophic landslides and impulse waves induced by them in fjords, lakes and reservoirs. *Eng. Geol.* **2009**, *109*, 124–134. [CrossRef]
21. Liu, C.; Pollard, D.D.; Shi, B. Analytical solutions and numerical tests of elastic and failure behaviors of close-packed lattice for brittle rocks and crystals. *J. Geophys. Res. Solid Earth* **2013**, *118*, 71–82. [CrossRef]
22. Itasca, Consulting Group Inc. *PFC3D Particle Flow Code in 3 Dimensions. User'sGuide*; Itasca, Consulting Group Inc.: Minneapolis, MN, USA, 2008.
23. Hungr, O. A model for the runout analysis of rapid flow slides, debris flows, and avalanches. *Can. Geotech. J.* **1995**, *32*, 610–623. [CrossRef]
24. Iverson, R.M.; George, D.L. A depth-averaged debris-flow model that includes the effects of evolving dilatancy. I. Physical basis. Proceedings of the Royal Society A: Mathematical. *Phys. Eng. Sci.* **2014**, *470*, 20130819. [CrossRef]
25. Jeong, S.W.; Wu, Y.H.; Cho, Y.C.; Ji, S.W. Flow behavior and mobility of contaminated waste rock materials in the abandoned Imgi mine in Korea. *Geomorphology* **2018**, *301*, 79–91. [CrossRef]
26. Liu, K.F.; Wei, S.C.; Wu, Y.H. The influence of accumulated precipitation on debris flow hazard area. In *Landslide Science for a Safer Geoenvironment*; Springer: Cham, Switzerland, 2014; pp. 45–50.
27. Han, Z.; Wang, W.; Li, Y.; Huang, J.; Su, B.; Tang, C.; Chen, G.; Qu, X. An integrated method for rapid estimation of the valley incision by debris flows. *Eng. Geol.* **2018**, *232*, 34–45. [CrossRef]
28. Pastor, M.; Haddad, B.; Sorbino, G.; Cuomo, S.; Drempetic, V. A depth-integrated, coupled SPH model for flow-like landslides and related phenomena. *Int. J. Numer. Anal. Methods Geomech.* **2009**, *33*, 143–172. [CrossRef]
29. Wang, F.W.; Sassa, K. A modified geotechnical simulation model for the areal prediction of landslide motion. In Proceedings of the 1st European Conference on Landslides, Prague, Czech Republic, 24–26 June 2002; pp. 735–740.
30. *FLO-2D User's Manual*; Version 2007; FLO-2D Software Inc.: Nutrioso, AZ, USA, 2007.
31. Hutter, K.; Szidarovsky, F.; Yakowitz, S. Plane steady shear flow of a cohesionless granular material down an inclined plane: A model for flow avalanches Part I: Theory. *Acta Mech.* **1986**, *63*, 87–112. [CrossRef]
32. Havenith, H.B.; Torgoev, A.; Schlögel, R.; Braun, A.; Torgoev, I.; Ischuk, A. Tien Shan Geohazards database: Landslide susceptibility analysis. *Geomorphology* **2015**, *249*, 32–43. [CrossRef]
33. Sun, J.Z. Environmental Geology in Loess Areas of China. *Environ. Geol. Water Sci.* **1988**, *12*, 49–61.
34. Peng, J.B.; Lin, H.Z.; Wang, Q.Y.; Zhuang, J.Q.; Cheng, Y.X.; Zhu, X.H. The critical issues and creative concepts in mitigation research of loess geological hazards. *Eng. Geol.* **2014**, *22*, 684–691. (In Chinese)
35. Qi, X.; Xu, Q.; Liu, F.Z. Analysis of retrogressive loess flowslides in Heifangtai, China. *Eng. Geol.* **2018**, *236*, 119–128. [CrossRef]
36. Leng, Y.Q.; Peng, J.B.; Wang, Q.Y.; Meng, Z.J. A fluidized landslide occurred in the Loess Plateau: A study on loess landslide in South Jingyang tableland. *Eng. Geol.* **2018**, *236*, 129–136. [CrossRef]
37. Bai, J.; Li, J.; Shi, H.; Liu, T.; Zhong, R. Snowmelt Water Alters the Regime of Runoff in the Arid Region of Northwest China. *Water* **2018**, *10*, 902. [CrossRef]
38. Derbyshire, E. Geological hazards in loess terrain, with particular reference to the loess regions of China. *Earth Sci. Rev.* **2001**, *54*, 231–260. [CrossRef]
39. Gao, G.R. Formation and development of the structure of collapsing loess in China. *Eng. Geol.* **1988**, *25*, 235–245.
40. Zhang, M.S.; Liu, J. Controlling factors of loess landslides in western China. *Environ. Earth Sci.* **2010**, *59*, 1671–1680. [CrossRef]
41. Pei, X.J.; Zhang, X.C.; Guo, B.; Zhang, F.Y. Experimental case study of seismically induced loess liquefaction and landslide. *Eng. Geol.* **2017**, *105*, 23–30. [CrossRef]
42. Tu, X.B.; Kwong, A.K.L.; Dai, F.C.; Tham, L.G.; Min, H. Field monitoring of rainfall infiltration in a loess slope and analysis of failure mechanism of rainfall-induced landslides. *Eng. Geol.* **2009**, *105*, 134–150. [CrossRef]
43. Peng, J.B.; Fan, Z.J.; Wu, D.; Zhuang, J.Q.; Da, I.F.C.; Chen, W.W. Heavy rainfall triggered loess-mudstone landslide and subsequent debris flow in Tianshui, China. *Eng. Geol.* **2015**, *186*, 79–90. [CrossRef]
44. Hu, W. Acoustic Emissions and Microseismicity in Granular Slopes Prior to Failure and Flow-Like Motion: The Potential for Early Warning. *Geophys. Res. Lett.* **2018**, *45*, 10406–10415. [CrossRef]

45. Zhang, M.; Yin, Y.P. Dynamics, mobility-controlling factors and transport mechanisms of rapid long runout rock avalanches in China. *Eng. Geol.* **2013**, *167*, 37–58. [CrossRef]
46. Zhang, D.; Wang, G.; Luo, C.; Chen, J.; Zhou, Y. A rapid loess flowslide triggered by irrigation in China. *Landslides* **2009**, *6*, 55–60. [CrossRef]
47. Kong, R.; Zhang, F.; Wang, G.; Peng, J. Stabilization of Loess Using Nano-SiO_2. *Materials* **2018**, *11*, 1014. [CrossRef] [PubMed]
48. Liu, L.N.; Li, S.D.; Jiang, Y.; Bai, Y.H.; Luo, Y. Failure mechanism of loess landslides due to saturated-unsaturated seepage—Case study of Gallente landslide in ILI, Xinjiang. *J. Eng. Geol.* **2017**, *5*, 1230–1237. (In Chinese)
49. Hungr, O.; Evans, S.G. Entertainment of debris in rock avalanches: An analysis of a long runout mechanism. *Geol. Soc. Am. Bull.* **2004**, *116*, 1240–1252. [CrossRef]
50. Hungr, O. Analysis of debris flow surges using the theory of uniformly progressive flow. *Earth Surf. Process. Landf.* **2000**, *25*, 483–495. [CrossRef]
51. Buchholtz, V.; Pöschel, T. Numerical Investigation of the Evolution of Sandpiles. *Phys. A Stat. Mech. Its Appl.* **1993**, 202. [CrossRef]
52. Geotechnical Engineering Office (GEO). *Guidelineson Enhanced Approach for Natural Terrain Hazard Studies: GEO Technical Guidance Note No. 36 (TGN 36)*; Geotechnical Engineering Office, Civil Engineering and Development Department, Hong Kong Government: Hong Kong, China, 2013.

MDPI
St. Alban-Anlage 66
4052 Basel
Switzerland
Tel. +41 61 683 77 34
Fax +41 61 302 89 18
www.mdpi.com

Water Editorial Office
E-mail: water@mdpi.com
www.mdpi.com/journal/water

www.ingramcontent.com/pod-product-compliance
Lightning Source LLC
LaVergne TN
LVHW071246170726
843515LV00006BA/1342

* 9 7 8 3 0 3 6 5 0 9 3 0 3 *